The Next Step

Dedicated to those who strive to preserve.

The Next Step

Interpreting Animal Tracks, Trails and Sign

by

David Brown

The McDonald & Woodward Publishing Company
Newark, Ohio

The McDonald & Woodward Publishing Company
Newark, Ohio 43055
www.mwpubco.com

The Next Step
Interpreting Animal Tracks, Trails and Sign

Printed in the United States of America by
McNaughton & Gunn, Inc., Saline, Michigan, on paper that meets
the minimum requirements of permanence for printed library
materials.

First printing October 2015
10 9 8 7 6 5 4 3 2 1
24 23 22 21 20 19 18 17 16 15

Library of Congress Cataloging-in-Publication Data
Brown, David, 1941-
  The next step : interpreting animal tracks, trails, and sign / by
  David Brown.
    pages cm
  Includes bibliographical references and index.
  ISBN 978-1-935778-25-7 (perfect bound : alk. paper)
  1. Animal tracks—North America. 2. Tracking and trailing—
  North America. I. Title.
  QL768.B753 2014
  591.47′9—dc23

                                                  2014037948

Contents

Preface

When I was a boy there were only a few books on animal tracking available, and most were of questionable value. Really, the only one that provided at all accurate information was the classic work of Olaus Murie, produced in the early part of the 20th Century. Not only was Murie a scientist and, as such, unwilling to invent what he had not actually seen, he also had the opportunity to roam the West before so much of it got transformed by the aggressive exploitation of its natural resources. I think many trackers today wish they had been able to lead Murie's life at the time he did and been able to experience the places that, as a biologist for the Fish and Wildlife Service, he got to visit. Off-hand references to his many adventures, such as dog-sledding down the Yukon River when that was the only way to traverse it in the winter, fire the imagination of the latter-day naturalist.

It seems that few nature writers up to Murie's time had taken the accurate depiction of animal tracks and sign as a serious subject and so presented to the public what sometimes were fanciful and bizarrely inaccurate images and related information. Since Murie, others, intent on original work of their own, may have lacked his eye for significant detail as well as for perceiving fundamental shape. The result has been that, until recently, the literature available to the reader has presented a hodge-podge of images that mix the accurate with the fantastical.

Today the availability of reliable tracking information is much better. Trackers like Rezendes and Elbroch, picking up the thread of Murie's dedication to accuracy, have produced much improved modern guides to track identification. I like to think that I am part of this new age, as well, with my own *Trackards for North American Mammals* and *The Companion Guide to Trackards for North American Mammals*, which I self-produced in small quantities for many years and which have recently been published by McDonald and Woodward.

With these resources, Rezendes and Elbroch on the desk and mine in the field, it should be possible today for a tracker of any experience to identify much of what he or she finds in the woods, fields and swamps that they roam, better informed than ever before.

I have always wanted to discover secrets, whether the secrets are hidden in a code or hidden in the symbols of a foreign language or hidden in the night or behind foliage. This ultimately has been the motivation for my study of animal tracking. The Nature that surrounds us teems with invisible life, living by its own rules and following its own roadmap that only occasionally coincides with our own. Finding several whitened droppings on the forest floor and then tracking them to an inconspicuous nest in a pine where a female goshawk incubates her eggs is a thrill for me, not because I have never seen a goshawk before but because I have discovered, hidden in a pine tree, a secret of Nature that I was not supposed to find. Thus it is not the sign itself that is as exciting as the interpretation of it in order to find the nest.

One should not, I think, aim at the finish line in one's study of animal tracking, that is, at the point where you hope to be able to say, "There, I've learned it all." But, like learning a language, one should look at the process as a pleasurable, life-long pursuit. If gaining insight into the lives of wild animals had a finite end, I expect that after all these years I would be bored with it. Instead, I cheer myself with the fact that I know I will never know it all, never get to the end of it. Rather than feeling frustrated at the prospect of an endless journey, I can expect the surprises and delights of discovery to go on as long as I am alive and able to see or, at least, think about what I have seen.

This book, then, is intended to encourage "the next step" to the many who have persisted through the confusions caused by the initial and necessary oversimplifications of an art and feel ready to begin *interpreting* what they have already learned to *identify*. Although the difficulties are many, the rewards of insight into a hidden world of life are generous compensation. I hope and trust that in progressing into the subtleties and complexities presented in this book, paradoxically some of the initial confusion will clear. This is the promise of persistence in this difficult and rewarding pursuit.

I will not pretend to complete or perfect expertness on the subject of animal track interpretation. All I can hope is that some information contained in this book, gained from my experiences over several decades, will be helpful to the reader. And all I can promise is that the information is honestly arrived at. Despite efforts to explain as lucidly as I can, no doubt some parts of this guide will seem confusing, especially to the less-experienced tracker. Mixing field experience with reading is the best way to reduce this confusion, using one to lever up the other in the climb to competence.

All of us who are engaged in rediscovering the lost art of animal tracking are somewhere on that climbing road. No one knows all of it and hopefully never will; the journey, as is often said, is much more satisfying than the arrival. Ultimately, the hope is that along the way we will gain valuable insights not only into the hidden lives of wild animals but also into that larger life upon which our own species depends, despite the civilized artifice behind which we hide from its realities.

The art of animal tracking is a work in progress. Lost to us is most of the information, little or none of it written down, passed by oral tradition from father to son in aboriginal societies in North America. We are, in Rezendes' phrase, "relearning an ancient art" or perhaps reinventing it, since the original largely has been lost. But we are obliged to reinvent it with modern Western minds, selected for traits other than those of the ancient native hunter.

Taking the next step in the art of animal tracking has many rewards as well as many difficulties. It seems most things in life that give us abiding satisfaction must have both. This book is intended to provide a hint of the former, pointing developing trackers in what I hope is the right direction and then setting them loose on their own personal journey of discovery. It also guides the trackers through some of the difficulties in mastering the tracking art, helping him or her to overcome obstacles to clear reconstructions of natural events encountered in the field. To those, I extend my very best wishes for as many happy and exciting days as I have been fortunate to have in my nearly three decades of applying myself to that art.

Acknowledgements

I wish to thank my friend, Joseph Choiniere, for his invaluable help in editing the original draft of the manuscript for this book. He is the best general naturalist I know and has served here to keep me out of trouble with the peripheral facts. He has also been a frequent companion on tracking excursions in central Massachusetts where his insights as a life-long naturalist have always added richness to these walks. Donna Nothe-Choiniere has also been a dear friend over many years. Her enthusiasm for Nature, not to mention her way around a roast, has been a great support during doubtful times.

Since the mid-1980s my friend Kevin Harding has been a frequent companion on many tracking excursions in New England and Arizona. Whether paddling along the muddy banks of the Megalloway or trudging the snowy Quabbin woods, his independent insights have been a great help in refining my own as they appear in this book. During those years I have had the privilege of being a frequent guest in his and his wife, Rita's, various homes in Massachusetts, Maine and Arizona from which many tracking adventures have emanated. Emerging from mosquito-infested bogs to a shower and one of Rita's excellent dinners has been a continually restoring experience.

In two decades of presenting tracking and other wildlife programs I have had the honor of meeting hundreds of eager trackers whose enthusiasm has buoyed my own. Their questions and observations have been invaluable in helping me entertain perspectives different from my own. Their questions and observations have led to insights into better ways of explaining what are often complex processes of identification and interpretation of tracks, trails and other sign. I am indebted to them for their questions and perceptual insights.

Tracker and photographer Paul Rezendes provided early guidance way back in the 1980s and early 90s. He was the first person I had ever met who had the competence to confirm or deny authoritatively my intuitions about tracks and sign. I might never have gotten started on this long journey of discovery without his help.

While I have not quoted other authors in this book directly, their collective knowledge contained in the bibliography has contributed greatly to what I have written here. No one starts from scratch in any art, and I have the advantage of an awareness of their past works in producing this new one. I am greatly indebted to these authors who have gone before and shown the way.

Finally, I would like to thank Dr. Jerry McDonald, of McDonald and Woodward Publishing Company, for taking on the risk of publishing not only this book but also *Trackards for North American Mammals* and *The Companion Guide to Trackards for North American Mammals*. His editorial insights have gone a long way to improving the quality of all these projects beyond the rough original manuscripts. Thanks as well to Trish Newcomb, also of McDonald and Woodward, for her marketing efforts in bringing these projects into public view. I hope the future for these books will reward their confidence and efforts.

The Next Step

Chapter I

Introduction to *The Next Step*

The Next Step seeks to extend the art of animal tracking beyond the scope of other books currently available that concentrate mainly on the identification of wild animal evidence. Animal tracking is composed of three complementary and often-overlapping phases. The first is finding the sign in the first place. This is an area that has largely been overlooked in current publications, and so I have used some space at the beginning to deal with this often difficult and neglected aspect.

The second phase is the identification of the found evidence, the area where most of the better tracking books concentrate their information. The third phase is what I have for many years called "eco-tracking," that is, interpreting what has been found and identified in order to answer the questions: What was the animal doing at the time it left the evidence, and furthermore, why was it present or passing there at that particular time? This step, interpreting the evidence, is often touched upon only lightly in other books and so it is the main occupation of the body of this book.

The art of animal tracking has many rewards, but few things that are worthwhile are ever mastered without first surmounting difficulties. It may be good, then, to begin in this introductory chapter with a discussion of some of them, followed by the rewards that one may expect to find by persistence in developing the skills of a wildlife tracker.

The Difficulties and the Rewards

The Difficulties

Although *The Next Step* concentrates on interpretation, one must first identify the animal that left the discovered sign. In the case of tracks and trails any manual must, in the finite space within its covers, involve some simplification of what may be found in the field by the

3

reader. Although subtle variations of the appearance of an animal's track may be many, a book is confined to, at best, a few illustrations and in many cases to a single representation. This, in turn, may lead the beginning tracker to regard an animal's foot as if it were made of wood — a wooden stamp that leaves behind a print of unvarying dimensions and characteristics in every medium in which it is impressed. An animal's foot, however, is a living entity connected to the creature above and to the surface over which the creature is traveling below. It can be distorted by muscular contraction and disposition of weight so that even the stiffest foot, that of a deer for instance, can be contracted or expanded laterally and articulated to some degree. Thus the foot of a dead animal will not always, or perhaps even often, reflect the life that formerly energized it. A dead opossum's front foot, to cite an extreme example, will show toes contracted laterally; the weight-bearing foot of a live opossum, on the other hand, will leave a front print splayed like a star. The rubbery toe pads of a mammal distort under load, even in a low-impact gait like a walk. But in a photograph of the sole of a dead animal's foot, the pads are unloaded and thus not appropriately distorted from their relaxed shape. This is especially the case with mammals such as cats that are adapted to jumping down from heights. Being thick and rubbery, their pads deform into many variations depending on gait and surface.

A mammal's foot exists in three dimensions while representations of tracks in a book, including this one, must necessarily be two-dimensional. The lobes of the secondary pad of many mammals, gray fox and coyote for instance, may be held in different planes, sometimes registering, sometimes not, depending on the softness of the substrate, the weight of the part of the animal's body above it, the angle of the pad-plane relative to the surface and the impact of the gait. However, a single drawing or photograph of a track can only show one plane, a limitation that may confuse the novice who expects an illustration to look like the animal's track in all circumstances. The prints of a bear are instructive on this point. The 5th medial toe, the "little toe," is often carried on a slightly higher plane than the other four so that in shallow prints it may not register. Should an illustration of a bear's prints show this 5th toe or not? If it is presented, a novice is unlikely to confuse it with anything else. But if the illustration depicts a shallower plane and removes this toe imprint, then the novice is left with the impression that a bear has only four toes. This is how bear prints, improbably, get identified as cougar prints. Finally, if you add the fact that a photo or drawing can only be viewed from a fixed perspective and in the lighting available at the time, the confusion compounds.

The issues above involve the animal itself from the pads upward. But from the pads downward more difficulties await. The medium in which the print is impressed can vary widely and, therefore, may present to the observer widely different impressions. While many of these impressions will be in clean white snow, snow itself is variable, and each variation will alter to a greater or lesser degree any print impressed in it. Furthermore, mammals may alter their foot configuration for different snow surfaces. More than once I have been presented with a photo of a bobcat track by someone who suspects it to be mountain lion or lynx. In soft, wet spring snow bobcats spread their toes for support as wide as 4 inches, giving the impression of a much larger animal or, in the case of confusion with a lynx, a much larger foot. Finally, while prints in snow are generally gray against white, prints in dirt are usually brown on brown, and neither may bear a precise likeness to the chiaroscuro drawings in black ink on a white page.

Tracking books, including my own, usually give dimensions for prints as a help in identifying the species that made them. However, such measurements must be regarded loosely since there are many differences among prints of the same species depending on sex, age and other factors. Although the human animals are all recognizably human, we all have more or less different feet, and our leg lengths will provide different stride lengths as well. The same is true of other mammals. Rather than being rigidly homogeneous in size and shape, the footprints of any mammal should be regarded as variations on a theme. I find that my own track identifications in the field depend more on perceiving a fundamental form within expected variations than they do on measurement. Some people are happier with this "gestalt" method than others. If you need measurements, then use them but do so while taking into account their limitations.

Up to this point I have discussed only the difficulties of identification. But identification is only one step in learning the art of animal tracking. I elaborate upon it because it is a difficult step. Moving on to the main focus of this book, the tracks and trails of an animal are in a sense a diary of its daily (or more likely nightly) activities. This diary can be read for a tremendous amount of information about the everyday life of wild animals that goes on all around us but is largely hidden from our view. However, that diary is written in a foreign language. One cannot expect to learn another language by simply looking up words in a dictionary, in a sense identifying them. Those words must be organized into a recognizable syntax in order for them to communicate anything worthwhile. In a sense tracks are the words, and the patterns and trails into which they are organized are their sentences. In order to interpret, as well as identify, we need to study the whole language. And as anyone knows who has tried it, learning a foreign language can be challenging indeed. After studying a half-dozen foreign tongues during my lifetime, I have found that motivation is the key. The language will not simplify itself for my benefit. Throwing up my hands at every obstacle and saying "I can't learn this!" will result only in not learning it. On the other hand, if one psyches oneself up instead, saying, "Isn't it interesting that they express this thing differently from the way we do?" then the language flows into one's consciousness more readily. This is another way of saying that the learning should be a pleasure, not a chore, and so one must actively seek out the potential pleasure in it.

The Rewards

We are surrounded by Nature. What we can see directly, however, is very limited; we see its surface but little more. For many people, Nature is scenery and that is enough for them. For others who are inclined to active searching, Nature is a bird singing in a tree to be identified and listed before passing on. Theirs, I think, is more a hunter's frame of mind, perhaps excited by some remnant of our hunter-gatherer past still floating around in our genetic material. In this book we will try to extend this process into interpretation of what is seen in order to gain more of an inside perspective, a look into the animal's hidden life.

Interpretive animal tracking, then, proceeds from finding to identifying and then to determining behavior and its relationship to habitat. Since much animal sign accumulates and lasts over time, it may be pondered at leisure. One might sit in the woods for days without seeing a live wild animal pass. But if an animal deposits its scat on a cut stump, the sign of that animal can be found continually through time by the careful searcher until weathering absorbs it into the background. This is the "cloud chamber effect," where the sign of passage of an atomic particle can be observed by its smoke trail even though the particle itself is invisible and long gone. Furthermore, unlike the smoke trail, the sign of wild animals accumulates over time so that its density is far greater than the density of the animals themselves. One can seldom see wild animals directly, but with enough skill and effort one can always find their sign

From the found evidence, the tracker may be able to recreate a past event in imagined detail that increases with skill. The event thus recreated has the advantage over other more intrusive methods in that what is imagined is candid and unadulterated information. Whether a wild animal knows that its tracks and sign can be followed is difficult to say. Not many animals beside the human one are smart enough to lie. I have seen a bear heading toward hibernation in early winter snow dissemble in its trail, but I think that few other animals suspect that a human can come after them and discover secrets about them from the evidence they leave behind. So, rather than actually seeing no more than the rear end of the animal retreating at maximum speed from our view, the interpretive tracker gets to "see" the same animal when it was unaware of observation, behaving in its normal way.

Eco-tracking

This approach to interpretive animal tracking is what I have for some time called "eco-tracking." Ecology is a buzz-word that has been so loosely used that for some it has come to be little more than a synonym for concern for the natural environment. Properly speaking, however, ecology is the study of the relationship of an organism to its environment, or for our purposes, how a wild animal relates to its habitat. Traditionally, science has approached this study quantitatively. For example, to see where an animal goes, a radio collar is attached to an anesthetized animal and then it is followed electronically as it moves around the landscape; to find out what an animal eats, scat is collected and chemically analyzed in a laboratory; or to see how an animal moves, it is captured, harnessed with electronic sensors and obliged to run on a treadmill. All of these measurements result in quantifiable data that can be used to produce theories that can, in turn, be defended against the critical judgment of other academic minds.

While this process results in much useful data, it has certain limitations. First of all, it is very slow to produce useful results. Secondly, much data is collected during experiments that may alter the natural responses of the animal. Field biology is difficult in that if you study the animal in Nature, you may find the study exhausting, difficult to control and expensive in time and money. If you move the behavior into the laboratory for better control, you may create conditions that artificially affect the result.

The difficulties of field biology often mean that researchers tend to concentrate their efforts on what can be studied easily with the tools available today and to leave the more difficult topics to await future technological innovation. The fact that papers written from current methods of data collection will be scrutinized rigorously by other scientists in the process of peer review tends to make researchers careful not to bite off more than they can chew.

Science would say that what cannot be quantified, manipulated mathematically and reproduced by others is useless, that until those criteria are met, one does not really know something in the scientific sense. I certainly don't advocate abandoning scientific rigor. However, this conservative approach also means that one's overview of animal behavior is confined to very narrow sight-lines, sight-lines that might be broadened by a less-well-defined supplementary approach.

Ideally, the scientific process starts with a theory, thought up by the scientist, who then sets about collecting data to support or disprove it. It is in postulating theories that I think the interpretive tracking approach may be useful. By following and interpreting an animal's trail after the fact of its creation, the tracker acquires information about an animal behaving in a natural way, free of collars or sensors or other paraphernalia of data collection. In no way does the tracker interfere with the animal's normal life. (I sometimes wonder how a female reacts when a prospective mate shows up with a man-made goiter around its neck.) Hopefully, advances in the technology of miniaturization will eventually yield sensor units that will be far less intrusive. At the moment, however, most that are in regular use are large and cumbersome.

While an electronic tracking collar will tell the data-collector where the animal is at any given time, it does a rather poor job of telling what the animal is doing at that moment. Once again, advances in videocamera miniaturization may eventually allow the electronic tracker not only to record the animal's location but also to see what the animal is seeing. Other sensor innovations eventually may yield data about what the animal is hearing and smelling as well as its physical condition, the sort of pheromonal data that an animal gives off in its breath and sweat. But such useful technologies are at best somewhere in the future. At the moment the only way of experiencing a little of what the animal experienced is to follow after and sense the surroundings that it encountered.

By itself, eco-tracking will not satisfy scientific scrutiny since what one finds by physically following an animal is seldom effectively quantifiable. Tracking, after all, is more art than science. However, intimate experience with the actual life of the animal may help generate the right questions, may point the scientist in profitable directions, and may generate the theories that data collection by conventional technological methods can then prove or disprove. Furthermore, broadening one's experience beyond the narrow sight lines of one's study may reveal parallels of behavior among different species that also can also provoke testable theories.

Organization of the Book

After this introduction, the next chapter of this book, "The Inner Game," deals with how the tracker thinks. It tries to reveal what habits of mind and eye lead to the discovery of sign and then how the interpretive tracker translates and processes the raw information that lies before him in the snow, mud and sand. What common habits of the human mind should be capitalized upon and which should be countered? How dependent upon measurement should the tracker's efforts at identification be? How helpful are categories and rules and when should the rules be broken? What sort of characteristics and habits of mind allowed the native hunter of ages past to excel at track, trail and sign interpretation? Finally, to what extent can we use that exclusively human talent, the ability to imagine, to enter the mind of the tracked animal, seeing the world as it does and internalizing the feeling of moving through it with a quadruped's body? "The Inner

Game," then, treats tracking from the eyes backward to the mind that actually does the seeing.

The following chapter, "The Outer Game," begins by considering various contexts in which tracking is performed. This section includes a discussion of habitats and surface conditions as well as a brief history of the eastern forest. The reader is then ready to begin interpreting animal tracks, trails and other sign. This phase begins with a discussion of track patterns and the gaits they represent. In conditions where clear footprints are unavailable for identification, the pattern of prints in the snow or mud often can be used instead, since many mammals characteristically use specific gaits in particular situations. Those patterns also can serve to help us visualize the animal in the gaits that those patterns represent. By imagining the animal in motion in its habitat we gain valuable insights about its secret life, insights that supplement and in some cases may contradict the information even in respectable monographs on these animals. In this sense, the advanced tracker is on the cutting edge of knowledge since little attention has been paid in the past to this avenue of investigation.

The process of track and sign identification alone is anything but easy, and every book on the market, even the good ones, must necessarily present some simplification of the process in order to give the person bent on learning to track a start that will not totally discourage him at the outset. This may be a matter of presenting sign in idealized circumstances, free of the many potential confusing details that weather, ground conditions, the action of other animals and so forth may introduce into the frame of reference. This book assumes some prior experience with these difficulties and at least some success at surviving them.

In the fifth chapter, "Putting It Together," we depart from the idealizations, take off the gloves and deal with the messiness not only of some real, complex identifications but, more importantly, with the interpretations that follow. Here the reader is walked through several incidents gleaned from my ancient and tattered field notebook in order to demonstrate the process and show the kinds of often subtle information that can be derived from evidence on the ground. In order to understand this evidence, the chapter includes an explanation of the process of predation, that is, of the brutal but indispensable relationship between predator and prey.

In the sixth chapter of the book, "Species Notes," the reader is provided with additional information of the sort that identification and life history guides have little room for. This is the information that would confuse the beginner by cluttering his clear view of tracks and sign. Here will be found many exceptions to the rules that one learned early on in one's experience about various species of wild animals in North America. No effort has been made here toward encyclopedic coverage, but rather toward including interesting details of evidence that can translate into the interpretation of the sign of the animals generally, including those not specifically dealt with at one point or another elsewhere in the book. The hope is that, in thinking about these interpretations, a sophisticated mental approach toward analyzing evidence will filter though, instilling an habitual nagging doubt about conclusions too easily arrived at.

The seventh chapter, "Applications," provides a number of tracking tips that I have found useful over the years in analyzing what I find in the woods. It then invites the tracker to try his hand at interpreting information found within the frames of a series of photographs selected from my collections. These latter are arranged with increasing difficulty from start to finish and are intended as a sort of graduation exercise where the tracker can test his or her level of skill.

The third part of "Applications" is a mix of information and incident. Three short accounts on tracking mountain lions in Arizona investigate the process of finding and interpreting the sign of this fascinating, almost mythical animal. The book concludes with an effort to look into the mind of one species, the red fox, by the evidence one of its kind left in its trail, and through that evidence to the millennia of experience and ruthless selection that produced that mind. It is, after all, the spirit of the animal that we are really after. All of our efforts are an attempt to experience the world as the animal does, to inhabit its skin for a glimpse or two of an orientation superficially foreign to our own but, in some deeper and perhaps very important way, the same.

By way of whetting your appetite for what is to come, I end this "Introduction" with an account of an incident I experienced in the woods of central Massachusetts a few years ago. Although each imaginative encounter in animal tracking is unique, this one embodies many of the difficulties and rewards that I have found in pursuing the tracking art. It also involves many of the aspects of eco-tracking that I have discussed and so I include it here at the outset.

An Engagement

The French have a military term for the moment when two opposing forces that have been probing and feinting with one another finally come into an irrevocable and lethal embrace from which either or both might wish to escape but cannot. The struggle, once fully engaged, must continue until one or the other gives up. The term is *engage* and it reminds me of a strange incident a few years ago.

On a crisp mid-January day, Kevin Harding and I were tracking in the Federated Forest on a few inches of snow when we cut the trail of a deer bounding and galloping downhill. As this was unusual behavior for a deer in mid-winter, we tracked the animal downhill until we came to a pair of commotions a few yards apart in the snow involving both the tracks of the deer and those of a bobcat. Clearly, the bobcat had made an attempt at taking the deer down but, since the deer's trail led off walking back up the hill it had just descended, the cat obviously was unsuccessful (Figure 1.1). A look around showed the trail of the

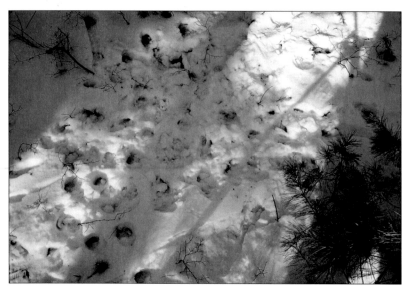

Figure 1.1. A commotion of tracks shows where the cat tried to take down the deer.

Figure 1.2. Between bouts, the cat simply sat and rested while the deer browsed nearby.

bobcat parallel to that of the running deer but off to one side by perhaps 10 meters. The cat too had been bounding and galloping to a confluence of its trail and that of the deer.

Following the deer's walking trail back uphill, we came, in less than a hundred yards, to several more commotions in the snow. Next to each turmoil was a place where the bobcat had sat down, oriented toward the deer (Figure 1.2). The deer for its part had been uninterested in running off but, instead, had taken only a few steps to some stump sprouts where it began browsing. In one of the commotions we could make out the impression of the entire hind foot of the deer, that is, the part of the leg from the hooves up to the "elbow," showing that the bobcat had at one point very nearly brought its victim down. Farther up the hill we backtracked the bounding deer trail to a low promontory with a lone pine tree where there was a deer bed and, around the back of this rise, we once again came upon the tracks of the bobcat, here walking toward the bed.

As we had not discovered all of this in proper sequence, we took a moment to piece it together. The bobcat, a young one by track size, was prowling the hill randomly, searching for something live to eat, when it approached the rise. There a deer, also an undersized one perhaps two years old, had been bedded under the pine, perhaps absorbing radiated warmth from the dark tree. As the cat came over the rise the deer detected its presence and spooked, bounding off down the slope below. This apparently triggered the chase response in the young bobcat. An older animal, we reasoned, would have seen such a pursuit as a waste of energy given the doubtful reward of taking down a healthy deer. In any event, our young bobcat raced after the deer on a parallel course until it closed on the animal at the bottom of the hill and attacked it. There, first one struggle and then a second took place within a few yards of one another. One of the sites had a little blood in the snow, showing that the bobcat had inflicted some injury. From the second altercation the deer headed back up the slope toward its bedding site, but not in the escape gait that one might expect. Rather it simply walked, with the bobcat doing the same behind. Half-way up the slope the bobcat made a couple more attempts to bring down the deer and on one of the two nearly succeeded. After each struggle the bobcat, exhausted at the effort, sat down to rest within a few feet of the deer, which continued to browse the low growth around it. A little farther up the slope, parallel with the original bed, we found one last commotion in the snow from which the deer trail led off one way and the bobcat, having had enough, led off in another. The bobcat was so tired by this time that it neglected to climb up onto and walk along a fallen log as bobcats habitually do (Figure 1.3). Its trail led off over the top of a hill and down while that of the deer headed for some dense pine regrowth. Its trail patterns showed a limping animal, wounded from its several struggles (Figure 1.4). Among the thick pine seedlings it had laid down to lick its wounds (Figure 1.5). More blood was in both of the beds that we discovered in this dense, protective cover. Sometime later the deer rose and walked off, its trail entering the trails of several other deer passing nearby where we lost it.

What about the strange behavior of the deer? Why, after each encounter, had it not tried to flee, instead of calmly turning to browsing any twig end that was handy? Was it daring the bobcat to take its best shot? Or was it, in some strange dance of death, cooperating in its own demise? Does the deer know that it is prey and that in the right order of things it eventually must succumb? A little stunned by the profundity of what we had witnessed after the fact, we reviewed what we had found.

Figure 1.3. The cat was so tired from the unsuccessful attacks that it didn't bother to walk a log along its path.

The deer, once the initial surprise and reflexive attempt at escape was over, no longer tried to flee. Such flight would only have provoked another chase, or perhaps it was already so weakened by its wounds that it was unable to flee effectively. Either way it instead just stood there and waited, attempting to restore its energy with browse. This restoration was not for the moment; it would have to rely on energy already stored for that. Was it instinctively taking the long view and browsing for its later survival?

As far as the behavior of the bobcat was concerned, when an adult bobcat makes a try at a deer, it must either catch it sleeping in a

Figure 1.4. The deer limped away from the scene in an irregular walk with foot drag.

bed or otherwise descend on it from above, from a tree limb on which the cat was crouched for instance. From such a position it can land on the back of the deer where it quickly sinks its teeth into the neck of the animal before it can react and either separates its vertebrae, severs a major blood vessel or suffocates the animals by choking. Only from such an advantageous position would an adult, with the wisdom of several winters, make the attempt. For this reason we decided that this had to be the work of young bobcat that didn't know any better. Trying to take down a healthy deer that is fully aware of your presence and taking it from below, no less, would look like a waste of

Figure 1.5. The deer found some dense pine regeneration in which to lie down and lick its wounds.

energy to an adult, perhaps a lethal waste of energy. Winter is a testing period; all animals are starving, and only the fittest will survive, winnowed down to those with a set of genes selected for appropriateness to the current environmental situation. The chase response had triggered the encounter and once it had begun, the fight had continued long past usefulness or instinctive wisdom. It is as if the bobcat, once engaged, could not cut its losses and let go. Like a poker player who has started betting on a weak hand, the cat had allowed itself to get bid up to potential disaster. It had sat and rested and tried again and again, unwilling to call off the effort. As it had walked off, too exhausted even to mount a log in the woods, I reflected that it may have experienced not only its first, but perhaps its last, winter.

⌒

This is the sort of amazing story written in the snow that keeps me coming back to our evolving art. No human was present to witness this life or death altercation, but it could be reconstructed with just a little skill by the human mind arriving after the fact. And from the reconstruction comes enlightenment or, at the very least, some insightful questions. This melodramatic moment, hidden from direct view, provided a candid look into the natural behavior of the wild animals involved, behavior unaffected by our knowing. This, it seems to me, is far better than "outsight" that one gets by trying to observe the animal directly or by watching a wildlife documentary made with trained or pet animals that are aware of or even habituated to human presence and thus affected by it. The insight of the tracker is the real thing; it is knowledge of the inner lives of wild animals expressed in their outer movements recorded in the snow — the ultimate reward of interpreting the tracks, trails and sign of these creatures.

Chapter II

The Inner Game

Perception: Looking and Seeing

"Tracking is seeing; seeing is done with the mind."

The English language regards "looking" and "seeing" as nearly synonymous. Other languages, however, draw a greater distinction between them and so will I here: looking is done with the eyes, seeing is done with the mind. That is, "looking" is what happens when the color and shape of outside stimuli excite nerve endings in the retina of the eye while "seeing" is what happens after that. Once those nerve endings send electrical impulses to various parts of our brain, an image is formed, more or less corresponding to the reality before our eyes. The operative part of that sentence is "more or less." Once various regions of the brain receive these impulses, they are collated with all the information that is already stored there. Since we all have a more or less different store of pre-received knowledge, this collation of the new image occurs somewhat differently in each of us. We "see" things differently. Much the same can be said of the differences between listening and hearing. But tracking is for the most part a visual art and so we will concentrate on that sense with a nod to the sense of smell.

We are "relearning an ancient art," to use Paul Rezendes' expression, but we are trying to do it with modern means and modes of thought. American Indians, who as Asian immigrants crossed the Bering Land Bridge in ancient times, brought their habits of mind with them. With our Western minds applied to animal tracking, we are trying to figure out today what they knew centuries ago. Indians did not have calipers or computer screens, did not collect and sift data, could not then manipulate groups of similar data in mathematical ways as we do almost reflexively even in our daily lives. They were obliged, by circumstances or inclination, to rely on intuition, the sum of what they had been told and shown by elders and what they had observed and experienced themselves. With this approach they may have made many mistakes in the modern view, but they were also able to do and find out things, perhaps central things, that may not be knowable by modern investigative practices alone. Few of us have the Indians' ethnic background, and none of us have their wise elders to inculcate their practiced skills. But perhaps some of those skills can be rediscovered and blended with our own Western habits of thought to achieve new insights not available to either mode alone.

Tracking is seeing, and visual perception is complicated. It has two basic parts: what happens forward of the retina and what happens behind it. Forward of the eye in the track or trail itself we have to deal with differences among individual animals of a species as well as, with variability in gait and behavior of the animal and, with almost infinite variability of the light in which the sign is observed. Behind the eye we have to deal with the human brain, the organ which actually does the seeing, the eye itself merely serving as the conduit to the mind. Visual acuity and the editing function of the brain presented with too many details as well as the tendency to see what we want or expect or fear to see, these and other factors influence the inner part of perception. Traditionally, seeing is thought of as believing, but any serious study of the more sophisticated aspects of tracking, or any other perception-based activity, must disabuse the reader of this notion at the outset.

Perception: Looking

Animal Perception

If the ultimate goal of eco-tracking is to perceive the world as the animal does and so come to understand the animal's relationship with the habitat in which it lives, then a little space should be devoted to investigating just how animal perception differs from our own.

Animals seem both to "look" and "see" differently from the way we do, that is, not only do their eyes and the arrangements of those eyes differ from ours, but their minds seem to process the information that passes through their eyes differently as well. Many prey animals have eyes positioned on the sides or top of their head. This sacrifices depth perception for a panoramic view of their surroundings. For them it is more important to detect a predator quickly than to judge its distance away. Once the animal has detected something suspicious, it may flee immediately to a nearby burrow, or in other cases it may turn its head to center the threat in whatever narrow cone of binocular vision its eye arrangement affords and then freeze until it identifies the intruder.

Most predators, on the other hand, have eyes with a more substantial area of binocular vision. Being fierce, they are less likely to be preyed upon themselves, and so a panorama of two-dimensional vision is far less important. Thus a face-on portrait of a felid or canid shows both eyes simultaneously. It is more important for them to judge the distance needed for a sudden charge or pounce than to apprehend the area around them that is extraneous to their immediate task.

Binocular vision is useful for range-finding because the predator's mind calculates the difference between the picture from each eye and uses that difference to create a three-dimensional perception. Animals whose eyes are on the side of their head range-find differently. For example, if you watch a kestrel perched on a telephone wire, you will see it bob its head frequently. By so doing it gets to compare the image at the top of the bob with that at the bottom, essentially taking two pictures separated by an instant in time, and then comparing them mentally. Another predator, the eyes of which are arranged farther forward on the face, may do the same thing simultaneously, creating a binocular, 3-dimensional image without moving its head. Owls, which have similar binocular vision to our own, hunt in such low light that they need to abet the binocular range-finding with a circular motion of the head. We actually do a similar thing, although not so noticeably, by moving our head slightly to determine the distance to a far-off object, say a tree standing out from a treeline. I have read that a mountain lion can widen its pupils just before pouncing. Analogous to a camera, the wider the lens opening, the narrower the depth of field, that is, the distance near to far in which an object is in focus. This may seem counterintuitive and the explanation is long, but it works. By widening its pupils while it focuses on its target, the cat can perceive the exact distance to the target since the prey is in sharp focus only in a very narrow zone while the foreground and background are blurred.

Various species of wild animals may have different eye arrangements from us but may also process what they see differently from us as well as from other species. Many mammals, even the clever ones like coyotes, seem to have difficulty distinguishing humans visually from our containers. You are much more likely to see a coyote or other predator out the back window of your house than you will while walking in the woods. Not only does the house serve as a scent shield, but animals seem to have difficulty separating us from an anomalous background, for example, the house itself or the car that we ride in. Many times a bird or mammal will pause and look at me in a stopped vehicle, not fleeing until I step outside and create a separation between myself and the container. Once it sees me as a discreet entity, it recognizes an ancient enemy and flees.

Our form at any moment may also confuse animals that are watching us if it is not the usual one they anticipate. Many times while riding a bike I have been able to approach a wild animal quite closely. Once, on a back road to Eastman, New Hampshire, I saw what appeared to be a cat crouched on the asphalt ahead, looking intently at me. I stopped pedaling and glided closer until the animal decided that, whatever I was, I was coming too close. At that point it turned sideways so that I could see that it hadn't been crouched at all, it simply had the short legs of the weasel family. As it loped off into the roadside brush, I realized it was a female fisher. Friends tell me of a time when they were bicycling in Nova Scotia and rolled up on a bobcat in exactly the same way. It simply sat and looked at them with the curiosity of all cats until one of my friends got off his bike to go through his panniers for a camera. Suddenly seeing my friend as human, the cat was gone in a flash.

I also have approached deer in this way often enough that it has led me to speculate on what finally triggers flight in prey animals. As I roll toward a deer, its first action is to center me in its cone of binocular vision and hold completely still until something about me confirms identification. Two cues seem to be needed before it will flee: any combination of sight and hearing, sight and smell, hearing and smell, etc. Why should an animal wait for a second cue? Wouldn't it be safer to flee at once? Perhaps the answers to these questions lie in energetics. Try to imagine feeding effectively if you were obliged to bound off at every rustle of foliage. All animals have an energy budget in which they must judge whether an expenditure of effort, as in bounding away, is justified by the potential danger. A flighty deer would soon weaken from hunger and become easy prey for the predator, whether that predator was detected at once or not.

Human Perception

If we were to order our senses according to their perceived importance, most of us, I think, would rate sight at the top of the list, followed by hearing, touch and, last of all, smell. In my tracking classes I have noticed that the last thing people think to do when confronted with animal sign is to smell it. Civilized man is for the most part a sight-dependent species, although hearing runs a close second, especially for urban dwellers required to negotiate traffic-clogged streets. As trackers we should be wary of assuming that wild animals are as sight-dependent as ourselves. Actually since much of their hunting is done at best in crepuscular light, the actions of predators are more likely initiated by something smelled or heard rather than seen. The fox doesn't see the vole under the snow and neither does the owl.

Human eyes are arranged in our heads so that we can, without moving our head or eyes, "look" approximately 180 degrees from right to left and about 160 degrees from top to bottom. However, the cone of our binocular vision within this 180 degrees is somewhat smaller. A simple exercise will demonstrate the limit of your binocular vision. With your head fixed directly forward, extend your right arm out to the side. Now close your right eye. You shouldn't be able to see your right hand with your left eye due to interference from the bridge of your nose. Gradually move your right hand forward until you can just see it with your left eye and note its position. Now do the same with your left hand and right eye. The field of view where you can see both hands simultaneously with either eye is your arc of binocular vision, for most humans between 80 and 90 degrees.

Perhaps from the beginning it was important for us to judge the distance needed to cast a weapon accurately, be it a stone, a spear or an arrow. Each of us has a slightly greater or lesser ability in this regard, partly depending on the distance between our eyes and partly on the prominence of the bridge of our nose. The wider the distance between our eyes, the better our perception of distance should be, and the more deeply indented our nose bridge, the wider our binocular field.

Narrower yet than the cone of binocular vision is the part upon which we can acutely focus the mind, perhaps 15 degrees outside of which our mind "sees" with increasing vagueness to the periphery. It is within this smaller cone that the human hunter centers prey and all of us center work that must be done precisely with our hands. Few animals other than humans are required to judge the anticipated flight of a thrown missile and none to perform work as precise, let's say, as surgery.

The design of our eyes, then, would seem to be ideal for the business of tracking wild animals. It may have evolved in early humans partly for that reason, not only to judge accurately the distance and force needed to hurl a projectile but also to follow quarry in the first place to the point where a weapon could be used. Without moving his head the primitive hunter could combine a vague apprehension of nearly half of his surroundings in order to warn him of approaching danger from the side (an individual human is not as fierce as other predators and so needed to be careful about being attacked himself), a narrower cone of binocular vision with which to see objects such as tracks and sign in relief so that they stand out and can be recognized from the background, and a narrower cone yet with which to isolate objects for identification. These three cones (not to be confused with the cones and rods in the eye's retina), moreover, can be shifted around by moving our head to look in different directions, allowing the aboriginal hunter as well as ourselves, although not at one moment, to see three-quarters of the surroundings with one degree of acuity or another. We may not be able to magnify our vision like the hawk, nor can we see especially well in low light as can an owl or a flying squirrel, but as compensation for those handicaps we are able to distinguish color with infinitely more discrimination than they can — witness the 64-bit color range of the modern computer. Furthermore, as I point out here and often elsewhere in this book, we can at least partly make up for debilities incurred by centuries of civilized life by using the preeminent skill that belongs to us alone in the animal kingdom — the ability to imagine.

The Butterfly Search

I often think of the advantages and limitations of my modern mind when I am walking in the woods searching for animal sign. Try as I might to concentrate on the here and now, I often catch myself wandering off mentally and failing to see what is right around me. It is clear that I lack the mind of the aboriginal, and this can be a handicap in attempting to approach the skill of that ancient being in "hunting" for animal sign. To remedy this I have tried various methods in an effort keep myself focused.

As I leave the asphalt, it can take a little while for my mind to leave the concerns of daily civilized life behind and join my body in the woods. I find that it takes me about 15 minutes to a half hour to block out the pending bills, arguments, perceived slights and inconsiderate drivers and to slow my mind down enough to take in the woods around me, to begin seeing, hearing, smelling my new habitat and to begin to tap into that deeper, obscure reality that surrounds me. After that transition period, I am in the right frame of mind for tracking.

As you walk along a trail, you may not realize how much of your attention is focused on the ground in front of you, searching for stable positions for your feet so that you won't stumble or turn an ankle. This wouldn't be a problem for you as a tracker if all you wanted to find was sign right in front of you. But animals leave their sign in other places as well, and keeping your eyes strictly on the ground in front causes you to miss a great deal to either side and above the surface.

I have evolved a search method that addresses both of these problems. Rather than allow my eyes to roam at random over my surroundings, a permission that soon drags my mind back to the civilization I just left behind, I have developed a pattern of looking that requires conscious application to my surroundings. I call this the "butterfly search." First I run my eyes forward and back for 10 yards or so in front of me. This is a track and scat search of the trail ahead. It also allows me subconsciously to memorize the ground ahead for obstacles over which I might stumble so that I do not have to look at it again for the next ten yards. As I move forward, I rotate head and eyes in a broad circle to the left, beginning far to the side and up, then across the canopy to perhaps 15 degrees of the mid-line of my course and then back down the ground at my feet. This is a sign-search in which I am looking for evidence of animal activity above the ground such as browse-nipping, scrape marks on tree bark, nests and natural cavities higher up, anything that fits one or another of the catalogue of search images collected from past experiences and stored in my brain.

If you try this, you will notice that your brain resists processing what passes through your eyes as a continuous sweep. The brain, it seems, is not accustomed to analog motion. It prefers to see digitally, in a sense. So as your eyes sweep through the loops of the search, your mind will insist on seeing a series of still images in rapid succession. During the sweep, your brain is not interested in what is in sharp focus in the central cone of perception, but rather on the total picture including the periphery. Thus, your mind prefers to process this wide field in the same way that the brain of a small bird does. If you watch a chickadee at your bird feeder, you will see that it is constantly jerking its head about in a series of short motions with brief pauses between. During those pauses, it takes a still picture of its surroundings. Most of its vision is monocular because it does not need sharp visual acuity except in the very narrow space just in front of its bill with which to judge the distance to the seam in the sunflower seed. Instead, its wide-angle monocular vision is designed to alert it to the approach of the sharp-shinned hawk while it is feeding on that seed, and a rapid succession of wide-angle, still images provides more safety than a narrower field of binocular vision.

Unless we are in grizzly or cougar country (or in the inner city!), you and I are not searching for the swift approach of a predator but rather for anything in our field of view that might prove interesting. For that purpose, we can adapt our own mode of vision to that of the small bird by moving our head and eyes to increase the field of view and doing so incrementally with the same jerky motion of the bird, allowing time for our mind to process each of the still images in the sweep, comparing it to our mental catalogue of search images or for what other information it may contain. This process, too, will help to slow us down to woods-pace and maintain our attention to our surroundings.

At this point our head and eye motion has described the body and the left wing of a butterfly. Now, once I have finished a wide sweep of these jerky motions to the left, I repeat the up and back trail search and then move head and eyes in a broad circle to the right to complete the shape of the butterfly. Because I must consciously move head and eyes according to a prescribed pattern, my mind tends to better maintain its attention on the surroundings, rather than wandering off to imaginary discussions with potential publishers or what I should have said at the last selectmen's meeting.

Because the butterfly search forces us to look away from our feet, we are apt to be uncomfortable with this method at first since we fear stumbling over tree roots and stones while our eyes are sweeping across the canopy on either side. This reluctancc can be countered in several ways. At first we may restrict these butterfly searches to level, even terrain to gain confidence in allowing our eyes to leave the ground. Secondly, bend the knees. Our normal walking stride tends to lock the knee straight as it passes under us; this is efficient since the muscles rest during this bone-support phase of our stride cycle. However, this locked-knee posture gives no way of countering unexpected irregularities underfoot. By bending our knees we can counter rises and dips as well as unexpected obstacles without looking directly at them. This is the posture of the skier, whose neutral bent-knee posture allows his legs to compress for bumps and extend for dips. At first, walking on the forest floor in this posture will be tiring, but after a while the muscles of the thighs strengthen and allow one to continue in this posture for longer and longer periods. This was probably the usual posture of the primitive human while hunting, one that allowed him to search his surroundings constantly while moving through productive or dangerous habitat.

Another aid to this searching method is the use of hiking poles. With one in each hand, ready to catch a stumble or fall to either side, you may find yourself more willing to allow your eyes to leave the trail for longer periods. In rough terrain I find the poles or at least a hiking staff (I often use a monopod for this as it serves also to steady photos of what I find.) to be indispensable in freeing my mind and eyes from what is immediately in front of me.

I often time my incremental head-eye movements according to my pace. Walking along a smooth dirt road fairly rapidly, I may coordinate each motion with every two to four steps. If I am on a hiking trail, where I am likely to move more slowly, I may time the stops for every two steps and for off-trail movement every step. There, where I am moving much more slowly, I use the bent-knee posture, raising my feet higher and sometimes interrupting a visual sweep in order to look at obstacles in front of me before going back to the sweep at where I left off.

If I still find my mind wandering, I may mentally talk to myself about what I am seeing during the sweeps. I may verbalize the dead tree with pileated cavities off to the right, the red fox berry scat on the trail in front, the raptor nest high in the canopy on the left, and so forth.

27

These verbalizations crowd out the imagined conversations that are ready to creep in and take over my consciousness. Even at that, there are days when I still wander off to reverie. I excuse myself by insisting that such failures are the mark of an imaginative mind and then bring myself back to the present, continuing the verbalized search.

A Catalogue of Search Images

Another cause of the wandering mind is the lack of a pertinent search image or images, that is, a mental picture or pictures that the mind has been pre-conditioned to notice in a confusing background to the exclusion of other detail. Confronted with a surface of infinite complexity, the mind may be daunted by all that detail and may escape from it by slipping into reverie. This is where experience is critical. The veteran tracker has spent so much time tracking and finding, that he has assembled an experiential library of search images in the back of his head that he can apply to any new scene. He may have thousands of these images available for instant recall. When he surveys a scene that bears a resemblance to some past experience, it pops the key elements of the scene into view in his mind's eye. The vast, dull monochrome of green suddenly acquires texture as some things stand out to the mind while other things recede. This is how the experienced tracker finds sign that the amazed novice looked right past or through. As is often said, there is no substitute for experience.

Or is there? The most dedicated among us spend much less time "tracking" than any wild animal, which is constantly racing away from imminent, violent destruction and toward food and reproduction. Busy people with careers and family to attend to do not have unlimited time to spend in the woods, time that is needed to assemble in the back of their mind a library of search images with which to compare future scenes. However, we can use our large brain, and specifically the remarkable human ability to imagine. At home before bed, or during a long automobile trip (with someone else driving!) or train/plane ride, pore over the books and try to imagine the described scenes as vividly as possible. The resulting images, stored in your brain, may not be as vivid as field experience, but they are better than nothing, and vivid or not, they are there, ready for use on the occasions when you need them. Don't worry about remembering them; once you imagine something, it is stored in your subconscious whether you want it there or not. Remember that intuition is the *unconscious* application of everything we know to the scene at hand. Imagine and then allow yourself to "forget;" the image will still be there, ready to inform your intuitions.

Whatever the source, real or depicted, an accumulation of more and more of these mental images traces the development from novice to expert. It may seem amazing to the novice that an experienced tracker can see what almost isn't there, that is, sign so subtle that he himself walks past it without noticing. And it may appear even more magical that the expert can then identify the animal and perhaps tell what it was doing from the slightest traces of evidence. Much of this skill is simply a matter of having seen it and solved its puzzles before and having, imbedded in a large fund of memories, something enough like the matter at hand that the similarity jumps out of those memories to his attention, where without such a "search image" he also might notice nothing.

While a single obscure print in snow or dirt may not result in identification, a succession of obscure images may. The tracker may take from each a detail or two and so build up a composite that exists only in his mind. This composite in the ether of his imagination can then be compared to his mental catalogue of search images, resulting in an identification that could not have been made from any one of the actual prints.

The Other Senses

Wild animals are much more dependent on the sense of smell than are we. Just look at the paucity of words in the English language for odors. We must resort to elaborate circumlocutions to describe an odd smell, usually comparing it to something more familiar but not quite the same. When I kneel down to smell a scent mark on a fallen log lying beside a road, I may say that it smells like "sweet skunk" or for a coyote "mild skunk mixed with burnt hair." Or I may resort to a metaphor across the senses. For the urine mark of a fisher or gray fox I use the phrase "chemically treated paperboard" and explain that it is an odor related to a taste I have preserved from my childhood when I apparently chewed, as children will, on the back of a clipboard or some such. Generally all of these descriptions are greeted with incomprehension by my associates. This is not the case with either sight or hearing for which we are richly endowed with language that can fix the character of something seen or heard with a single word and communicate it fairly reliably to the mind of others.

The ability of a wild animal to use scent to detect prey, danger or a potential mate seems phenomenal to us. Not only can many animals identify using odor, but can also evaluate what they identify, whether it is toxicity in a plant, the breeding condition of a potential mate or physical vulnerability of prey due to illness or injury.

We are animals, of course, but it is often helpful in understanding the behavior of wildlife to try to imagine the differences between them and us. This may be a matter of imagining ourselves as a quadruped with a very different body and form of locomotion. It may also be a matter of imagining what it would be like to have preternatural senses of hearing and smell. If we perceived the world as wild animals do, as a world of odors more than visual images, then the wisdom of the Indian sweat lodge makes perfect sense. Natives who prepared for the hunt with a ritual bath of this sort cleansed themselves of odors that prey animals might detect. Success in the hunt may have given the ritual religious or spiritual significance to the Indian, but the mechanics are forthright. Exactly like the deep cleansing of a Finnish sauna, a sweat lodge removed, with the flow of sweat, bacterially active oil from deep within the skin, and with it the odor that would help to identify the hunter as human. In effect he cleansed himself from the inside out rather than, as we do, simply washing off the surface oils with soap in the shower. Back in the days when I trained with a Finnish ski club, we always adjourned to the sauna for a good sweat at the end of a workout. Afterwards in the woods I discovered that I could walk up on deer without being identified. The deer might start and run a short distance, but rather than wasting any more energy, it would stop and look back, trying to get a second cue of scent, sound or movement to identify the disturbance.

Wild animals have scent glands all over their bodies, even between their toes, with which their passing is recorded. For most of us these odors are locked away, beyond our ability to apprehend them. Just as we can see neither into the ultra-violet or infra-red zones of light, we cannot hear or smell into the ranges that are available to most wild animals. However, even with our limited range of smells, we can still tell instantly the difference between the disgusting odor of dog poop and the mild, almost sweet smell of wild carnivore scat. And we can certainly detect the herbal odor of deer and moose urine in a winter trail as well as the turpentine reek of a porcupine's den. Positioning our nose close to the ground in a recently vacated lay site may yield the odor of wet fur from its recent occupant, and crushing bear scat in late summer will often give up the winey and recognizable smell of congealed berry juice. For most of the rest we must surrender to our human limitations. Just as an otter or a mink sees poorly above water and relies on acute hearing and smell to detect danger, we must generally face up to the fact that we are a sight-dependent species, using our other senses where we can.

Measurements

In seeking to describe what we find in the woods, the western mind tries to quantify. Measuring things like track width, step length, pattern width and straddle not only help to identify an animal but also to gain insight into its gait and behavior. To this end we will spend a few pages discussing both the advantages of certain kinds of measuring and, more importantly, the pitfalls associated with an over-reliance on them.

Most things in tracking have soft edges, including the tracks of many animals themselves. Rather than standing out discreetly, the footprints of soft-padded animals blend into the background in which they were impressed. In a sense, such a track is shaped like an old-fashioned bathtub, and where should you measure a bathtub? At the bottom? If so, exactly where at the bottom — at the beginning of the upward curve or at its top? Or perhaps we should measure it at the rim of the tub. But again, exactly where on the curve of the rim? An animal track blends into the background with just such imprecision. In preparing the illustrations to my identification guide, I initially (this was before the age of digitization) projected a photograph of the track on a sheet of paper taped to a wall. Then I would draw the outline of the track on the paper. Again and again I had to pause and step back to decide where the edge of the track "really" was, or better, where the human eye of the tracker using the illustration would perceive it to be. The process of science requires precise measurement, but precise measurement, in this case at least, is what Nature denies. Added to the vagueness of tracks are the difficulties of measuring accurately in the field, encumbered with gear and with snow melt trickling down the back of one's neck. Often I have taken measurements of a track and then cast it with plaster. Later in my study, in comfort and with better light and tools, I would find the initial measurements in my field notebook to be off by perhaps an eighth of an inch or more from the cast impression.

The bottom line in this is — be wary of precise measurements given in books and of your own measurements in the field. Track identifications are more often made on the sum of evidence or on a mental impression than on a single imprecise measurement. And while measuring a track, watch yourself. Do you want it to measure to bobcat? And is that wish affecting your choice as to where to delineate the track? My old friend, Paul, was a brilliant tracker with excellent intuitions but with, in my opinion, an excess of faith in measurement. Many times I would watch him unwittingly measure to his expectation. He knew from long experience what he was looking at, but he always insisted on measuring in order to provide some quantifiable evidence for his judgment. In measuring something small and with soft edges, there was lots of room for measuring to the dimensions of the animal that he already knew it was.

I certainly don't mean to suggest that measuring is never useful. Indeed, a tape measure is a well-used item in my own tracking kit. In the case of tracking in deep, soft snow, for instance, or for old, melted-out tracks, measurement of step length and straddle may be most of the evidence available. And a step length of 15 inches in a walking gait of direct registrations with a very narrow straddle remains the best way to identify a walking red fox trail when track details are absent. I only suggest that over-dependence on it is unwise. Identifications are made on the sum of the evidence and measurements are only one part of that. If you must rely solely on measurement, you will probably get an ambiguous answer. A recent incident illustrates.

I was training a group of docents for a land trust in the North Country in nearly two feet of powdery, unconsolidated snow. Most animals were plunging right to the bottom, leaving only the entry/exit holes of their feet in their trails. Early on, the participants learned to identify fishers by their diagonal bound patterns and deer by step length, visibly wide straddle and key-hole appearance of the holes in the snow. After that we developed ways to distinguish deer from coyote trails by the latter's narrow straddle, consistent step length and reliance on the snowshoe trails left by previous human visitors. Then we came upon a line of holes that measured between 15 and 19 inches with a wide straddle. Was this a deer taking abnormally small steps? Certainly a deer could do this just as a human might in deep snow. The group set to work. They noticed that the longer step length tended to be associated with places where the animal was descending a slope, so that the 15 inch steps were more representative. Still measurement could not rule out either a small deer or one that may have been shortening its steps in deep snow. Finally, one of the group took a mental step backward and looked not at the snow but above it. At one point the animal had passed beneath a branch that was only a foot and a half off the surface. No walking deer could do that without the patterns showing an obvious break in the stride. Here a combination of measurement and a contextual clue identified the animal as a bobcat where measurement alone provided only ambiguity.

"Measurements are a beginner's training wheels," so says my friend and fellow tracker Kevin Harding. Without them the novice might never get started on the road to competence. But there comes a time when they must be questioned and our dependence on them at least partially abandoned in a transition to educated intuition. Intuition, as I suggested in the previous section, is the simultaneous and unconscious application of everything one knows to the problem at hand. The more experience, the better the intuitions. Selectively abandoning a dependence on measurement means unlearning some of the things that the novice was taught at the outset and of which he had hitherto felt certain. This is the natural process of education, the process of being "led out" of one frame of mind and into another. Training amounts to repeating what one is told; true education means acquiring the experience and mental tools to evaluate what one is told for oneself.

Statistical Tricks

Another aspect of measuring deals with relating statistical measurements to the real world. For example, gray foxes tend to be dimorphic, with the male somewhat larger than the female. You may read that the average step length for this species is 11 inches. But this is a useless and misleading statistic. Because of their dimorphism, the walking step lengths for this fox tend to surround either 9 or 13 inches with a narrow distribution of about an inch around each. An average of the two will give you the 11 inch measurement, but this is an unlikely step length for this species, at least on a firm, even surface. Thus, one should be suspicious of averages that do not have information about the distribution around that average.

Another statistic I often hear is that weasels have an extremely short life-span averaging about a year and a half. This might suggest that all weasels are in a desperate race to procreate before they die, which is partly true, and furthermore that this is a statistic peculiar to weasels. Actually, the highest period of mortality for any wild animal, including weasels, occurs in the first few months of life. On the other hand, if a young animal manages to survive its first year, as a few do, it has enough time to gain lessons in life that may sustain it for, in the case of weasels, several more years. As it matures, it becomes a dominant member of its species, out-competing younger rivals and so becoming responsible for much of the procreation of that species in its area. In this way the statistics are misleading, being heavily skewed by early mortality while minimizing the important survival of the few strongest or shrewdest of the species.

Over the course of many years of measuring animal tracks and gaits, I have found that outside parameters tend to spread as I find more and more exceptions to the rule. Eventually they get so wide that they become useless for distinguishing one species from another similar one. Accumulation of such statistics may be admirable scholarship, but not very useful for the tracker in the field. Instead, averages that cite the amount of distribution around the average are much more helpful. And if the distribution is very wide, so wide as to be useless for either identifying or interpreting the animal's sign, then it might do well to reduce statistical clutter by leaving them out of accounts altogether.

Light and Memory

Since tracking is seeing (and remembering where we saw it), it may be worth our while to explore both of these facets of the art.

What we "see" is light reflected from objects, with various colors representing different frequencies of light waves. In the field, the source of that light is usually direct sunlight. It may also be reflected from the energy of the sun indirectly, through that which is stored in a flashlight battery or re-reflected from the surface of the moon.

Whatever the source, as light strikes an object, certain frequencies of color are absorbed and others are reflected to our eyes. Ironically, although we relate the color green to life, this is the frequency that the live leaf rejects, reflecting it back to our eyes. Different amplitudes of light are also either absorbed or reflected, creating color shade, highlight and shadow. Between its origin in combustion on the surface of our star many things can happen to light before it bounces, finally, off the surface of the track or sign and enters the aperture in our eye. Among them is the modification of light by atmospheric interference such as weather and time of day.

Changes of light both reveal and conceal. The low angle of light in the early morning or late afternoon may cast into relief a shallow track that was invisible at midday with sun directly overhead. Any sunlight that creates shadow may make any sign so obvious that the tracker may be fooled into thinking that he can return at leisure to photograph or otherwise collect it. Returning later in the dull light of an overcast sky, he may find to his chagrin that the sign has "disappeared." Sometimes changes of light may even alter the appearance of the site so that the tracker can't even find the approximate location of the sign. A scene recorded in memory under gray skies may look remarkably different from its appearance at another time, for instance under the green dapple of filtered sunlight.

Memory itself is suspect. Several winters ago I returned to a boggy seep intending to photograph a collection of fisher scat that Kevin and I had found two days earlier. Two days is not a long time, but as I retraced our footprints in the snow, the order in which I encountered things seemed out of the sequence in my memory. Backtracking our exit trail I found that it left at the location of a peculiar dig the female fisher had made at the base of a live tree. But I remembered this dig as having been encountered earlier in the afternoon before our extensive search of the seep had really begun. Around and around the seep I tromped, tracking our old footsteps, looking for a commotion of human prints where we had examined the scat two days before. After a couple of hours all I could find were my own fresh prints; apparently I had followed every trail but the right one, found everything of note that we had found before, except the scat pile. A streamer or GPS should always be part of one's tracking kit.

Point of View

What we see is not only governed by our mental perspective, but also by our physical one. Point of view can mean literally the physical point from which we view a scene. On one occasion years ago I pointed out a deer track to a client and asked her to identify it. I was standing behind the track facing in the direction of the animal's travel while she stood opposite me. Where I saw a deer, she saw a duck! For a moment I was astonished at her mistake until I stepped around to her perspective and looked again at the track. From her point of view it did, indeed, look like that of a duck.

Figures 2.1 and 2.2 present the opposite confusion. In this case a duck's print that could easily be mistaken for that of a deer depending on where the observer stands.

Varying one's point of view by circling to view at different angles is also helpful for sorting out an obscure commotion of tracks (see page 399). For some probably anthropological reason we tend to distinguish

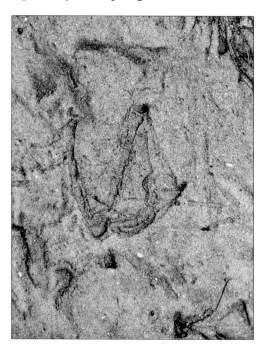

Figure 2.1. In this photo of a duck print the point of view is from behind the bird.

an animal's prints from the background more readily if we are facing in the direction in which the animal was heading. This can be startling the first time you experience it. I recall my own first experience many years ago looking at a commotion in the mud down below Gate 37 at Quabbin. I was examining a seasonal isthmus between the shore and an island where I suspected wild animals might cross. At first all I could see was roiled mud. Moving around it, however, I got to the point where I was facing in the same direction as the coyote had while crossing the isthmus and there, suddenly, two of its tracks in a trot pattern jumped out of the background

The sudden appearance of animal sign where one might have been convinced there was nothing is something a tracker experiences often. Usually this sign only appears after one ponders a particular scene for some time. I can remember many times suddenly realizing that I was looking right at a scat that I had at first not seen at all. This seems almost magical and reminds me of something I read years ago. A photographer who ventured west to the frontier to document the

Figure 2.2. Viewed from the opposite side the same print might easily be mistaken for a deer track.

Indians tribes before they were contaminated by civilization described his surprise when an Indian silently and suddenly appeared before him seemingly out of thin air. I had a similar experience years ago while sitting in a tree stand with a camera. After a couple of hours I noticed what looked like a brown barrel in a field only a hundred feet from my position. Why hadn't I noticed it before? And then the barrel moved, and I realized that it was the torso of a deer. How could it have got there without my notice? The answer is that we see with our mind, not our eyes, and our mind is a complex, crowded and clouded emulsion of perceptions that we must see through as much as with. Sign may appear out of the background if only we look carefully and with a naked mind, as free of expectations as we can make it. If the mind decides that there is nothing there, it will see nothing. And careful examination even of found sign is a virtue for the tracker. Evidence as tiny as carrying notches on an acorn that one must detect with a loupe can result in identification of its transporter and interpretation of its behavior at the time.

Perception: Feet and Tracks

It may be useful at this point between discussions of "Looking and Seeing" to spend a few pages on what we are looking at and trying to see with our mind, specifically the relationship between the foot of a wild animal and the track it leaves behind. I think it is helpful for any tracker, in preparing to "see" animal sign, to carefully consider the relationship between an animal's foot and how that animal lives, that is, how the foot is tailored to the animal's lifestyle. A foot cannot be all things to all animals. Its shape, its "morphology" if you will, both enables and limits the activities of the animal. What helps the animal in one respect may necessarily exclude facility in another. The relatively shallow but tough and insensitive pads of a coyote's foot permit it to travel long distances daily over rough, abrasive ground, but the necessary insensitivity of those same pads limits the animal's stalking ability, and their shallowness limits its ability to climb up to and jump down from heights. The deep, rubbery pads of a bobcat, on the other hand, are designed to absorb the impact of jumping down, and their relative sensitivity allows the cat to detect the twig that will break underfoot during a stalk. However, those same characteristics make the bobcat a less mobile hunter than a coyote, which spends less time stalking than traveling, creating coincidences between itself and potential prey.

The structure of a predator's foot can predict how it hunts and kills. In a chase a bobcat snags its prey with its claws and pulls it back into its mouth for the mortal bite. To do this it must be able to rotate its foot (remembering that the cat's foot extends all the way up to what we may think of as the animal's elbow) in order to snag a fleeing animal by the fur on its flank. The feet of a wolf, by contrast, are relatively rigid, at least rotationally, forcing the animal to catch and kill with its mouth, its feet being used chiefly for getting to the prey and then, with small prey at least, pinning it for the killing bite.

Mammals that swim for a living, like otters and minks, show five well-expressed pads on their feet, arranged in a shallow arc at the anterior of the track. Such an arrangement provides more paddling surface than is the case for animals for which swimming is less important. The latter usually have their fifth medial toe recessed farther up the foot where it is useless for propulsion in water. For a bobcat, with only four registering toes, this structural simplicity favors stealth — a smaller foot is quieter than a large one because it encounters less terrain resistance as it travels forward and makes less noise as it is placed down.

In comparing rabbits to squirrels you may note the tendency for the rabbit's feet to register relentlessly parallel to the direction of travel while the squirrel's often splay outward. The former reflects a stiff foot-structure adapted to straight-ahead speed, while the latter shows an animal with the flexibility needed for climbing up and down tree trunks. Indeed, a squirrel, like other tree-climbers, can rotate its hind feet backwards for grip when climbing down.

Long, fully exposed claws on the front feet of skunks and grizzly bears indicate an animal that digs for a living, while the plantigrade prints they leave behind show creatures whose defenses are so fearsome, each in its own way, that they have little need for a sprinter's speed.

The hard border of a deer's hoof combined with its rubbery central area shows a foot designed for traction on all surfaces. The hard edge allows the animal to dig into soft surfaces as well as to "edge," as alpinists call it, on steep rock. The rubbery center, on the other hand, provides grip through friction on firm, dry surfaces as well as cushioning on impact as the deer descends from its impossibly high, graceful bounds.

Fur-covered pads on an animal's feet are an adaptation to life on both cold and hot surfaces. The red fox has such insulation whether it is found in the sub-Arctic or, as is its close relative, the kit fox, in the Southwestern desert. Lagomorphs, that is, rabbits and hares, also have fur-covered feet, the better to stay still in one spot in the snow to feed for a length of time without being noticed. Another adaptation to travel on snow is large feet. The lynx and its principal prey, the snowshoe hare, both have enormous feet for their size, the one to pursue over snow and the other to flee that pursuit. Even the red fox has larger feet for its size and weight than does its cousin, the coyote. This relative difference between the two supports a boreal origin for one and a temperate origin for the other. Those large fox feet not only function as snowshoes but also help in pinning a vole under snow at the end of a pounce where its exact location under the surface may be uncertain.

Pad Deformation: The Protean Bobcat

Many times in my life as a tracker I have had the experience of trailing a small bobcat only to have it turn magically into a large one or vice versa. How does this happen? Over the years I have developed a theory that seems to hold up to application. A cat has deep globular pads that, uncompressed, form a cross-sectional shape as shown in the left hand figure in Diagram 2.1A. Being deep and rubbery this shape can be distorted by variations both in the animal's gait and the surface in which it is impressed. High-impact gaits such as jumps, bounds and lope-bounds tend to flatten the pad, making it appear large, while at the other extreme a flat gait such as a walk, having little impact, allows the pad to retain more of its unloaded shape. The impact of a high-angle gait also tends to spread the toes so that not only do the pads seem larger but the width of the entire print also expands.

However, the story doesn't end there. A track has an external component that also affects its shape: the surface in which it is expressed. If the surface over which a high-angle gait is used is firm, the impact of contact allows the pad to spread more than it would in a softer surface. In this case there is nothing to prevent horizontal expansion of the rubbery pad as in Diagram 2.1B. However, on a soft surface another effect is involved. As the ball of the pad descends into soft mud, for instance, it initially meets little resistance and so there is little deformation. By the time the animal's full weight is brought to a

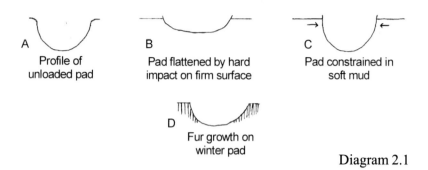

A — Profile of unloaded pad

B — Pad flattened by hard impact on firm surface

C — Pad constrained in soft mud

D — Fur growth on winter pad

Diagram 2.1

stop, deeper in the mud, the walls of the track impression prevent a lot of horizontal spreading, as shown in Diagram 2.1C, so the print seems relatively smaller than would be the case if the surface were firm, even though the animal is performing the same action.

The genesis for this theory was an incident on the Saugus River one dry September. A bobcat had jumped down from the high bank onto a mud flat that was six feet wide, exposed by the recession of the stream. Having landed high up on the flat where the mud was firmer, the cat's pads flattened and spread (Figure 2.3). The animal then took a couple of steps to the water's edge to drink. Disliking mud on its feet (it is a cat, after all!), it did not walk back up to the firmer mud before leaping back up onto the bank but rather vaulted back up forcefully from its position on the softer mud near the water. One could hardly believe that the tracks from jumping down on a firm surface and those from jumping back up off a soft surface were made by the same animal. Both were strong applications of downward force, but in the jump down the pads had been allowed to spread, absorbing impact, while on the vault back up from a soft surface, the mud forming the sides of each pad-mark as well as of the boundary of the entire track had confined their spread, making the tracks (and the cat that made them) seem much smaller. None of this was immediately apparent in the field while looking down into the tracks themselves, but only became so when I examined the three-dimensional plaster cast that I made of each of the tracks. The larger tracks showed the pads as shallow, flat and wide, while the smaller ones from the softer mud are closer to the deep, globular shape of an unloaded foot. The photos show both top and side views (Figure 2.4) of these prints. Note that claws that registered in the smaller track made it look even less like

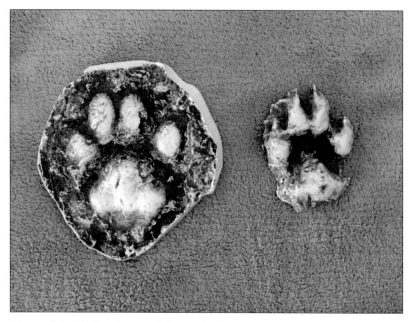

Figure 2.3. A bobcat's spread front print is on the left; it's compact hind print is on the right.

the larger one. These claws were extended for traction to support the vault up and forward.

A last detail that can affect the size of a bobcat's track is the fur on its pads. Examination of a road-killed bobcat showed that the fur on its foot doesn't just grow out of the inter-digital spaces but out of the pads themselves. In an inverted pad such fur gets sparser the higher up the pad you go until at the top the skin is bare (Diagram 2.1D). Think of a rocky outcrop: the trees down low around its base are tall while the farther up the slope you go, the smaller they become until finally at the summit there is a small area devoid of growth. Now revert the foot and press it down into various surfaces. Especially in snow this growth around the bare area disguises the exact outline of the pad, making the furless space seem like the impression of the entire pad and the four of them plus the secondary pad seem like the impression of the whole foot. The disturbed snow from the fur growing out of and around the pad is easily dismissed as peripheral fur rather than part of the "track," and so, in what the mind sees, the actual size of the foot and its pads is underestimated.

Figure 2.4. The two casts in profile show how impact on hard mud spread and flattened the pads while soft mud contained their spread.

Another example of the advantages of casting (page 500) has to do with a bobcat's hypothetical "little toe." In a photograph of a track, the lateral toe of a bobcat may appear to be smaller than the others. Indeed, with a micrometer and a dead animal's foot you may detect some small difference in size between this pad and the medial toe. An examination of several three-dimensional track casts, however, shows that this seeming smallness, visible in some photographs, may be more a function of weight distribution than actual pad size. Because bobcats have more straddle in their walking gait than foxes or coyotes, the medial side of the foot tends to be more heavily weighted, being closer to the centerline of the animal's body, and so tends to spread horizontally more than the lateral toe, which is well off this line and thus relatively lightly weighted. The deeper impressions of the medial toe in several of my bobcat casts tend to support this conclusion.

I have used some space discussing this relationship of feet to tracks in order to show the subtleties that can be part of interpretation. It is worth keeping in mind the forces that apply to the tracks of all animals, from the shallow-padded canids to the more deformable pads of cats. One should also keep these in mind when viewing illustrations of animal tracks made from the feet of dead animals, either by casting them or "foot-printing" them. A live, weight-bearing foot under muscular control and subject to the vagaries of natural surfaces can have both obvious and subtle differences from those of a dead and unloaded one. A foot is not a track.

Perception: Seeing

It may be helpful at this point to extend the discussion of seeing with which the chapter began into exploring some of the differences between our perceptive ability and that of the aboriginal hunter who was much better in many ways at our art than we can hope to be. Perhaps there are ways in which we can use our own talents to compensate for the deficiencies of our over-civilized brains. We will follow that with a discussion of some habits of mind that are typical of our modern species, how to combat them and how to capitalize on them in pursuing the tracker's art.

The Primitive Mind

I think it likely that the primitive mind was selected for qualities other than those of our own. Aboriginal man lived in a close and brutal relationship with raw Nature that demanded close attention to his surrounding at all times by way of finding food and avoiding danger. He could not afford to allow his mind to wander. But the wandering of the human mind is precisely the quality that resulted in civilization. The leisure to indulge one's curiosity about how things worked led to invention. Invention led, in turn, to more leisure time for thinking. All of this was permitted by the division of labor and the accumulation of wealth. (Could Darwin have thought up the theory of evolution without a country estate and a lack of need for a paycheck?) The western mind looks at a process, imagines it differently and then invents a substitute. What then of the North American Indian who failed even to imagine the benefits of a wheel? It was not a lack of intelligence or laziness as was often assumed by the early colonists. Rather, the tendency to allow the mind to wander to imagining easier ways may have been ruthlessly selected out of his genes. If Nanook had been in the habit of reverie, he might not have noticed the slight movement of the feather stretched across the tangs of a fish bone set in the snow next to the seal's breathe hole and in front of which he had been standing for hours in the cold. He would have failed in the hunt; he and his family would have starved and the tendency toward reverie would have died with him. Instead, he and his fellows had to be constantly "right there," constantly focused in the present. The result would be the avoidance of danger and success in the hunt. He and his family would survive, and the tendency of immediate focus would be preserved in the genetics of the line at the expense of the wandering mind.

In contrast, most of us have very modern brains, selected and conditioned by several centuries of life since the onset of the Industrial Revolution. We cannot realistically hope to match the focus and attention to the detail of our surroundings that characterize the primitive mind. But for the modern human who is intent on becoming a competent tracker, some of the habits and tools of the modern approach can at least partially compensate for the loss of primitive attention. The aboriginal, for all his skills, did not have a computer or even a tape measure, could not write down what he observed for later contemplation and correlation with other data since he had no pencil, paper or even the ability to write. He lacked our mobility and the breadth of experience that modern mobility makes available for comparison. His knowledge of the past was restricted to oral tradition; he had no books, bookstores or libraries that represent the collective memory of one's culture. His constant fear of ghosts and evil spirits closed off avenues of knowledge that might otherwise have opened. And he seemed more than ready to attribute anything untoward to the activities of those spirits rather than to seek plausible explanations. Our hope, then, is to recover enough of the aboriginal's skill that, blended with our own modes of thought and resources, will result in opening a way not otherwise available into the natural world and into the secret lives of wild animals that live all around us but hide their lives from us in foliage and the night.

Selective Attention: The Editing Function of the Brain

A person who is new to tracking when asked to look at a natural scene may be overwhelmed by it. Having no editing principle with which to sift through the myriad detail, his mind reduces what is viewed to a few large features and then blacks out the rest. So he may notice a tree perhaps, some low brush, a boulder. But after some experience has tutored his perceptions, he may see in the same scene scratch marks on the bark of the tree, a vague animal trail through the brush, and a small scat deposited on top of the boulder. In order to notice these things, however, he may have to miss others. The intense concentration on visual detail may cause his mind to ignore the song of a scarlet tanager singing in the canopy and only slightly notice the delicate bloom of a wood anemone on the forest floor. Unfortunately, that is the price to be paid by a mind only capable of entertaining so much detail at one time. Our perception of Nature grows both deeper but also at once more restrictive, altering our view of it. Mark Twain, in *Life on the Mississippi*, noted with some regret that in training to be a riverboat pilot his perception of the river changed. No longer did the bony snag protruding from the river lend itself to picturesque metaphor; now it was a warning of a deadly sand bar hidden under the surface. So it must be, apparently; something is gained but also something is lost.

In any track there are virtually infinite details. To deal with them all, the mind edits them, sorts through them, emphasizing the "significant" ones and ignoring the rest, so that finally we arrive at a manageable set of details. However, each of us has his own idea about what is significant. My mind may edit out one set of details that I subconsciously regard as unimportant, while your mind may ignore a different set. The result is that the same track will look different to you than it does to me.

Furthermore, we have to remind ourselves that since we are seeing the track with the mind, not the eye, and that the mind may have de-emphasized many potentially significant details, we see what we expect to see, we see what we want to see, we see what our previous experience has conditioned us to see, and the rest is neglected. Knowing this about ourselves at least makes us aware that we may be missing important details that we can go back and find, pushing our minds "out of the box," as the currently popular expression goes, to look at things in new ways. It also reminds us that there is no track beyond the perception of it, and that there are as many versions of it as there are perceiving minds.

The education of a tracker is the process of artificially tutoring the mind as to what details are personally more important than others. There is nothing very precise about this. Some details that the expert tracker views as significant may simply not work especially well for the journeyman, or some "gestalt," some assemblage of details, creates an appearance that only his mind perceives or is capable of perceiving. To the expert this may seem like a mistake, but it may not be, at least to the mind of another observer. In other words, if it works reliably, use it, regardless of whether it works for other people.

Suggestibility

I recall once coming upon some canid prints on a road down in the Federated Forest. "Look at the gray fox tracks!" my friend announced. As suggestible as the next person, I looked at them and saw gray fox tracks. A few minutes later a man came around the bend returning on his walk with two small dogs and embarrassed us both. Because perception has the two parts described above, it is not enough merely to look at a track, trail or other sign of an animal. Part of our mind must always watch ourselves watching, so to speak, vigilant to subjective influences on what we see. So much of our attention is directed toward the site, the act of observing requires so much concentration on what is before our eyes that we can easily forget this need for self-perspective. As in other complex activities, the process of learning to track is often the process of learning ourselves. I do not mean this in any spiritual or mystical way. This is simply the practical business of learning our own habits of mind. Where do we often make mistakes, and what sort of mistakes are they? Perhaps we are suggestible, or tend too rapidly to jump to conclusions, or tend to view sign as having been made in the same conditions in which it is observed. Learning these habits allows us to watch ourselves more carefully from the back of our mind.

Experience and Practice

Experience it the best teacher, it is often said, but, I would add, only if we pay attention to our experience. Unexamined experience is useless. And a great deal of learning from experience is realizing how we arrived in the past at right and wrong decisions so that we can gradually assemble our own unique kit of evaluative tools.

To stay on top of your tracking game, you need a lot of experience, but you also need practice. Self-knowledge fades with disuse. Keep mental notes as you watch yourself watching, and remind yourself of your tendencies from time to time. I learned this lesson years ago after puzzling over some tracks in the dry sand of the Moore Island channel. My first impression was that they were bobcat, a small female that I had tracked before in the same area. Even though there was very little detail in the prints, the sinuous S-trail is one that anyone who tracks bobcats quickly becomes familiar with. So I followed this trail, confident of my identification until I came to a small spot of liquid under an overhanging rock. I stirred it and detected the odor of "chemically treated paperboard," as I mentioned before, an odor from my childhood. Bobcat urine in my experience had always smelled of ammonia, the common odor of housecat urine. But this was an odor I associated with fishers and gray foxes. Could I have been tracking a large gray fox all along? These particular canids often behaved more like cats than dogs. Had I made a mistake? Could this species, too, leave a sinuous trail pattern? I went back over the trail, reexamining its clearer tracks. They measured to gray fox; the toe arrangement was what the fox might have used for traction in loose sand; and the step length of 11-12 inches in the few places where the animal direct-register walked was short for bobcat but at least marginal for a male gray fox on this surface. I looked at a single print in sand in the shadow of a rock where the surface was a little stiffer than elsewhere. The scalloping of the trailing edge of the secondary pad, which shows in many bobcat tracks, could also be the indentations the gray fox shows in the "winged ball," as the shape of its secondary pad is often called. Granted that the lobes of that ball were wider than they should have been, but impact, I thought, often spreads the pads of animals whose feet are adapted for jumping down from things, and sand has a tendency to spread features. (My confidence had been shaken and I ignored the fact that nothing in a walking or trotting trail on a flat beach would have resulted in such impact.) Foretracking, I found another spot of urine under another overhanging boulder. Above it on the boulder itself I found the spot from which it had dripped. Ammonia! Suddenly it was a cat again. Then I remembered the single registration way back up the beach where I first found the sinuous pattern; it had had asymmetrically arranged toe pads with the medial of the middle two extended, a sure sign of cat. And then I remembered the winter I had tracked what may have been this same small cat over in the Federated Forest in dry loose snow where I couldn't get a well-defined track. I had puzzled there as well, vacillating back and forth between small bobcat and gray fox, until the animal sat down long enough for the warmth from its pads to melt defined prints into the snow. Then I saw the asymmetry of the front prints and the wide outer lobes of the secondary pad. Here on the beach I had overlooked the same evidence because of the odor of a single spot of urine. The ironic postscript to the story is that when I went home that night, I read over my journal for that same date several years earlier and found myself on the same beach, puzzling over the trail of the same small bobcat.

The message is obvious; review your own experience. And the more experience you accumulate over time, the more time you need to devote to reviewing it. To do that you may wish to keep a journal, and in that journal to keep not only measurements and drawings, but also notes on yourself watching yourself, admitting for later review the mistakes that you made.

And what about that single spot of paperboard urine? Wild animals, as we will see later, often mark their own scent in places where they sense the passage of other competing animals. I suspect that a gray fox at some point may have left its mark along this commonly used bobcat trail that passed through its hunting range.

Intuition

Something needs to be said about intuition with regard to the above incident, as well. A habit of mind that I have identified about myself is that my first intuition is usually correct. Then I start to think over what I've observed, applying consciously remembered details of experience. This piecemeal application sometimes results in my talking myself out of the initial identification. As I consciously recall more and more of these past details while re-imagining the event, I am bent back by the accumulating sum of the evidence to my original identification. Because these details from past experience were recalled individually, rather than comprehensively as when intuition is applied, it takes a while for enough of them to be dredged up to consciousness in order to come to the same, original conclusion.

An incident from years ago illustrates. A group of local trackers had found "bear" sign in the woods at Great Brook nearing the far western suburbs of Boston. Since in those days bears were uncommon in those parts, I joined the trackers at the site. The sign consisted of claw marks on a beech tree with a total width of slightly over 2½ inches. The assembled trackers thought the tracks had been made by a cub, perhaps shimmying up the tree for safety. My initial sensing based on intuition, however, was that the marks had more likely been made by a porcupine. Aware of my tendency to get swept up in suggested misidentifications, I examined the sign at length and began rummaging after conscious details of experience to support my contrary view. I remembered in my collection a photo of Paul Rezendes holding up a ruler to bear claw marks on a beech. The measurement showed them to be 10 inches across five toes, twice the width usual for New England bears and the result of 10-15 years of spreading beech bark. As the bark of this tree ages, it stretches, spreading any indentations on its surface as well. From this conscious recollection, I realized that this 2½-inch spread at Great Brook was what a bear cub would make only if the marks had been made yesterday. The individual claw marks before me were very wide and therefore very old; by the spread of the individual marks they looked to have been made at least 10 years earlier, probably more. Halving the track width, then, gave ¾ of and inch as the original paw width and made the tree, which was nearly saw-log size, into a sapling. I defended my intuition from the evidence of a now too-narrow breadth by suggesting to myself that it was just a young porcupine climbing after buds or tree flowers in the spring. But doubt was creeping in, not helped by the fact that the distance between the sets of marks as the animal climbed was rather long.

Aware of my doubts, someone suggested gray fox or bobcat, and for a time I aligned myself with that theory until, thinking it over later and consciously recalling more details of experience, I realized that cats, at least, have very sharp claws, normally retracted from dulling wear so that they will be sharp for climbing and snagging prey. When they climb bark, their sharp claws don't slip very much; instead they usually leave tiny points. The claw marks on this beech showed scrapes where the animal had slid backwards about an inch with each move. Since gray foxes are able to climb like cats and have retractile claws, I reasoned that they too would have sharp claws and leave shorter marks. This brought me back to porcupine. I imagined the animal hugging what was then a sapling with its front paws, scraping backward slightly, and energetically launching itself upward with its hind feet, to account for the distance between the sets of marks. Thus, my mind was bent back to my original identification by process of elimination. It was not a perfect fit with the evidence, the spacing was still a little long for a porky, but the conclusion, such as it was, follows the pattern my identifications often take: first an intuition based on consciously forgotten or half-remembered experience (porcupine claw marks on beech that I had seen and initially misidentified out at Minns several years earlier), then a tendency to go along with a suggested identification, then consciously recalling and accumulating past experiences, then imagining the event, returning in the end more often than not to my original intuition. In this case the processes of both identification and interpretation levered each other to a more or less satisfactory conclusion.

Reflection

Further thought on the issue of the marks on the beech tree followed a few days later while I was hiking down from the top of Mount Wachusett. This is a time when I often think out tracking problems. The beech at Great Brook, I recalled, only showed marks for the front feet. It occurred to me that when a bear climbs a tree it is apt to show marks of its hind claws somewhere below those of the front, as it digs in its hind nails to launch itself upward. A porcupine, on the other hand, hugs the tree with its front legs and places its thick, knobby hind plantar pads on the bark near the centerline of its body and vaults upward on the traction they provide, leaving little or no hind nail marking. This is more consistent with the evidence on the beech tree. Secondly, on the issue of whether beech bark stretches both vertically as well as horizontally, it seemed to me that it must expand both ways. Otherwise small marks in its surface would over time spread into narrow oblongs with their long axis horizontal when actually they spread as circles. This might account for the fairly long distance between sets of marks on the beech in question. If the marks spread vertically the same amount that they clearly spread laterally, the extra distance, which would seem unmanageable for a short-legged animal like a porcupine, would be explained.

The fact that this occurred to me only days later rather than on-site suggests that at this stage of my experience I still couldn't summon up this sort of analysis on the spot but only after the evidence had been digested for quite a while.

A return to the area sometime later generated even more thought on the issue. Although I was unable to locate the original tree in a forest with many beeches, I examined the marks on several trees and decided that it appeared that there was in fact more lateral spread than vertical. Even so it still appeared that there was some stretch upwards, perhaps enough to account for the spread of the nail marks on the original tree.

However, this problem wasn't done with me yet. As a postscript to the story, several years later while studying bear sign in the White Mountains I watched one of these animals descend a cherry tree. From that, I realized that long nail marks in bark might be made by an animal climbing down rather than up, as it lets its claws slip a little with each placement in order to speed its descent. This seemingly simple realization threw the conclusions about the evidence at Great Brook back into turmoil.

Expectation

"Tracking is seeing. Seeing is done with the mind." Another way to express this epigraph cited both elsewhere here and on several pages of my website is "A track is the perception of it." This, of course, refers not just to impressions in the snow and mud but extends to all animal sign.

We see what we expect to see, what we fear to see, what we want to see. The eye is simply the conduit to the mind. It is sensitive but as insensible as the camera's lens. Just as the film records on its surface the light impulses delivered to it in different ways depending on the camera settings, so one's mind records according to its own settings of anticipation.

Here are two examples. Twice in my experience I have seen people mistake the tracks of a gray squirrel for those of a fisher! Neither of these people were fools. One was a fairly experienced tracker, the other an animal control officer. In the first case, a line of squirrel bounding patterns had been worked on by sun and rain to the point that each pattern melded into a single impression something like an inverted trapezoid. This shape could look to an appropriately expectant mind like a line of fisher tracks in a walking gait. However, the line of impressions was direct with no hint of zigzag straddle, a fact that should have given it away as something other than a walking animal. Why didn't the observer notice this directness? He was leading a group in the woods, looking for "charismatic" species. No one in the group would be interested in the sign of the lowly gray squirrel. He wanted to show them fisher tracks, and so his mind saw what it wanted to see, ignoring contradictory evidence.

The second incident is more complex and interesting. A dog officer in Saugus was called to the home of a town resident who claimed to have seen a fisher in the backyard. The homeowner feared the fisher had been there hunting her cat. The officer went into the back yard and discovered under a bird feeder the following "track," the spread of which was about 4 inches.

Diagram 2.2A

This he took to be the impressions of the toes of the cat-marauding fisher and photographed them. When I was shown the photos, I inverted them so that the arrangement looked like this:

Diagram 2.2B

Seen from this angle the photograph showed the foraging pattern of a squirrel. The "toeprints" were actually four footprints arranged in the way they appear when a squirrel levers itself slowly forward looking for food beneath its nose. How could such a mistake have been made? The officer's expectations had been aroused by the testimony of the homeowner who had seen a fisher in the back yard "hunting her cat." This suggestion combined with the vagueness of the somewhat melted out prints was enough to cloud his mind to the reality. He saw a fisher track because he had been led to expect to see one. Of course, the fisher she had seen was not in the backyard in the first place to hunt the woman's cat. It was there because she had a bird feeder, the fallen seeds of which had attracted squirrels, the main prey of fishers in the Boston suburbs.

Both of the individuals above were embarrassed by their mistake, but they didn't need to be. All of us, from the expert tracker to the novice, see what the mind anticipates. The trick is to know this about oneself, to understand that tracking is done with mind and the mind is easily deceived. This self-knowledge makes one wary of the certainty of one's identifications and such wariness provides at least some protection from error.

This latter case perfectly illustrates the wisdom of the tracking tip you will encounter later in this book: When identifying vague tracks, walk around them so that you view them from a variety of angles. The results can be almost magically revealing.

Cause and Effect

There is no doubt a great deal of valuable information in "hunter lore," given that at least some of it results from the long experience of real woodsman. One of the problems with this lore, however, is that some of it results from editorial attempts to enliven dull pieces submitted to sportsmen's magazines by these woodsmen. Once this misinformation, invented in the magazine publishers' offices, gets mixed in with real experience, it is very hard to tease out the gold from the dross. Furthermore, once it is in print, this information acquires a life of its own, taken up and repeated as fact from mouth to mouth around many campfires. The power of print can be so malicious that misinformation, generated in this way may even invade respectable journals, framing discussions among legitimate researchers and wildlife managers.

A common defect often found in such pages is the error of proximate cause. I discuss this error later in Chapter V on Predation (page 156) with respect to a dogged insistence that, because fishers prey on snowshoe hares, they are the cause of the decline of hares. This is a linear error in that it results simply from a failure to trace a visible effect back beyond an immediate, visible and thus apparent cause. This error is so frequent in the "lore" that it bears some additional elaboration here in our discussion of habits of mind. The lengthy syllogism goes as follows: there is a crash in the hare populations, fishers prey on hares, there are a lot of fishers in the woods, hare fur is showing up in fisher scat, therefore fishers are the cause of the hare decline. This reasoning also illustrates another common error of material logic in which two nearby events are taken as a cause and its effect when both may in fact be effects of some unrecognized cause. The hidden cause in the case of the fisher-hare story, too far back in the causal chain to be easily perceived, is the periodic increase in anti-browsing toxins in common food plants of hares. These elevations occur cyclically every 7-10 years. During periods when these toxins are elevated, the hares weaken from starvation and fishers, along with other predators, concentrate on taking off the weakened animals. Thus the elevation of plant toxin is the hidden cause and the rest of the associated phenomena are both sequential and parallel effects. While all of this is not directly related to tracking, it is important in "eco-tracking," the payoff at the far end of identification and interpretation of what the tracker finds in the woods.

Timing and Placing the Event

Another habit of mind that the bear/porcupine incident at Great Brook illustrates is the tendency to see a track or trail as having been made in the same conditions in which it is viewed. If you find a bobcat trail in the snow at noon on a sunny day, there can be a tendency to assume that the animal made it on a sunny day at noon. If claw marks are found on a beech tree a foot in diameter, it is also natural to assume that the tree was the same size when the marks were made. If we find a cut high up on a plant while poking around in brush in the spring or summer, we may carelessly assume that the cut was made in the same snowless conditions and thus be led into mistaking the sign as that of a large browsing animal. Careful examination of the cut, however, along with a moment's thought may reveal that the work was done by a hare standing on two feet of snow.

If we are trying to recreate the event in the mind's eye, then we must recreate the conditions under which the event occurred. If, while standing on bare ground at noon, we wish to see what the bobcat saw in his hunt or what the hare sensed while it was feeding, we have to "see" the situation before us in the dark, on the snow.

A track or sign is an expression of behavior as well as an expression of relationship to habitat. When you find sign, don't get so absorbed in it that you fail to step back and look at its context. Think of habitat and terrain as functions of species content and behavior. Considering the content of the habitat, what species has a reason to be here? An incident a few years ago with a tracking class serves as an example. The group found a scat in some brush about 20 meters off a road and were encouraged to speculate at length about what might have left it. Much discussion concentrated on its shape, tubular, and its size, a little over a half inch maximum diameter, as well as its composition, fur without large bone chips. After a few minutes the consensus was that it had been left by a coyote.

Suggestibility may have played a part here because we had just been discussing a canid print out on the muddy road that showed some coyote characteristics. However, the context of the scat we were examining at the moment suggested otherwise. It was pointed out that coyotes are not usually stealth hunters that creep carefully through brush looking to make a kill. Instead, they tend to put in a lot of miles often over human infrastructure (trails, roads) looking for scavenge or targets of opportunity that they can take in a rush. Such would account for the apparent coyote tracks out on the road. But how likely was it that a coyote scat would be found in dense cover? Furthermore, during the group's intense concentration on the characteristics of the scat, they had missed what was directly under the scat — a vague prepared scrape in the ground cover. The two features, habitat and local context, suggested that the depositor was a bobcat rather than a coyote.

A related issue deals with counter-marking. This is the tendency, especially among predators, to deposit their own scat on top of or next to the scat of others (page 406). Because we find both scats at the same time, should we assume that they were also deposited at the same time? And by the same animal? A predator prowling a river bank, for instance, may come upon the deposit of an otter, mink or raccoon, all of which habitually leave scat at such sites. The predator will then deposit its own scat, often on top of the deposits of the other animals, presumably as an assertion of its presence in the area and perhaps its exclusive ownership of the range.

A Snake, a Coyote and a Mink

A friend sent me a photo a while back that had led him to wonder whether snakes can crawl out over snow in mid-winter! The photo showed the remains of a small water snake lying at the edge of a beaver flowage on snow roiled by a lot of coyote tracks. As he reconstructed the event from what he saw, the snake had crawled out of the pond over the snow, and had been discovered by a coyote or coyotes patrolling the bank who had attacked and killed it. The snake was partially eaten, lying mostly on its back and with a scrap of stained crust stuck to its belly.

Snakes are homeotherms, animals that must absorb ambient heat in order to move about. As it is unlikely that a such an animal would be able to, or have any reason to, move out of water onto snow where the little heat it had been able to absorb from the water would quickly dissipate, I began to search for another explanation. The key detail that I noted in the photo was the stained crust stuck to the inverted snake's belly. This suggested that the snake had been lying dead on the snow for at least a few minutes while its body warmth drained by conduction into the surface along with its body fluids, melting the snow just under its belly. As the snake's body cooled, this melted snow re-froze, sticking to its underside. Since the snake in the photo was in-verted and the chunk of the frozen snow was lying above it, I sur-mised that the commotion by the coyotes came later after the dead snake had lain in place for a while and cooled to the point where it had stuck to the underlying snow. Sometime later the coyotes arrived, per-haps drawn by the scent of the dead animal, and pawed at the now frozen snake. In the process they inverted it from its normal position, breaking loose a piece of the crusted snow along with the snake. If the coyote had been the actual predator rather than a later scavenger, the frozen snow should have been found under the snake, not on top of it.

But if the coyotes had not made the kill on a snake crawling over the snow, what had actually happened? Here's my version. I suspect that the snake had been swimming in the flowage, the water of which under the ice was above freezing. A mink hunting the brook below the dam spotted and killed it in the water, then dragged it ashore to feed. Perhaps disturbed during the feeding, the predator abandoned the snake before consuming it completely. The coyotes arrived after the snake had lain in place for awhile and pawed up the snow around it into such a turmoil that sign of the original predator was erased. In the process of all this pawing they inverted the carcass and its attached chunk of crust before leaving.

As I was not present at the scene, I could not gather the kind of detail needed to prove this revision of my friend's account, and cer-tainly there are other possibilities. However, the incident illustrates an important tracking principle — it is natural for our minds to assume that what we discover in an instant took place in an instant. Without some discipline to resist this impulse, our minds are likely to compress an extended event into a moment. This relates to the natural assump-tion that a discovered incident took place in the same physical condi-tions in which it is discovered. It takes a strenuous mental leap to consider that what we find in a single place did not take place at a single time. A dead snake and coyote prints suggested that the snake was killed by the coyote. Only a careful search of the details, in this case the frozen snow stuck to the inverted snake's belly, showed that the coyotes and the demise of the snake occurred at different times. Once more this is an example of a common error of material logic that I paraphrase in my fractured high school Latin as *Proxima hoc, ergo propter hoc,* and this relates to time as well as place.

Perception: A Tricky Bird

Down at Barre Falls several winters ago Joe Choiniere and I came upon an interesting scene. On a narrow tote road in the woods on about 4 inches of fresh snow, we found the wing marks of a bird that were about a foot and a half across. Between them were a couple of punch holes in the snow. A line of grouse prints led into and out of some sapling growth on the side of the road. Since hawks and owls regularly use such tote roads as hunting lanes, I jumped to the conclusion that this was a predation attempt with the raptor leaving its fist marks in the snow. As the wing marks were rounded, I took them to be owl prints and speculated that an owl had flown down the road and made an attempt at a grouse. But Joe followed the grouse trail into the woods a few yards where he found another shallow impression in the snow after which the grouse trail returned to the road and joined the wing marks. Then I noticed that the orientation of the wing marks was not quite perpendicular to the road as it should have been with my theory. Joe said he didn't think the marks had been made by a predator, but by the grouse itself. If that was the case, then, what about the deceptive punch marks in the snow, typical of the "fist" marks of a striking owl?

After falling back and mentally regrouping we finally realized that a grouse had been looking for snow deep enough to roost in for the night. It had landed on the road and tried to plunge its breast into the snow, but finding it too shallow, had moved off into the woods in search of deeper snow. Not finding it deep enough there either, the bird returned to the road and took off in search of a better roosting location, leaving the wing marks in the snow.

I had seen grouse wing prints mistaken for hawk or owl marks many times before and should have been ready to read the scene correctly. However, a detail, the "punch" marks, had overwhelmed my insight and led me to totally misread the scene. Also I didn't realize that, although I could tell the snow was only a couple of inches deep because I was standing in it and walking through it, a grouse could not tell its depth from its aerial view as it flew in.

There are several lessons in this incident. First, it may be helpful to try to see or sense our surroundings as the animal does, filling in the deficiencies in our own sensory ability with our imagination. Secondly, we need to widen the circle both physically and mentally, to look around for more evidence and to keep an open mind, ready to abandon a neat theory if further evidence shows it to be wrong-headed. It also helps to have another experienced mind look at the same scene and come to its own conclusion.

Categories

In Western music, we create scales of discreet notes and place great store in a singer hitting them precisely without any slurring in between. Eastern music, on the other hand, uses sliding scales, deliberately merging one note with the ones above and below. Tracking, it seems, is more like Eastern music than what we are used to. As we will see in Chapter V (pages 162-164), measurements, those indispensable features of modern scientific thought, tend to blend together in real-world animal tracking. This is true as well for categories, the end result of all that measuring. At what point in a progressively changing gait and the track pattern that results does a bound become a lope? Our western modes of thought would like to know and to assign precise limits. But such precision becomes arbitrary in the real world and unhelpful to the tracker who is trying to visualize an animal in analog motion with his mind's eye. These arbitrary limits are difficult to apply to an animal that is mixing and matching gaits or is in transition from one to another. The fox didn't go to school and presumably feels little need to restrict its movements to the appropriate boxes.

These attempts at precision may also lead to the arbitrary misappropriation of language. Words in common use, like "gallop" and "lope," get borrowed into a quasi-scientific lexicon with arbitrary meanings attached to them to describe the precise limits of categories of pattern or gait. For instance, the word "lope" in popular use suggests an easy run with a rocking-horse motion. I think it is not helpful to include it, as some books do, in the category of a "gallop" which in popular use is the energetic gait of a race horse or a pronghorn. If, in the course of loping or low-energy running, a fox encounters deep soft snow, it may increase the verticality of its gait and blend it into a "bound," a series of short leaps with a tighter cluster of prints. In between the two gaits may be a set or a series of prints that is somewhere in between the two. Precisely where one gait ends and the other begins is arbitrary and of no interest to the fox, nor should it be to the tracker trying to recreate the scene in his imagination.

Rules

"Rules are made to be broken." So goes the old saw with its concession to human nature. We seem to be habitual rule-makers, and rules certainly are needed to assure orderly governance of both societies and studies. One must know, however, when to break them, and this applies to tracking. Some books on the subject give many rules to guide the novice in his identifications. Eventually, he may come to realize that some of those rules are unhelpful, or even downright wrong, and need to be abandoned. This shouldn't be a moment of regret and self-doubt but rather of self-celebration. With this realization, the novice begins to graduate to the next plane of awareness, a plane that will open new doors to insight and knowledge. Any information that truly educates must be internalized, ransacked, examined and re-expressed as one's own. And this necessarily involves rejection of explanations that were good enough at the time but do not hold up in the long run to careful scrutiny and increased experience.

Along the way in that direction, be suspicious of anything that you read containing absolute statements. The words "always" or "never," stated or implied, are seldom appropriate to tracking explanations. Wild animal behavior is so versatile and so poorly understood that exceptions inevitably crop up. And one cannot be sure that the evidence before one's eyes corresponds to the rule or the exception.

Derivative Information

As I have suggested before, the power of print is amazing. Almost any notion that would be dismissed if heard in conversation tends to gain credibility once it is committed to type. Be prepared to discover that even information in respectable sources may have been tainted by misinformation first committed to print in the pages of an adventure magazine. Once in print, the notion may get repeated in the work of others until it gains the status of a canon.

Otherwise, more accurate information in print may simply be misinterpretation. A study by the Forest Service decades ago noted that sign of long-tailed weasel tended to show up in wetland habitats somewhat more often than that of short-tailed weasel. A look at the data showed, however, that both species could be found at any given time in both upland and wetland habitats. Many times since then I have seen derivatives of this information in books and articles, suggesting that sign of short-tailed weasels will only be found in uplands. (Note the absolute "only," reason enough for suspicion.) For the tracker, the original information is next to useless, once again, because he has no way of knowing whether the trail he encounters follows the rule or is the exception.

On the subject of derivative information, there is an easy way to test whether the tracking guide in a bookstore is original work by the author or is derived from earlier work. The illustrations that appear in many of these guides have their source in the important pioneering work of Olaus Murie in the early part of the 20th Century. Murie led an admirable life that has inspired many after him, including myself. His *A Field Guide to Animal Tracks* was the first really authoritative book on animal tracks and sign.

As careful as Murie was, however, he did make a few mistakes in his book One of these was the porcupine error. In drawing porcupine tracks, he apparently mistook two hind prints of different sized animals for a front and a hind print of a single animal. His representation, then, shows two prints, one larger than the other but both with five claw marks. The problem is that porcupines have only four toes on their front feet. In thumbing through a tracking guide in a bookstore, flip to the porcupine plate and see if the front print is depicted as having five toes. Then, turn to the muskrat page. Murie apparently mistook part of the nails of the muskrat hind print for part of its toes, thus representing the animal as having long toes and short nails. Muskrats actually have medium length toes on the hind foot and very long nails. If either of these mistakes is repeated in the guide you are browsing, then you may have reason to suspect that at least some of the information in the book may not be original with its author.

Chapter III

The Outer Game

Here we move our concentration forward of the eye and the mind behind it. The first thing we will consider is context in which animal sign is embedded. Identifying and interpreting wildlife sign often depend as much on the circumstances in which it is found as they do on the characteristics of the sign itself. An intimate knowledge of the natural background, then, is critical to successful tracking. Context is discussed both on the macro level, the various habitats within which tracking can take place, and on the closer level, the various media and environments in which sign is placed and discovered — earth, snow, sand and mud — and the effects of weather on the formation and duration of the sign. Next we look at specific habitats that are often fruitful for the tracker to explore. Lastly, we consider the differences in sign that the tracker might expect or encounter from season to season.

Contexts

If one compares the difficulties in distinguishing the eight-hundred or so species of birds in North America to the much smaller number of mammals, on might think that identifying a track, trail or other sign of the latter should not be so difficult. However, an animal's footprint is subject to the conditions of the surface in which it is imprinted, the so-called "substrate," and then to the many effects of weather that accumulate as time passes after the registration. The track in snow may be affected by additional snow of various depths, by rain into that snow, by the effects of wind and sun and of thaw and refreezing. A track in earth or sand, on the other hand, can be affected by wind, drying and crumbling of its walls, accumulation of falling debris or formation of frost pillars rising from dampness in the ground overnight. All of these factors can make a track protean indeed, and greatly increase the challenge to the tracker just in identification alone, a necessary step toward accurate interpretation. This section, then, deals with the various contexts in which a footprint is imbedded.

In the effort to attribute tracks and other sign to the animal that made the sign, it is easy to get so caught up in the minute observation of details of print morphology, the shape and distribution of pads, the measurement of scat and so forth, that we miss these various distorting factors, including those that occur within a track as well as in its background. Contexts should be considered not only spatial, location and habitat, but also temporal, night or day, time lapse since origination and, as mentioned above, action of intervening weather — sun, wind, overnight refreezing, rain and snow-melt, etc.

When we are attempting an identification, we unconsciously try to recreate the scenario in which the sign was produced, to "see" the animal performing the action in the mind's eye, embedding it in its various contexts. Let's say we are walking down a dirt road in a pine-oak woodland in New England next to a beaver pond. On a boulder about a foot across and a few inches high by the side of the road we discover a scat. We go through our usual procedures of measuring for maximum diameter, looking at the surface for bone chips and, at the "tail," for excreted fur. Deciding it is indeed a carnivore scat deposited by one predator or another, we are left to puzzle — is it coyote, bobcat, red fox, gray fox, fisher? Let's say the size and consistency of the scat is intermediate among all these species. The next step is to back away from the scat itself and look at the large spatial context. All the above species range into New England, all are denizens of the pine-oak woodland, and all hunt the forested edges of beaver ponds. But one or two become less likely. Fishers rarely hunt along roads and do not typically prowl the open edges of beaver ponds. Coyotes and red foxes do, however, and both use the human road networks to get around, especially in the winter. Bobcat trails usually cross roads rather than follow them, except in winter when the animals will walk down plowed dirt roads as the easier course, overruling their habitual secrecy for convenience and savings of energy. The same applies to gray foxes, which normally prefer overhead cover. Was this road plowed in the winter? Does the apparent age of the scat suggest that it may have been deposited in the previous winter? Investigating this possibility, look for the fraying that often occurs with fur scat during spring thaw as the softer material in the scat is leached away.

All of the remaining species in our consideration use scat for marking, choosing conspicuous locations along their hunting routes for placement. These may be boulders, cut stumps, gaps in stone walls, raised portions of a road or trail, trail intersections and so forth. Some are more usual than others. For instance, gaps in stone walls and the raised stone wall itself are more often used by bobcats than by the others. Coyotes leave their scat at trail intersections more than the rest.

As we move in from geographical region to habitat type and then to closer contexts we begin refining our mental "video" of the event. We picture each of these predators in turn, walking down the dirt road in the season in which we suspect the scat was deposited, and we begin evaluating likelihoods. Is one a more credible version of the event than another as we home in on closer contexts?

Now let's move in closer yet to look at the boulder on which the scat was deposited and picture each animal in the actual process of dropping the scat. Is the boulder large enough that we can imagine each in turn perched atop while it squatted to defecate? Is the boulder stable? We give it a wiggle to see and decide that it is or isn't. Would it have been more stable in the winter, covered by a large mound of snow? Again we picture each of these scenarios in turn, evaluating them for relative likelihood. Was it possible that the animal was not actually standing on the boulder to squat, but was long-legged enough to have stood on the ground and defecated on the boulder? The latter would suggest the lankier coyote or bobcat rather than fox. We look at the disposition of the scat itself. Was it arranged linearly or was it curved or looped? If the former, it may have been deposited by a height about equal to the length of the largest segment. If it is curved or looped, it would have been deposited with the anus closer to the landing spot.

The more contextual information we accumulate as we ponder, the clearer the relative likelihood becomes, until ultimately we make our identification. We should be humble, however. Some identifications made in this way are easy to refine. Others may have more than one likely contender, and we may not be able to say which with perfect certainty. In tracking classes, I usually qualify my identifications with a percentage of certainty attached. For instance, I may say the identification is 80 percent certain. Given the tendency of the sign of different species to resemble that of others and given the myriad factors that affect animal sign over time, one does not often get to assign 100 percent certainty to anything. We are trying to reconstruct in our imagination a past event on very partial evidence. The best we can do is to see it through the proverbial glass darkly, and so we may be wrong. Don't worry about it. Take what you can learn, make your best estimate and move on, confident that with more experience, the accumulation of which never ends, your batting average will inevitably improve. Or it will as long as you are willing to learn and not become prematurely dogmatic about what you encounter in the woods.

Snow

It might be hard to believe but it is possible to confuse turkey tracks with those of a coyote! One winter during a thaw I came upon a trail of narrow tracks in the snow. From a distance I attributed the trail to that of a wild turkey. It did have a bird-like gestalt: narrow middle with a flaring tip. Upon closer inspection I realized that this was actually the trail of an eastern coyote! Damp snow has an elastic quality that allows it to deform under its own weight. The coyote had walked along, direct registering in snow to a depth of about 8 inches. Afterwards the sides of each track deformed, slowly bending inward to create the bird-like surface impression.

Snow is amazing stuff. Although it is essentially ice crystals, paradoxically it can insulate creatures below its surface, especially if it is fresh, fluffy snow that traps a lot of air. Furthermore it comes in many different consistencies, each of which can affect the appearance of tracks impressed in it. A red fox track on a dry, firm snow surface can look quite delicate; the same track in soft, damp snow can look robust. Bobcat tracks, as we have seen, are notorious in this regard. Because their pads, like those of other cats, are quite deep, rounded vertically, they can be easily deformed so that they can look quite different on different surfaces (pages 40-43). The actions of time, sun and changing temperatures on snow can also strongly affect the appearance of a track. The "yeti tracks" found in snow by alpinists in the Himalayas are a good example of this. Although these mountains are quite high and cold, they are located at fairly low latitude where the sun is quite strong even in winter. The action of direct sunlight on tracks in the snow causes them to magnify quite rapidly and to lose their fine detail. In this way, bear tracks, which can look rather human in the best of circumstances, come to look like the gigantic footprints of some hairy pre-human with snowshoe feet. I call this distortion the "yeti effect," noting that it can occur to a somewhat lesser degree in any snow-covered area of the globe where the sun has a chance to work on a track for a while.

Not only does the action of direct sunlight melt out a track, but the shadows in its contours ironically do so even more strongly. If a track in snow were perfectly white, we couldn't see it. Instead it is a mix of white, gray and black. Since fresh snow lying perpendicular to the sun is a nearly perfect reflector of sunlight, those areas of a track that are pure white resist melting. However the gentler gradients of a track that lay at an angle to the sun are less white, assuming a range of gray shades depending on the steepness of the angle. These areas reflect sunlight a little more poorly than the pure white areas and so begin to melt. At the edges of the track closest to the low winter sun, dark shadows form, shifting gradually as the sun crosses the sky. These darkened areas, then, although shielded from direct sunlight, paradoxically absorb ambient reflected light faster than the exposed areas and as a result melt the fastest. Over time all these characteristics of snow reflection can strongly affect and distort the appearance of a footprint.

Not only sunlight but also temperature can affect a track's appearance. A typical 24-hour-day in winter is a mix of temperatures, warmer at mid-day and coldest in the early morning before dawn. While pure, white snow is a good reflector of short-wave solar radiation — sunlight — it is also a good absorber of long-wave infra-red radiation — heat. A thaw-freeze cycle that lasts for several days with the snow absorbing heat at mid-day progressively distorts a track, softening its contours and widening it. This adds to the distorting effects of direct sunlight on the "yeti tracks" cited above, which are also exposed to warm Himalayan days and bitterly cold nights.

I recall an incident tracking with a friend in which he believed he had found a bobcat track in shallow snow over a pebbled roadway. The secondary pad impression in the original print had fallen on a spot where dark pebbles exposed in the old asphalt were just under the surface of the compacted snow. After the sun climbed in the sky and worked on this for while, the slightly more darkly shaded snow above the raised pebbles absorbed light differentially and melted away, exposing these pebbles to sun. The dark pebbles then absorbed the warmth of mid-day and melted even more of the nearby snow, exposing what appeared to be the large secondary pad typical of a cat. However, combining imagination to create a history for the track with knowledge of the effects of sunlight on snow resulted in a reconstruction of the original appearance of what turned out to be a coyote print.

In the species notes on bobcats later on in this book I describe the experience of following the faint trail of a bobcat over a heavy crust. This experience illustrates another characteristic of snow. At mid-winter any settled snowpack develops a temperature and moisture gradient from the ground to the surface, warmest and dampest at the bottom, drier and colder near the surface. Since warmth moves to cold and moisture moves to dry, over a cold winter night the moist warmth deep in the pack migrated to the surface, forming crystals. In the case of the example in the species notes, these crystals were needle-shaped. The bobcat had walked over this recrystallized surface leaving no impression in the crust other than the crushing of these tiny, delicate structures. It was hands-and-knees tracking as I searched for the slight yellowing caused by ambient light reflecting from the flattened areas, but it was enough to follow the trail to its conclusion.

Not only may snow consistency distort the tracks of an animal, it may also affect the patterns the animal leaves behind. Red foxes, for instance, normally leave beautiful walking trails in snow, with one track directly ahead of the next like stitches in a garment. However, in soft spring crust that threatens to collapse at any moment, a fox will widen its straddle so that if the snow collapses, the animal will fall to its medial side rather than laterally. A fall medially is easily caught by the feet on the other side, while a fall laterally would tumble the fox over into the snow. A 15-inch step length and wider straddle in a trail with poor track rendition can easily be mistaken for a bobcat, especially since the two animals have similarly sized feet.

Soft, deep and unconsolidated snow may also affect the appearance of an animal's trail, distorting it out of the expected parameters. I have seen foxes and deer, struggling with deep snow, lose patience with the heavy going and begin bounding through it toward firmer footing. Such a fruitless expenditure of energy would otherwise be inexplicable in the depth of winter when survival demands efficient movement. Foxes are mostly solitary animals, hunting in pairs only during mating season, and thus must shift for themselves in deep snow.

Coyotes, on the other hand, often band to together in loose packs in the winter for several reasons, one of which is to apply teamwork to breaking out trails in unconsolidated snow.

Snow and Predation

When snow first falls it is usually composed of feathery crystals. Like feathers in a pillow, these accumulated crystals muffle sound. However, their loose arrangement allows air and carbon dioxide within the snowpack to move to the surface. As the snow ages, the crystals lose their fine detail, taking on the form of progressively larger pellets. These pellets pack together and may actually fuse into layers of ice. At this point the snowpack is less gas-permeable — but it does conduct sound better.

The significance of these changes for the predator moving on the surface and trying to detect the presence of prey underfoot is major. While the snow is soft, the odors of a vole, for instance, made by its breath, sweat, urine and scat drift to the surface where they can be detected by the nose of the fox. Although this will tell the predator that there is vole activity under the surface, the same soft, loose snow also muffles sound. And timing is everything. The scent alone cannot tell the fox where the vole is at the moment the predator passes overhead. Perhaps this is one of the reasons that predators like foxes and bobcats tend to lay up for awhile after a snowstorm until a combination of hunger and a firming surface provokes them to hunt. All predators are on a strict energy budget. The fox or bobcat must calculate accurately whether expending energy by active hunting is a better idea at the moment than laying up and conserving calories until conditions change in its favor. For time changes the snowpack, settling it into a firmer surface and granulating it into packed and ultimately fused pellets that better transmit sounds made deep within to the outside world. It is this change that allows the fox to move easily over the surface and to replace its sense of smell with its sense of hearing, a much more timely sense with which to hunt a vole scurrying through a runway a foot beneath the surface.

The tracker, too, may be on a budget, not of calories but of time. It may be a better idea for him, as well, to lay up for a couple of days after a snowstorm until the predators start to move again and the page begins to fill up with the cyclical and short-lived phenomena that we seek.

I highly recommend the lucid and readable explication of the properties of snow presented in Peter Marchand's *Life in the Cold*, an excellent book that also presents interesting information and theories about many other aspects of winter that affect wild animals.

Sand

Sand can be a good medium in which to track wildlife. For one thing, it often is located along water. And water, being one of the necessary ingredients for life on Earth, attracts animals. Many predators, for example, move along beaches, using them as highways to get around their hunting ranges. Other animals such as otters and bears have to cross the beaches to get from woods to water to drink or hunt their prey.

Such a beach, wetted by either boat surge or rain, is a fresh tablet on which to write the nightly diary. The damp consistency of sand keeps the grains stuck together and preserves for a short period fairly fine detail in tracks impressed in it. While it is still damp, the sand also remains firm so that the trails of wild animals retain their normal appearance. As the sand dries out, however, it loosens and the tracks and trails impressed in it can change. After the foot leaves the print, the steep sides of the track begin to collapse toward the "angle of repose," the normal gradient to which loose material devolves if it is not held back. This angle is somewhere around 35 degrees. When this happens, the normal bathtub impression left by a soft-padded animal like a coyote begins to transform: it widens as the grains of sand at the rim fall into the track and the sliding grains falling in from all sides create a point or crease at the bottom of the impression as the original roundness is filled in.

Loose sand, of course, also can be acted upon by wind. Here, unlike gravity, the distortion is horizontally directional. Figure 3.1 shows a gray fox front track masquerading as that of a red fox. Both wind and gravity caused the familiar central pyramid in the print, typical of many canids, to flow into the secondary pad impression from the middle toward the posterior of the print. As it did so, it filled in the familiar winged ball of the gray fox's secondary pad, replacing it with a deceptive bar, roughly similar to the transverse chevron in the secondary pad of a red fox. Note, however, the catlike roundness of the print and the lack of nail marks. Consideration of the macro-context is important

Figure 3.1. The edges of this gray fox sand print slid downward over time, giving the false impression of a bar at the deepest part of the secondary pad and making the print look superficially like that of a red fox.

here as well: red foxes do not range into central Arizona where this photo was taken.

Just as you and I are likely to shorten our step in soft sand, so do wild animals. This shortening outside normal parameters should be taken into account in making identifications that depend largely on step length. Furthermore, the mounding of material in front of a track in snow where the animal is moving in a low-energy gait also appears behind the track in loose sand. Thus, in an identification that depends on print shape alone, the tracker must orient his mind to the direction of travel.

Mud

While soft mud tends to retain the appearance of tracks quite well, there are still some distorting factors. A normal bobcat track, for instance, shows no nail marks, as the animal keeps them sharp and protected in sheaths at the top of its toes. As described in the previous chapter (page 42), a bobcat can extend its claws for traction and support on soft, slippery surfaces.

Raccoons are one of the great foolers in the animal world. They can spread or close their toes both laterally and longitudinally, providing wide variation to their track appearance. On soft mud they may extend their club-like toes and their nails for support. Such a track looks very different from that resulting from closed and contracted toes on winter crust where the animal is conserving heat by minimizing contact with the snow.

The normal appearance of coyote tracks shows closed toes, held tightly together. This appearance, along with frequent absence of nails on the medial and lateral pad impressions, is diagnostic for this animal and serves to distinguish it from the tracks of domestic dog. However, on soft, slippery mud, coyotes will spread their toes for support and extend their nails for traction, leaving behind tracks that look more like those of a domestic dog.

There are two problems with mud as a medium for tracking. First is uniting the wild animals with the surface, and the second is uniting the tracker with the tracks. There is rarely a problem finding mud. The problem is getting wild animals to go where the mud is and to leave their prints in it. At Brownfield, Maine, there is a dirt road across a man-made (and more recently beaver-improved) impoundment (Figure 3.2). Such a road is a funnel for any mammal that wishes to get from one side of the bog to the other. Rather than swim or plod across the muddy surface of the bog that is exposed at mid-summer, these animals naturally choose to move down the raised road above the mud. Fortunately for me, the road fell into neglect and developed a lot of depressions that filled with rainwater. On weekends, pickup trucks would drive down the road and roil the puddles, grinding the sandy soil into a nicely homogenized fluid that dried out to perfect mud. Timing is significant, once again. The puddles had to fill before a weekend so that enough vehicles would traverse the bog to roil the road soil into mud. Then the puddles had to dry to expose the mud during the week

Figure 3.2. Bog Road, Brownfield, Maine. Lightly used dirt roads that cross wetlands create natural funnels for animal activity. This one has a fine silt roadbed that churns to nicely homogenized mud after rain.

when few other vehicles would ruin the mud tablets they had just created. If the puddles were too wet, then fastidious animals like bobcats and foxes would simply walk around them rather than get their feet wet. If the puddles got too dry, then the soft mud would harden and become impervious to track registration. And of course, all of this had to take place before the return of traffic on the following weekend. It could get a little maddening, but persistence paid off and rewarded me with a great many of the plaster casts I have in my large collection.

The second problem is getting the tracker to the tracks. A drying bog, marsh or swamp provides a lot of mud, and certainly many wild animals will cross these features when the mud firms. But how is the tracker, who weighs a lot more than a mink or a muskrat, to get out on that same mud to examine and record the tracks? Having lost several boots to the sucking mud over the years, I admit I have not yet found a perfect solution. I have privately considered snowshoes as a way of distributing my weight over a large enough surface to stay on top. So far, however, the mental image of myself wearing snowshoes on a muddy bog in mid-summer seems too ridiculous to attempt, and then there is the problem of cleaning tenacious mud off the shoes afterwards. In the end, I have surrendered to the need to get dirty, really dirty.

Weather

Weather is a context all its own. Once a track or sign is created, it begins to age, and its appearance is increasingly affected by time and the weather around it. In the dry air of the Southwest, scat dries quickly and may appear older than it actually is, and older than it might appear if it had been deposited in the high humidity of Florida, to cite an extreme contrast. Conversely, old summer droppings of deer, moose and other ruminants can look quite fresh. These pellets, run through a multi-chambered stomach, have so little nutritive value to bacteria by the time they are excreted, that they lay on the ground until they weather into the background. At mid-day they look their age, but covered by dew in the early morning they can look deceptively fresh. Nightly frost on a damp beach can greatly alter the appearance of a track as the moisture in the sand freezes and expands. In any loose substrate such as dry sand, wind can directionally fill parts of a track, giving widely different appearances depending on its direction relative to the print. Ice storms and the absence of snowfall are two weather conditions that can have a profound significance on the behavior of the wildlife that we track.

Ice Storms

In a mature forest where the trees are old enough to have lost their supporting heartwood, live branches break off and fall to the ground under the combined influence of accumulating ice and wind. The fallen limbs bring energy back from the canopy to ground level (Figure 3.3). The resulting thinning of the canopy allows more sunlight to reach the ground as well. The buds on the live limbs provide critical browse for deer, which are starving over the winter, and the added sunlight will, in the following spring, provide the energy of the sun to the forest floor, sponsoring new growth at browsable height. Deer were presumed to be scarce in the pre-colonial forest, due to its supposed lack of browse at ground level. However, the last ice storm to cause widespread and prolonged blackouts in my part of the country has proved a boon to deer. Following the storm, deer trails vectored on

Figure 3.3. Ice and wind damage bring live limbs to ground level, providing food for terrestrial herbivores like deer.

every fallen limb, where the animals fed on the winter buds. Since ice storms are a common occurrence in New England and since, in the pre-colonial forest, there would have been many more old trees with weakened limbs vulnerable to ice and wind, it is very likely that deer were abundant in such woodlands, as well.

Lack of Snow

Lack of snow in early winter can result in a small animal die-off. Mice and voles have little body mass and so retain heat poorly. Under a blanket of 8 inches or more of dry snow these animals are insulated against the worst effects of winter. Such a blanket increases their likelihood of survival from the predations of all the carnivores that feed upon them from above: foxes, hawks, owls, coyotes, fishers, bobcats, weasels, etc. In drought winters, however, many of these small creatures of the prey base die of exposure and over-predation. While these animals may huddle in a converted bird nest or in a ball of grasses to pool their body heat, they must eventually leave this comfort and forage abroad where they may freeze before their return or get victimized in their exposed condition. Any significant die-off of this sort has ripple effects throughout the food web, as predators must either go without and themselves starve, switch to another food source or migrate to better areas.

Lack of snow also affects browsers like rabbits and snowshoe hares that depend on it for a rising platform to reach new forage as the winter progresses. Since lagomorphs reproduce quickly and cannot climb, in drought winters they rapidly exhaust browse at ground level, weaken and become easy marks for predators whose numbers may increase due to this boon. The plenty is short-lived, however, as the crash in lagomorph numbers following a snowless winter will be followed by a decline in predator numbers as well. Equilibrium is unnatural; Nature arranges itself in waves. The numbers of all species rise and fall in response to food availability and predation pressure.

Aging Tracks

I have already introduced the aging of animal tracks (page 50) but will expand upon the subject here.

Snow

As suggested earlier in this chapter, in a snowpack at mid-winter there is usually a temperature gradient from warmer near the ground to colder just under the surface of the snow. This gradient is exaggerated on a cold night during which warmer air laden with water vapor migrates upward and emerges from the surface where it encounters the cold night air. At this point it immediately crystallizes as frost on any surface available including the surface of the snow itself. The frost crystals themselves do not grow to the pretty Christmas ornament shapes of snow flake crystals, but rather grow into long needles. With subsequent exposure to sunlight with the dawn, these crystals "sublimate" — they evaporate directly from a solid state, ice, to a gas, water vapor. On the next night the water vapor again crystallizes on the surface. However, another attribute of snow comes into play: disturbed snow hardens. (Snow in an avalanche, to illustrate with an extreme, hardens into white concrete shortly after it stops moving.) Applied to an animal's track, the hardening of the surface under the previous night's tracks, disturbed and packed by the weight of the passing animal, prevents recrystallization within the outline of the track.

Thus, if you come upon the trail of an animal shortly after dawn on a cold morning and detect crushed crystals within the track, you can be sure the track was made on the night just past. If no needle crystals show in the track, then it is likely an older animal trail.

The sublimation of snow goes on continually, even in the depth of winter. And of course, with the intensification of direct sunlight from February onward toward spring, outright melting also takes place. This attribute also affects the appearance of animal tracks. Close examination of the edges of a track will usually show whether there is fine fracturing and crystal angulation along the rim. If it is there, the track is fresh. If it is not, the track is old. We tend to absorb this information intuitively when we look at a track, seeing age as a rounding of fine features, and so it doesn't need a lot of elaboration here. I might add the old hunter's trick of blowing into a track to see if snow crystals move. Once again, disturbed snow hardens. If the crystals at the edge of the track can be blown away, then it is fresh. However, it seems to me that this same information is available by just looking at the track edges without disturbing them. Furthermore, warm breath on the snow immediately rounds the crystals, making them look older than they are.

Mud

Aging tracks in mud and soft earth is harder than in snow since the medium changes less over time. A number of tracking books advise creating a test situation where a track, manufactured in mud or soft earth, is left to age. You then return to view the changes in the track every day and perhaps to photograph them as well so that the alterations can be reviewed over time. If this works for you, do it. However, what one is likely to learn by the end of the experiment is that tracks age over time, something you knew anyway. The rate at which they age depends on the surface and the weather, both of which vary infinitely. Thus, the changes that you see in the test tracks may never occur at the same rate again in your experience.

At best, a rough idea of the age of a mud/earth track can be gotten by looking for debris that has collected in the track. Trees, shrubs and grasses are continually shedding material that falls to the ground and gets moved around by wind. This debris collects in any depression, including animal tracks. The rate at which it falls, is moved and collects is variable and so your estimate of age from such evidence is likely, once again, to be inexact. It should be obvious, as well,

77

that raindrops pock the surface of soft substrate. Recent weather may also help. Not only will a recent windy period tend to cause debris to collect differentially under one rim of a track or another, but raindrops will also leave pock marks in the ground. To overstate the obvious, if those marks can be detected within the tracks as well as outside of them, then of course the trail was made before the rain.

Over time any depression in soil magically disappears as the substrate reduces to a plain. The depth of a track may make it seem fresh, but one has no way of knowing how deep the track was when it was made. However, pressure on the surface near the track may help. Press with you thumb on the ground to see how much it gives. Then extrapolate that to the approximate weight per square inch (your thumb) on the track by the animal in question. If the earth gives way less than the depth of the track, then it is an older print. I recall using this method once in an old log yard up in the Passaconaway intervale. A mother bear and cubs had walked through the clearing and left some castable prints. Pressure on the ground with my thumb caused it to yield very little while the depth of the cub tracks was about a quarter of an inch, showing that this was not a fresh trail.

There is a lot of print on the subject of aging tracks, but at best, in this tracker's experience, old tracks in mud or soft earth simply look old. I knew at once the trail of the mother bear and her cubs was old because it simply looked that way. My intuition was based on loose debris that had to be blown out of the tracks before I cast them, and an appearance of roundness to the details of each track that resembled a sort of muted version to what happens with snow tracks over time (Figure 3.4).

Aging Sign

Once again, old sign looks old. Browse on vegetation dries visibly over time. Scat also dries as its moisture evaporates and bacteria consume its mucus content. One caveat, however — as was the case just cited with deer scat, dew that collects on scat overnight may make it appear fresher than it actually is. Dirt and debris scraped up into a pile by paws dries and loses its coherence, with its fine upper points falling downward, giving the pile a more rounded appearance. The edges of scrapes begin to devolve toward the bottom of each shallow trench made by the scraping paw as moisture in loosened soil dries more quickly than that in the deeper earth in the center of a pad groove.

Figure 3.4. Old tracks look old. Drying over time has rounded the edges of this bear cub track, and wind has blown debris into the pad impressions.

It should be equally obvious that the odor of urine scent marks generally fades over time. Sue Morse recently pointed out at a trackers' conference that the sweaty sock odor of a tom bobcat's urine doesn't activate for a week or so after its deposition. However, the common sense rule still holds that tomcat scent as well as any other, once it is detectable by the human nose, fades over time. The initial pungency of the odor tends to vary with species and with the season so that, once again, you have no way of knowing at what stage of its fading you have encountered it. Red fox urine during the January mating period seems to be much more acrid than at other times of year, filling the woods on damp days with its skunk-like odor. At other seasons it is often detected as no more than a faint whiff as one walks down a path.

Habitats for Tracking

In this section I choose two habitats important to the tracker-naturalist, forests and beaver ponds, for more extensive discussion.

By knowing something of the natural history of each, the tracker can better prepare himself mentally before he enters the habitat in search of tracks, trails, and sign. I begin with the eastern forest, although much of the discussion is appropriate to any biome. Since more tracking is probably done in forested habitat than in any other, it will be helpful for the tracker to know what forest conditions likely will be more productive than others in the search for wildlife evidence and what sorts of evidence he or she should expect to find.

A Brief Recent History of the Eastern Forest

Historical accounts that speculate on the nature and appearance of the pre-colonial forest vary widely. The confusion arises because, of course, Indians had no written language with which to record the pre-history of the region. Secondly, during the early colonial period few literate Englishmen were interested in studying and writing about the nature of the forest. Mostly the colonists just wanted to get rid of it. Information available today to ecological historians is partial and scattered, and any reconstruction of that forest relies on extrapolation from accounts, recorded for other purposes and often of dubious accuracy.

The bill of goods that the colonists had been sold when they were beguiled by grant salesmen in England was that the forest of the New World resembled an English estate, a tall closed canopy of old trees over a shaded forest floor free of impediment to travel. By purchasing a piece of it, everyman could be a landed aristocrat himself. The fact that English estates only had that appearance through intense management by humans apparently didn't occur to the readers of these advertisements. On the other hand, a contrary and revisionist view of the New England forest presents it as so thoroughly burned over and stripped of wood by natives that that there was little woodland left to the view of the colonists from the ship's rail. What are we to make, then, of these widely disparate views? I would suggest that recent experience may illuminate the situation that actually confronted the new arrivals.

Most of the natives in eastern North America lived along the coast and along the banks of rivers inland within reach of the coast. Any scorching and removal of the forest by Indians would have occurred there and not so much in the interior. Along the inland rivers there also would have been small plots cleared by slash-and-burn

80

agriculture for squash, corn, melons and beans. However, the large-scale burning of the woods by Indians may not have been a tactic of primitive game management, as often cited by historians, but rather may have been largely accidental. Since Indians had no way of controlling wildfires, it is possible that some of this burning may simply have been caused by ground-creep of embers through the duff and peat from Indian campfires. I know from local experience that such underground fires can be very difficult to extinguish, even with modern fire-fighting tools like high-pressure hoses.

Confronted with a frightening landscape, not at all like the one they had left or had been told to expect, the colonists set about transforming New England into Old England by removing the forest. No one studied that forest the first time around; they simply cut it down or burned it to get rid of it. Today, however, as eastern woodlands slowly recover from repeated deforestations of the past 400 years, ecologists are watching. And some of the things they are witnessing are surprising, shedding some light on the nature of the pre-colonial forest.

In scattered locations where the eastern forest has been allowed to grow back for a century or so uninterrupted by periodic commercial cutting, we can see a couple of factors ignored or at least underestimated by past historians. First is the true nature of a "virgin," or even an "old growth," forest. (A virgin forest is one that has never been cut; an old-growth forest has not been cut for a couple of hundred years.) The second is the influence of a single mammal on the appearance of that forest. That animal is the beaver.

When the first religious colonists made their way to Concord in the Massachusetts Bay Colony, they reported their journey as arduous, a sort of Pilgrim's Progress through the "slough of despond." Later historians suggest that this may have been Nature imitating art and that the reporters were exaggerating their trials to make the trip resemble the death and rebirth of the soul in Christ. It is pointed out that trade had been carried on by white men from Boston with the local Indians for years and that trails that facilitated that business had to exist over which the new colonists may easily have traveled. The colonists in their accounts reported losing their uncertain way through a forest that tore their stockings to shreds, emerging scratched and bloody at the confluence of the Assabet and Sudbury rivers, the concordance or "Concord" of those streams. Having prowled those same woodlands often in the past 50 years or so, I hear more truth in the account than is often supposed. First of all, a series of smallpox epidemics in the early 1600s decimated the native populations. Under disuse, the Indian trails that were so well worn would quickly have been overwhelmed by shrubby growth. I have seen many instances where this has happened to unused trails in my own time. In as little as two years such paths can be become obscure, as bordering foliage, fighting for sunlight, impinges on the edge of the trail and then totally covers it. Indian trails, after all, were only single file affairs, much narrower than the double-tracked ruts that the white man would later develop in the service of the wheel.

The second factor that might have accelerated the obscurity of the trails to Concord was the nature of the surrounding forest itself. Rather than a relentless closed canopy of tall trees, the virgin forest in the interior of New England was actually a mixed-age forest composed of ancient trees with broad crowns interspersed with younger growth in plots where age, fire or weather had cause the demise of the old trees. In places where tree-mortality allowed sunlight to reach the ground there would be explosions of new growth ranging from dense ground cover in the more recently uncovered forest floor to young trees of all ages interspersed between. The resulting patchwork of aging would contribute to places where abundant sunlight would have been available to promote the erasing of an unused trail. Add to this the fall of aging trees and limbs across the trails and a journey over one of these routes could be arduous indeed.

The other factor that would have made the trip from Boston to the future Concord difficult was the beaver population. This factor, it seems to me, has consistently been underestimated by ecological historians in their reconstructions of the pre-colonial landscape. Only in my lifetime has this native mammal experienced a resurgence to something like its earlier numbers. And no other wild mammal has such a profound effect on the landscape. A compass march through these same woods today requires many detours around the impoundments of these animals. It can be expected that the same was true for the first colonists to move inland from the coast. In many places the trails that they were attempting to follow, in addition to being obstructed by tree-fall and being overgrown from disuse, would also have been flooded into boggy mires by beavers. Without natives to reroute trails around them, these beaver ponds would have presented a serious impediment to Englishmen inexperienced in this kind of landscape. Once the thread of trail was lost, the pilgrims would have been "bewildered," in the original sense of the term, with nothing more than the direction of the sun to help them navigate through a mixed-age forest of thickets, briars, tree-fall and more beaver bogs.

Wildlife of the Forest

A forest can begin in any of several ways. With the decline of agriculture in the Northeast in the mid-19th Century, old fields began to grow back to woodlands as farms were abandoned. In these plots, the dung of farm animals that had fertilized the soil for human use was available to the newly encroaching forest to use for natural growth. Fire could also renew a pre-existing woodland. As woody growth burns, it releases much of its energy to the atmosphere in the form of heat, but it also recontributes its chemical constituents to the soil, promoting abundant new growth in a sort of slash and burn silviculture. A third way in which a forest can begin is by cutting (Figure 3.5). In this case, lumbermen are careful to leave the slash by-product in place as fertilizer for a new generation of trees to grow from. The current experiment of the resulting regrowth as an eternally renewable resource, relying on this slash fertilizer and the energy of the sun, both directly and stored in the chemical bonds of soil components, is now about a century old but has not yet been proven conclusively. In this sort of forest renewal, a great deal of energy is removed from the system with removal of the valuable trunks of the trees. Over how many regenerations this removal can continue without impoverishing the system is open to debate.

As a forest, renewed by whatever method, begins to regrow, poplar and pin cherry along with various woody shrubs invade the site, growing densely and providing food and concealment for many species of prey animals. This growth, termed "early successional," is especially vigorous if the soil has had the advantage either of animal fertilizer or of fire-provided minerals. After a cut, on the other hand, several years are needed for the slash to decompose and recontribute its chemistry to the soil. During this initial period the cut area has very low wildlife value.

The early successional growth that follows this impoverishment harbors many bird species like chestnut-sided warblers, yellowthroats, black-throated blue warblers, ruffed grouse and many others. As for mammals, the dense early growth supports browsers like moose, deer, rabbits, hares and voles, all species that can reach the low growth without climbing. These "prey" animals, then, attract their predators — foxes, bobcats, coyotes, fishers and weasels. In western parts of the continent they are joined by wolves, wolverines and cougars and in northern latitudes by lynx.

Early successional tree species, such as aspen and some birches, are sun-lovers. As they grow, they provide shade for less sun-tolerant but more long-lived species like maple and oak that succeed them. Ironically, the shade that the sun-loving species provide eliminates them from the emerging mix of more durable forest cover that grows up under them and eventually overtops them. As the new tree species grow up to sapling stage, the energy in the system begins to rise with it and the increasingly shaded ground-cover starts to thin. The wildlife population changes accordingly, beginning on average a long decline as the energy of the woodland moves upward. The maturing growth

Figure 3.5. Clearcut in a managed forest. After several years a young successional forest will begin to regrow, providing habitat for many species of wildlife. After 7-10 years, however, it will mature into an even-age forest of less value to wild animals, and in 40-60 years it will be cut again.

of taller trees supports canopy bird species like tanagers and vireos along with bark and twig gleaners like chickadees and nuthatches. Tree-climbing squirrels become the most abundantly visible mammals, while tunnelers like shrews and pine voles replace other small mammals because of their ability to both feed and conceal themselves under leaf litter.

If the forest were left alone it would gradually take on some of the characteristics of an old growth forest, which would support most of the wildlife that was present in all its previous growth stages. However, woodlot forests are only allowed to grow to the point where the wood becomes marketable, at which time they are cut over again, the natural time scale is short-circuited and the cycle of succession begins once more. After its successional stage a maturing, closed canopy of more or less even-aged trees evolves, shading out the dense ground cover, after which its main energy contribution to the system is in the form of decaying leaf fall. Over 40-50 years it will grow this way, providing relatively little of use to wild animals until it is harvested once again before natural rotting has a chance to hollow out the heartwood of its trees.

However, if a managed forest were not cut over, it would begin to acquire old-growth characteristics. Contrary to the popular image of old-growth as a forest of towering trees with a closed canopy, actual old-growth is much more complex. As a forest ages toward this state, two things important to its wildlife content take place. First the older trees begin to lose their core wood, which is dead tissue, perfectly useless to the tree. The life of the tree, after all, is transmitted in the outer few inches of living tissue through which fluids move up and down the trunk from root to canopy leaf and back. The heartwood was only there in the first place to support the tree while it was young and relatively flimsy so that it could resist wind and storm. Once a tree gets to a certain girth, it can afford to lose this heartwood since the trunk achieves the mechanical strength of a cylinder. The tree thereafter is supported largely by the ring of peripheral wood, not by its core. As this core disintegrates, cavities from limb-fall or the work of woodpeckers begin to provide nesting, denning and hunting sites for a large number of animals (Figure 3.6). Barred owls nest in them, squirrels and raccoons den in them, fishers both hunt squirrels in them and use them as birthing dens. Furthermore, pileated woodpeckers dig into

Figure 3.6. A woodpecker hole in standing deadwood. Tree mortality in older woodlands provides denning and nesting opportunities for many species of wildlife.

the bark of old trees that support carpenter ant colonies and then use the holes themselves for nesting and roosting, abandoning them afterwards for use by other animals.

The second thing that happens to an older growth forest is that it starts acquiring greater vertical complexity (Figure 3.7). Where the woodlot forest is a stand of even-age trees with most of the energy in the single layer of the canopy, the old growth forest has energy and growth at all levels. As the woodland grows older, some trees succumb to age or disease and either lose their foliage, remaining as standing deadwood, or fall to the forest floor to decay. Either way, the chemical constituents of the entire tree are made available to new growth, rather than losing much of them to the mill. Furthermore, the sun space left by tree mortality promotes vigorous regrowth, fertilized naturally both by the sunlight and the decay of fallen tree trunks that are quickly hidden in the new undergrowth. The appearance of such an "older-growth" woodland, then, is a fine patchwork of giant old tuskers, middle-aged trees, saplings and underbrush, probably bearing a close resemblance to the pre-colonial forest. This mix of growth provides habitat at different levels for different animals, and the forest as a whole supports an improved diversity of wildlife, much greater than a woodlot forest or the even the early-successional growth in a clearcut.

True old-growth status is not achieved, by the various current definitions, until it is 200-300 years old. However, a grove of trees does not have to be left untouched that long to begin to take on some of the appearance and characteristics that are beneficial to wildlife. Instead it may start to develop vertical complexity in as little as eighty years, depending on the intensity of human use, as farmland, pasture or woodlot, prior to its reversion to forest. This may seem a long time compared to the mere three or four years needed for a clear-cut to succeed to young growth of benefit to wildlife, but once an older growth forest achieves this level of complexity it will continue to do so

Figure 3.7. Due to tree mortality and storm damage, an older growth forest develops an increasingly complex vertical structure that supports wildlife diversity.

indefinitely, that is if it is left undisturbed by harvesting. Even natural disturbances like ice and wind storms will only improve its complexity and value for wildlife. Successional growth, on the other hand, begins to decline as a wildlife resource in about a decade as it grows up to a young/middle-aged and even-aged forest.

There has been a concerted effort in New England to identify the scraps of old-growth forest that remain scattered through the forested landscape. What has been found is usually located in places where it was just too difficult and therefore uneconomical to cut it out. The tonsure of steep rocky hillside that surrounds the summits of the domed mountains typical of the Appalachians is one such location. Similar in inaccessibility are the steepest slopes of ridges in the Berkshires and Green Mountains. Another scrap I know of is on a plateau in the White Mountains that was so high, remote and boggy that it defeated even the avarice of the pulpwood companies whose mills fouled the air and water of the North Woods.

Most of these plots are remote and difficult to get to; that was, after all, what protected them. But on a smaller scale, patches of older growth can still be found on the steep portions of smaller hills. While perhaps not large or old enough to satisfy any of the current definitions of "old growth," these plots may still progressively acquire enough vertical complexity to support more and more diverse wild animals. This may also be true of many other woodlands, protected from the logger for one reason or another, that have begun to take on some of the desirable characteristics of old growth. These are sometimes wildlife sanctuaries when and if their staffs are aware of the benefits of such a forest condition.

I live in a wooden house and may be the last person around who still insists on wood skis. I certainly don't advocate against logging anywhere, but rather against logging everywhere, that is, against taking every last stick of marketable wood. I would like to see the creation of mini-wildernesses, areas within existing woodlot forests of perhaps ten acres or so that would be left uncut in perpetuity to support the diverse wildlife that gravitate to such woodlands.

Ironically, in my home state of Massachusetts some extensive woodlands that are coming to resemble the pre-colonial old-growth forest that the first white men found teeming with game are not the simple-structured commercial "wilderness" of western parts of the state but rather the woodlands of the Boston metropolitan park system.

Because the surrounding human population would not tolerate lumber trucks on their streets or cutting of their scenery, these woodlands have been left to age well beyond marketable value into something like the condition I have described above. It is true that dogs off-leash, brought to the park and released by the surrounding residents, repress many ground denning species like ovenbirds and hermit thrushes, and it is true that one of the main constituents of the pre-colonial forest, the American chestnut, is no longer present there or elsewhere, at least as anything much larger than stump sprouts. However, these woodlands are rapidly becoming older growth with many of characteristics beneficial to diverse wildlife that I have described.

Anyone looking for a variety of wild animals to observe in the woods, then, from the birder to the tracker intent on finding their sign in order to learn about their lives, should look not only for early successional growth but also for signs of older growth with complex vertical structure.

Tracking Beaver Ponds

Beavers have a bad reputation with foresters because the animals cut trees and a worse one with home-owners of low-lying property who fear flooding. However, the value of beaver ponds as a resource should not be overlooked by trackers interested in finding wildlife sign. Wild animals need food, water and cover. A beaver pond, which comprises both vertical and horizontal complexity, provides all three. As the forest gives way to the open area of the pond, the high tree canopy declines to dense shrubbery that provides browse at low height for deer and both browse and concealment for rabbits and other small herbivores (Figure 3.8). Closer yet to the edge of the water grow marsh grasses and sedges as well as various emergent plants. Red-winged blackbirds and rails nest in the cattails and leatherleaf while muskrats feed on the emergents. In summer out on the water itself beavers roll up lily pads like tortillas to feed them into their mouths. In the winter they swim under the ice to pull up the starchy tubers of these plants for the calories they store. In the summer as well moose use their long legs to wade out into the pond and then pull up with their long snouts the sodium-rich aquatic plants the salt from which they need to regulate their large muscle mass. The tracks and droppings of these animals can be found along the muddy margin of the pond where you may also commonly find the tracks of bears, raccoons, muskrats and herons (Figure 3.9). The point where the feeding brook enters a pond is a favorite location for otter haul-outs. These semi-aquatic mammals scratch up little piles of leaves and grass to anoint with urine and scat, the latter composed of fish bones and crayfish chitin. Dead snags that fall into the water are favorite locations for muskrat scat stations and mink often deposit scat on muskrat domes. Having spent a great amount of time in the last twenty years hiding with a videocamera on the edge of various beaver impoundments, I can testify to the amazingly lengthy list of macro-wildlife that can be directly observed in them along with the tracks and other sign that can be found along their margins.

Nature is struggle and change, and so it is with beaver ponds. Eventually, when the shoreline forage pays out, beavers temporarily abandon the pond and all their hard work. The dam breaches, the pond drains, and succession begins in its former bed (Figure 3.10). The muddy bottom of a pond is a repository for nutrients, leached out

Figure 3.8. Dense edge growth, stimulated by sunlight at the border of a beaver pond, provides food and cover for many mammals and birds.

Figure 3.9. A great blue heron wades in the shallows of a beaver pond hunting for fish and frogs.

of the surrounding hills. In impounding the water of the contributing brook, the beavers slowed the flow of water, allowing sediments to settle out into a nutritious matrix, now exposed to the stimulation of the sun. In the bed of the old pond a beaver meadow quickly appears, a dense growth of many varied plants that collectively provide food and concealment at ground-level for all of the animals described in the passage on successional growth, with the additional factor of better nutrition than is the case with cut-over land. These beaver meadows attract bears to succulent growth of skunk cabbage and other arums like jack-in-the-pulpit, both of which they relish. Deer feed on the succulent plants mixed into the low growth and, when the meadow begins to succeed to woody growth, they browse the foliage and buds as well. Rabbits, woodchucks, mice and other small "prey animals" hide and feed in the low growth in shaded seclusion. The foraging paths and runways of these animals await the careful and observant tracker.

Figure 3.10. After the dam breaches, the bed of the drained pond grows into a meadow rich with energy to feed browsing terrestrial mammals.

Tracking through the Seasons

Many people think of wildlife tracking as a winter activity. Certainly the tracks and trails of wild animals become much more apparent on a clean sheet of white snow than on the muddy margin of a beaver bog in late summer. However, the advanced tracker should be ready to challenge the other seasons for the information on animal life that they support. The other three seasons are the subtle seasons. Anyone can find and, with practice, identify sign on snow. Finding a brown track on brown mud and building up an ecological story from a discontinuous trail, however, requires a lot more practice. Here are a few of the expectations that these other seasons should arouse in the tracker.

Late Winter

Finding tracks and trails of mammals in late winter and early spring snowpack is not as easy as earlier in the winter. In the low light of early winter a track lasts until the next snowstorm. But from mid-February into early spring in the north temperate latitudes the sun, climbing ever higher each morning, melts out the shallower overnight trails of fox and bobcat by 10 or 11 a.m. Most of what one finds in the snow at this season are the trails of diurnal mammals like squirrels. Mammalian predators prefer to lay up during the day, curled up in a sunny spot where they will not be disturbed while waiting for the nightly freeze-up to stiffen the snow surface so that they can get around with a minimum expenditure of their precious energy.

Ironically, early spring that we associate with the rebirth of life is the season of highest mortality for prey animals. Deer that have been trying to stay alive by feeding on buds and conifer needles all winter succumb in numbers in March and April (Figure 3.11). This is a necessary testing period for their physical and genetic fitness. Those that are too weak to defend themselves against predators are killed and their weakness disappears from the herd. However, for predators and scavengers, which have themselves been tested by starvation all winter, this is the sudden season of plenty.

At this season, the tracker becomes increasingly attentive to the calls of ravens as they summon their fellows to a fallen deer. Ironically, the sudden disappearance of the many coyote trails he has seen distributed over the landscape all winter also signals a late winter kill. These scavengers and others localize their activities close to the sites of concentrated energy represented by the fat and protein of a deer carcass. Rapidly traversing the land on snowshoes or skis until he cuts a single coyote trail and then following that trail until it is paralleled by another and then another will often vector the tracker onto a kill site. At such sites he will find an abundance of sign of all the local scavengers. As the carcass is systematically reduced, it comes apart, allowing less dominant coyotes in the pack to remove a leg or two for concealment elsewhere where they will not have to surrender it to the more dominant members of the family. As the days go by the bones of the animal are exposed. Coyotes and other scavengers with strong teeth crunch these for the marrow and calcium. Bobcats, lacking strong carnassials, lick the bones clean of meat with their raspy tongue.

When most of the energy in the carcass has been recycled, the small mammals take over; mice, for example, reduce the bones by gnawing on them for their calcium. Coyotes go off in search of other carcasses or take advantage of seasonal vulnerability to create one of their own. If they are unsuccessful, they may return to old scavenging

Figure 3.11. Late winter deer mortality provides critical scavenge for many starving animals.

sites to gnaw on the bones as the mice do. The tracker may occasionally find a firm, well-formed coyote scat at this season that is solid and pale yellowish-white. This is what the scat of a starving coyote looks like when the scavenger is trying to glean a little energy out of nothing but bones.

Spring

As the snow thaws in early spring, tunnels that have been hidden below the surface all winter become apparent. First, the meandering tunnels of shrews close to the surface are revealed and then later in the thaw the deeper tunnels of red squirrels and voles at ground level are exposed (Figure 3.12). Scat that was found sparingly in the winter because it was periodically covered with new snowfall has been re-frigerated at different layers of the snow pack all winter. Now as the snow melts it all devolves to the single plane of the bare ground where it reveals the entire winter's history of feeding and predation.

Skunk cabbage is the earliest green growth at this season in the thawing seeps at the foot of hillsides. This arum is attractive to bears just out of hibernation that have little else to eat. The bears' saliva apparently contains an enzyme that coats the oxalate crystals in this

Figure 3.12. Meadow vole cores and runways. Spring thaw reveals the life that has been going on all winter under the snow.

and other arums that would act like ground-glass in the stomach of a human. Look for bitten leaves in the patches of green in these seeps for other bear sign.

Any early greening attracts starving prey animals. Deer, which technically are browsers rather than grazers, will graze new grass and other non-woody succulents in April before it hardens up and becomes abrasive to their teeth. Porcupines invade fields, as well, to graze on the grass shoots as they emerge from under the snow.

Many animals time their reproductive cycles so that they will drop their offspring as early in the spring as there is food available to sustain their growth. This gives them the entire summer to grow, fatten up and learn the survival lessons they will need in the next winter. As offspring emerge from dens, they often become vocal and draw the tracker to their location. In the spoil in front of a fox den one will find the remains of the larger animals that have been fed to them and get an idea of both the diet and the relative local abundance of various prey species. Young animals are dependent on their parents for more than food and training. They also depend on them for security, since they themselves don't know what to be afraid of and what not to. At this season, a fawn may approach you trustingly and fox kits may be summoned out of their dens by a whistle that piques their curiosity. Not until the fall will these animals, discharged by their parents into the cruel and testing world, develop a wariness of their own.

Summer

After the emergence of offspring from the den, they will in time begin to accompany adults to the food rather than waiting for the food to be brought to them. Even though the supervising adults try to corral them into hiding, the lack of wariness of these young animals as they insist on indulging their curiosity about everything they encounter in their new world makes them more visible directly than they will be through the rest of the year. This carelessness may result in mortality when the object of curiosity is a predator that the adult may be unable to ward off. That the highest rate of mortality among both predators and prey is within the first year of their lives is the reason many animals produce more offspring than are needed just to replace themselves. One animal's progeny is the means for another animal to sustain its own.

For the tracker, summer can be challenging, especially for finding tracks and trails. Late in dry spells he can visit the exposed mud on the margins of waterways where he will find beaver, otter, mink, muskrat and raccoon, muskrat trails perpendicular to the water and raccoon trails parallel to it. Because bears have a lot of body mass and have dense brown to black sun-absorptive fur, they need to drink a lot during the summer and so their tracks are often found near water (Figure 3.13). Oddly, one finds opossum tracks there as well at this season.

Sign is more frequently found in summer than are tracks. This sign is often in the form of feeding: browse on foliage, digs after invertebrates and turtle eggs, the resulting scat and so forth. Vague paths through grass can often be attributed to a particular species according to width, feeding destination and the surface of the path itself. Black bears pad down their trails into smooth paths while grizzlies, I'm told, often step so frequently in their own former footprints that they wear down a series of shallow basins in the soil. Such trails can be track-trapped by clearing and softening a patch of soil to record the footprints of animals using them. Trails made by hoofed animals tend to have a lumpy feel underfoot and show abrasions on roots and fallen

Figure 3.13. A search of the muddy margin of a beaver pond may turn up the tracks and sign of a thirsty bear.

logs where the animals nicked them with their cloves while stepping over. Predators leave scat at conspicuous locations such as trail junctions, gaps in stone walls and on top of stumps and trailside boulders.

As mid- to late-summer arrives, the berry crop ripens, attracting almost everything to the bounty of blackberry and blueberry patches. Remember the fable of the fox and the grapes? It's true that, as the wild grapes ripen, they too are exploited, especially by foxes and coyotes. The seeds as well as the odor and color of the resulting scat will tell right away what they have been eating.

Fall

This is the most difficult season for finding tracks since leaf fall is more or less continuous and quickly covers them. Even the muddy margins of ponds collect an obscuring cover of wet leaves. By now juvenile deer have molted into their gray winter coat and suddenly become wary independently of any accompanying adult. Adults and juveniles invade orchards as do black bears and coyotes, all after apples

97

and other fruit that has fallen to the ground. As hunters aim in their rifles on the day before the fall hunt opens, deer hear the reports of their weapons and have been trained by selection to know what is coming. On the day before opening day an orchard I know of has fresh deer sign everywhere, but on opening day the animals have disappeared, abandoning these fruitful locations for high ground or dense swamps where humans are unlikely to follow.

Beavers put the finishing touches on the mudding of lodges and dams as well as collecting cuttings and shrubbery to cache in the water next to their lodges for the coming winter (Figure 3.14). The energy of pond life recedes into tubers in the mud where the beavers can pull them up for their starch once the pond freezes over. Deer begin to feed on the buds of hardwoods as they will all winter. Jumping mice and woodchucks prepare their hibernacula for the winter and go into them quite early on as the cold approaches. Porcupines and deer dig after false truffles under hemlocks. Gray squirrels cache nuts individually, while red squirrels and chipmunks stock their larders in tunnels, inside stone walls and in the rotted root structure of cut stumps. In late fall the first light snowfalls begin to coat the ground, providing a readable slate for both the tracker and the hunter.

Early Winter

The timing of the first deep snows is important for the future year, for in a snowless early winter small mammals, denied an insulating blanket of snow and with little heat-retaining body mass, will die in numbers (Figure 3.15). This die-off can have a ripple effect through the food web: fewer mice and voles means a hard winter for foxes, hawks and owls. With the arrival of less than a foot of continual snow cover, voles and red squirrels begin tunneling through it next to the ground, using it for protection from predators as well as for insulation. Deep, soft snow in the first half of winter forces many predators to live mostly off accumulated body fat since such snow is energy-consuming to traverse and provides a muffling effect on the sounds of sub-nivean prey. Many predators find a windless spot near a rabbit or hare run and lay up, waiting for prey to run into their mouth as well as for the surface of the snow to firm enough to permit efficient travel. Weasels solve this problem by diving under the snow themselves, using their long, thin bodies to advantage in squeezing into the tunnels of their prey. Chipmunks, in turn, build a right angle into their subterranean

Figure 3.14. With a freshly mudded lodge and a cache of branches in place, the beavers are now ready for the winter.

Figure 3.15. White-footed mice recycle a bird's nest for insulation from the strengthening cold. Inside they can huddle together to conserve body heat.

tunnels so that the weasel with its long body will be unable to negotiate the corner. Deer gravitate to the south-facing edges of conifer groves where they can take advantage of shallower snow, back-radiation from the sun-warmed foliage and windless conditions. Every animal now goes on a diet, husbanding its energy in any way it can in order to stretch it until the following spring. For his part, the tracker, now, has a periodically refreshed page of varying snow conditions in which to improve his identification and interpretive skills.

Chapter IV

Patterns and Gaits

It will be helpful at this point to explain the general nature of diagrams of patterns and gaits introduced here and to elaborate on a few terms, found in the glossary, that will be used as verbal shortcuts in describing the relationship between an animal in motion and the patterns that its feet leave behind. We will also begin considering not only the various gaits that result in those patterns but also an animal's possible motives for choosing one over another.

Track patterns used in diagrams below are shown with the use of circles and ovals. Circles are used for (a) general patterns and (b) patterns of animals with relatively circular feet while ovals are used for (c) patterns of animals with relatively elongated feet, with the long axis of the oval aligned with the direction of movement shown in the pattern. In cases where a species has different sized front and hind feet, and where this is significant to the identification or interpretation of tracks and trails, I have presented the symbols at correspondingly different sizes as well.

Since an animal's trail is composed of a series of track patterns, let us begin this discussion of terms with the concept of "trailing," that is, following an animal for a distance in order to interpret its behavior. Since various species typically use certain gaits, trailing the animal may also be needed just to determine its identity in the first place.

Trailing

In coming upon an animal's trail the tracker has a choice. He or she can "foretrack" the animal by following it or "backtrack" it by proceeding in the opposite direction. For those of us whose main interest is in learning about wild animals rather than in actually finding them, we should foretrack and animal only if its trail is old. This is especially true in the colder months when the animals are under stress.

In winter animals are slowly starving. Some, with the best genetic and physical fitness for enduring this testing period, will barely survive to the lushness of spring. Others will not, ultimately to the benefit of the species if not the individual. Any additional disturbance, such as having to flee from a well-intentioned tracker, may be enough to tip the scale against that animal's survival. And one can learn as much from backtracking as foretracking, although the former requires a little mental gymnastic in translating the trail imaginatively into forward motion opposite to one's own.

Experience is the best teacher here, but book knowledge can help guide you in the right direction from any found sign of passage. From knowledge of the habits, habitat and preferred food of a species you can anticipate where the animal is likely to go next. This can be a great benefit in helping you re-find a momentarily lost trace or even find the trail of an animal in the first place. At least one prominent tracking book recommends this anticipation as a method of saving time in following the animal. That may be useful if you are a hunter or if you are trying to locate a man-eater in northern India. But for others the process of following the animal in detail is the reward itself, that is, in following and sensing the habitat through which it passes as it senses it. This is the intimacy-payoff, a feeling of inhabiting the wild animal's skin for a glimpse of its view of the world, in order to gain inside knowledge of the animal's psyche.

There are many examples of animal habits that may help you find an animal's trail or reconstruct a lost one. Bobcats, for example, often aim for and follow corridors along the north edge of powerline clearings, the southern exposure of which causes snow to crust quickly and melt early. Such a route supplies a double-benefit in that the man-made brushland in such cuts provides food and cover for just the sort of prey animals that bobcats like to feed upon: rabbits and hares. These cats also know where the gaps in stone walls are and regularly head for them, sometimes leaving a scat message in the gap. If you momentarily lose a bobcat's or a gray fox's trail in woods, look around in the general direction of travel for a downed log or other raised linear feature, since both species like to walk along them. Bobcat trails in woods often aim straight for thickets of laurel or other dense, shrubby growth where rabbits or hares may be hiding. Coyotes often follow the human infrastructure of trails and dirt roads, especially little-used ones. Gray foxes, on the other hand, usually avoid such infrastructure, preferring to stay in the woods. Bears typically cross dirt roads rather than follow them and equally often use dry streambeds for easy movement in their feeding range. Lynx have the unfortunate habit of walking on plowed roads in the winter, a practice that can decimate local populations when these cats are introduced to new areas more populated by humans than where they came from. For habitual passage on their circuits, fishers look for a downed log that forms a bridge across a stream, especially in the winter. Otters moving cross-country over the snow aim for southern exposures on their destination pond, especially shorelines that are overhung with conifer growth, where re-radiated long-wave radiation (infra-red heat) weakens the ice, permitting entry.

Once again, this sort of knowledge is acquired by combining book-time with woods-time. Over years of experience, for which this book is intended to help you prepare, you will develop a battery of knowledge, intuitions and search images to help you find or re-find an animal's trail so that you can continue to learn more about its hidden life.

For me, the ultimate objective of studying the track patterns and gaits is to visualize the wild animal in motion and to hook up the animal in motion with its environment. First, of course, is to identify the animal. But this only gets us to the starting line. A derivative and more interesting question is, "What did it look like in motion while it was leaving one or another track pattern?" Answering this second question and those that follow from it is one of the goals of "eco-tracking." Of course, we can't usually see the animal in motion directly, but we can determine what gait it was using from the pattern it left behind and so put the animal into motion in the mind's eye. From that visualization we can progress to the ultimate questions: "Why was it using this gait and not another?" "Was it traveling, hunting, attacking, fleeing?" "What characteristics of the landscape around it contributed to its decision to behave in one or another of these ways?" "Did the deer lope-bound through the woods because it was being pushed by coyotes?" "Or was it doing so to vault over debris on the forest floor?" "Is its zigzag bound when fleeing in the open an adaptation to confuse the aim of the predator that kills at a distance (with a spear, arrow or bullet)?" "Or is it a reaction to debris that the deer sees in the grass

103

before it?" From the narrow beginning of identification, then, the study of tracking can open up to broad insights into the otherwise hidden lives and motivations of wild animals.

Terms

Registration: Registration, which is the relationship between hind and front prints in a pattern, can be expressed in several ways. In some cases the hind "registers" within the outline of the front, in others it registers slightly outside of it, obviously outside of it, or completely separate from it. A different term is used to identify each of these relationships.

Single Registration: In this case, both the front print and hind print in a pattern register separately without their outlines touching. Most trotting patterns and the patterns of the faster gaits usually show single registrations. Also, a "wiggle walk" used by bobcats and bears shows this type of registration.

Direct Registration: Here the hind print registers totally within or totally covers the outline of the front so that the track appears to be composed of only a single registration. Which print, front or hind, dominates the appearance of the track varies with their relative size and the surface in which they were impressed. The large paddle-like hind foot of a beaver usually covers and erases the print of the front, while in species whose front and hind feet are nearly the same size, the front foot, bearing the extra weight of the head usually impresses more deeply and dominates the appearance. Direct registration is the pattern that appears in the walking gait of many predators that cover long distances such as red foxes and coyotes as well as bobcats. It is most useful in winter when pre-packing the snow with the front foot provides a firm platform for the hind, saving a small amount of energy with each stride. It is a habit predators pick up in their first winter, or that winter may be their last.

I also have seen video of a red fox leaving this pattern in an aligned trotting gait, in soft snow in which the animal appears to have been sinking down about 4-6 inches. We will explore this further in the section on trotting.

A number of sources refer to direct registration as "perfect stepping." Since this latter more properly describes the gait rather than the resulting pattern, I prefer the former, traditional term.

Figure 4.1. Typical trail of a domestic dog with sloppy double registrations and wide straddle.

Figure 4.2. Contrast the dog's walking trail with that of an eastern coyote, the efficiency of which has been honed by ruthless selection.

The habit of direct registration is one of the easiest ways to distinguish a wild predator from domestic dogs, which have lost the need for, and therefore the instinct for efficient direct registration (Figure 4.1). The red fox, an exquisitely winter-adapted animal, is the most relentless in direct foot placements while the gray fox, a more southern animal, is a little sloppy about this as, occasionally, is the coyote (Figure 4.2).

Indirect Registration: In this case the print of the hind foot impresses slightly off center from the front print, leaving the impression of a single, slightly distorted track. The distortion can be a slight understep or overstep, lengthening the appearance of the track, or it can be a lateral distortion as is sometimes seen in a line of otherwise perfect direct registrations of an eastern coyote.

Double Registration: There is some confusion in tracking literature about this term. Some authors use it to describe patterns of two single registrations and others use it synonymously with "indirect registration." Why waste a useful term? In this book, I use it to describe an indirect registration where the distortion is so obvious that it is possible at a glance to tell that there are two footprints involved. Among wild animals it is most often seen in the trails of prey animals that seldom need to travel long distances with any efficiency. It is usual in the trails of domestic dogs.

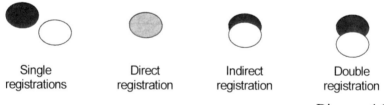

| Single
registrations | Direct
registration | Indirect
registration | Double
registration |

Diagram 4.1

Hereafter in diagrams throughout the book, front prints will be shown as black, hind as white and direct registrations as gray.

Straddle: Patterns of registrations can be measured in a number of different ways. Some of these may seem like quantification for its own sake, but one or two are indeed helpful in both identification and interpretation. The measurement of a pattern from one side to the other using either the front or hind feet is called "straddle." One prominent tracker insists that this measurement should be taken on the inside of each print, and another uses a center-to-center method. Most, however, measure from extreme outside to extreme outside, and this is the method that will be used in this book. For many animals moving in aligned gaits the terms "straddle" and "pattern width" can be used synonymously. These include mammals with more or less the same chest, hip and leg dimensions for their front and hind ends and are moving in an aligned gait. These include most of the larger predators as well as ungulates such as deer and moose. For other animals with less symmetrical forward and hind quarters, like mice, rabbits and squirrels, however, the terms "front straddle" and "hind straddle" are more useful as these can help decode the pattern into a mental image of the moving animal. At this point it is useful to know that in the faster gaits, straddle tends to narrow as an animal increases speed. This is not always so but is the case often enough to serve as an indicator in estimating the speed of an animal from its track pattern, particularly if the speed varies within a single gait — a slow lope of an individual animal, for instance, compared to its loping gait when approaching a gallop. The reason for this seems to be that lateral stability increases with speed. At slow speeds the bicycle rider who must steer radically to avoid falling over is in effect moving his center of gravity under the fall-off. But as his speed increases, the fall-off decreases as a function of distance and the action of the bicycle smoothes out, requiring only subtle corrections to keep it and its rider upright. As a wild animal increases speed within a general category of gait, its fall-off also decreases and the animal need only correct with subtle changes of foot pressure and straddle. This can be important in imaginatively recreating the event, the aftermath of which one is witnessing in observing and then translating an animal's trail.

Space, Time and Energy

Unlike the pattern it leaves behind, a gait is a temporal entity with three dimensions. Let's examine this statement in detail. A pattern in snow or dirt is a spatial entity, a thin slice of time translated to space, a photograph in two dimensions, if you will, which will last substantially unchanged until wind and weather soften and then erase its contours. The gait that made the pattern, on the other hand, is temporal rather than spatial, a continuous motion that existed before, during and after the pattern. Furthermore, a gait has three dimensions, whereas the pattern from which we must deduce the gait in order to recreate the event has only two. (It actually has three but the effects of track depth are irrelevant to the present discussion.) It is important to keep these differences between pattern and gait in mind when attempting to interpret an animal's trail.

A gait has two components that are largely invisible in the pattern it leaves behind. These are tempo and verticality. Two theoretical coyotes galloping side by side may have the same ground speed but one could create that speed by taking many short, rapid steps while the other could take fewer longer ones. If we measure the resulting patterns we may be deceived into believing that the trail with long patterns resulted from greater speed than the one with shorter ones. The difference in step lengths, then, represents a difference in applied power, not velocity.

The other component of gait that is not readily visible in the pattern is verticality. We cannot easily see in a two-dimensional pattern how high the animal's body arced during moments of suspension, with all four feet off the ground in between the patterns and, in the faster gaits, within it as well. An animal moving relatively slowly, with low energy input and a fairly flat arc of stride, can leave the exact same pattern as an animal using a lot of energy to launch itself high in between patterns. We may detect this additional energy indirectly in evidence of impact in the front prints, but the pattern itself will not show the difference. And that evidence of impact may or may not be present, depending on the looseness of the surface. We will discuss this in more detail later.

Most of us don't get to see coyotes running, but we can see marathoners on television. Note that runners moving in the same group, move at different tempos; some take many small steps while others, like the Kenyans, typically take long graceful strides. One might expect that the long stride would be less tiring than the quick short one and, indeed, the Kenyans seem very relaxed as they run. The short striders, on the other hand, look rushed and so we might expect them to tire faster. But the longer the stride, the necessarily higher the arc of stride must be. And verticality is wasted motion. So the shorter, flatter gait might seem to be more efficient. However, although each short stride has a short vertical vector, there are more of them over a given distance. Adding up all the factors evens out the comparison.

The Kenyans prevail so often probably because they are physically and genetically adapted for exertion in thin, hot air and also because their longer strides spread the exertion over larger areas and ranges of muscle, tiring the individual muscles less than those of the short-strider who uses fewer muscles with less range. Furthermore, they are so fit that they can afford to sacrifice a little efficiency for a little extra speed, even over a distance of 26 miles. That, presumably, is why they win and the short-striders do not.

This is the trade-off we see all the time in the trails of animals in the snow. A predator in the winter is on a strict energy budget, balanced between life on the one hand and death by starvation on the other. It is constantly making decisions about energy expenditure against the chances of hunting success. What are the odds that an energetic chase will result in a kill? Is it better just to keep on trotting in hopes of catching easier prey? Is it worth moving at all in fresh snow?

Imagining

Many analyses of track patterns bring the analyzer to his knees, both figuratively and literally. Moving on hands and knees to simulate the quadruped gaits of animals can be useful in gaining insight into the relationship between pattern and gait. One problem encountered at once is that the human on all fours cannot simulate verticality. On our hands and knees we are earth-bound, unable to duplicate the vertical dimension of the quadruped, that is, the moments of suspension in the course of any gait faster than a walk. And we are so clumsy on all fours that we cannot readily simulate differences in tempo either. Such simulations can only take place in our minds by imagining ourselves as wild animals, with bodies and capacities very different from those we were born with. A talent for this seems to vary with individuals, but don't overlook this possible avenue to insight simply because you find it hard at first.

Categories of Gaits

Using our Western modes of thought, the animal gaits that are expressed in the patterns they leave behind can be divided up into several categories. First of all, they can be viewed as "flat" or "bouncy." In a bouncy gait, a dot at the animal's center of gravity, when viewed from the side as the animal moves forward, follows a series of arcs, the height of which varies with the specific gait and the amount of energy introduced by the animal. In a flat gait, on the other hand, that dot rises and falls only slightly. As we will see, walking and certain kinds of trotting are flat gaits, whereas many of the faster gaits, such as loping, galloping and bounding are bouncy to varying degrees. A second division may be made between "aligned" and "displaced." In an aligned gait, the animal keeps its spine parallel to the direction of travel. In a displaced gait it slants its spine at an angle to that direction. Generally, displaced gaits have a lower arc of stride than aligned gaits and are more efficient on a firm, even surface. Finally the faster gaits, such as lope, gallop and bound, can also be divided into transverse or rotary, depending on the order of footfalls.

In any gait faster than a walk, a quadruped has the problem of how to get its hind feet past its front without tripping itself. This may require some explanation. Temporally, the front feet of a moving animal, being at the head of the torso, strike the ground first. Now comes the part that we bipedal humans sometimes find difficult to apprehend. In the case of an animal moving at a trot or faster, the hind feet cannot just catch up and register behind the front. To make any speed at all each hind foot must pass the front foot, or at least the print of the front foot of the same side, and register ahead. When we look from an overhead, following perspective at the resulting pattern of such a gait, the front print on each side registers at the bottom of our frame of reference or "below" the hind print of the same side, which registers at the top. This is just the opposite of what we might intuit. Here is a typical squirrel bounding pattern to illustrate (Diagram 4.2). The animal is moving toward the top of the page with the smaller, black ovals representing the front feet.

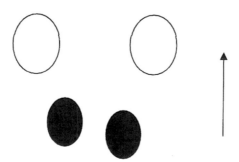

Diagram 4.2

Even though the squirrel is bounding toward the top of the page, its front feet appear at the bottom of the pattern and its hind at the top. How the squirrel physically accomplishes this magic can be hard to see directly since the animal's legs may move too fast to follow with the eye. Temporally, the right front foot is the first one to strike the ground, followed closely in time by the left front. This time difference is reflected in the fact the right front registers spatially a little below the front left in our frame of reference. The hind feet then pass by the planted front feet on the outside of each, landing simultaneously above the front in the pattern and, thus, spatially side by side.

The movement of hind past front that is required by the faster quadruped gaits is accomplished in one of two ways, depending on whether the gaits are aligned, as above, or displaced.

Aligned Gaits and Their Patterns

In an aligned gait, the animal holds its spine parallel to the direction of travel. In the faster gaits, however, it is faced with the problem, cited above, of getting its hind feet past its front without tripping itself. It solves the problem by elevating its body in a moment of total suspension so that all its feet are off the ground at the same time and then makes the switch, with the hind feet moving directly over the vacated prints of the front. In an aligned trot the pattern will look something like Diagram 4.3. Here, the animal is moving from left to right and the larger front foot (black oval) plants first (temporally) in each pattern, with the hind registering an instant later and farther ahead (spatially).

Diagram 4.3

An aligned trotting gait has a fairly high arc of stride. That is, there is quite a bit of vertical motion involved so that, viewed from the side, the gait looks bouncy. On a flat, even surface such a gait is inefficient, much energy being lost in the vertical vector. However, in the dense brushy cover frequented by bobcats, for instance, such a gait makes sense because it provides narrow frontage and also allows the animal to bounce over obstructions.

It is worth noting that this same pattern can be left by a particular walking gait that I call the "wiggle walk" often used by bears and bobcats whose flexible tree-climbing pelvises can twist in the direction of travel so that the hind foot passes over the front just as it leaves the ground and then plants ahead of the front print. In the course of the stride cycle, only one foot at a time leaves the ground, qualifying it as a walking gait. The resulting pattern is exactly the same as the aligned trot. Distinguishing between the two can be difficult and will be discussed later.

Displaced Gaits and Their Patterns

Because aligned gaits are inefficient, animals that habitually travel over flat, even surfaces use a different, more efficient technique. A fox rimming the edge of a snow-covered pond where the snow has been packed by the wind or a coyote using a dirt road to get around his hunting range will use a displaced gait. In this case the animal slants its spine at an angle relative to the direction of travel. In this posture the hind feet pass *by* rather than *over* the front prints. Not having to elevate its body in a moment of suspension in which to switch feet, the gait remains quite flat, greatly reducing its arc of stride and with it the wasted verticality typical of aligned gaits. The classic fox or coyote "displaced trot" pattern is shown in Diagram 4.4. with the animal once again moving from left to right and increasing its speed.

Diagram 4.4

Now all the front prints are lined up on one side of the trail and are "retarded" while all the hind prints are lined up on the other side and advanced. The resulting gait can be hard to imagine but if you think of one of those salvaged wrecks you occasionally see on the highway where the frame of the car has been twisted so that the rear wheels run off to one side of the track of the front wheels, you can get an idea of what this looks like. The gait can be observed directly by watching Rover moving over an even surface, perhaps jogging with his master. The dog isn't interested in exercise, but rather on minimal effort since it is only doing this out of obedience to its owner. So the dog uses the most efficient trotting gait it can — the displaced trot. Elbroch and some others refer to this particular gait as a "side trot."

These displaced patterns of two prints (2X displaced trot) reflect a gait where the animal is stressing the horizontal component over the vertical. Because this gait has a very low arc of stride and energy is not wasted pushing upwards, it is more efficient than many of the aligned strides. Its drawback is that, because it is not very fast, it is not useful for chasing or escaping. Furthermore, it can only be used on flat firm surfaces since its low arc of stride will not carry the animal over irregularities or softness in the surface. An animal attempting to use a displaced trot through untracked forest would trip over tree roots and fallen limbs while trying the same thing in soft snow would require a lot of plowing through and would quickly exhaust the animal. A displaced trot is most useful for medium speed traveling over a firm, flat surface such as a dirt road or wind-packed snow while the animal is on its way between denning and hunting areas, let's say, or just covering ground until it gets vectored onto some quarry by scent or sound. It is most often used by coyotes and foxes, especially red foxes, because both of these animals typically hunt by putting in a lot of miles. Eastern coyotes are especially clever at using man-made roads for getting from here to there when we are not around. Roads, of course, provide a perfect surface for this gait.

The choice of aligned versus displaced occurs in other gaits as well, especially in lopes and lope-bounding. These will be discussed later in the chapter. It is worth noting at this point that the speed of a displaced trot is variable and may be judged by the separation within and between each pattern of two. As the speed increases, the two-print patterns and the inter-pattern distance stretch out as in the illustration above.

Transverse and Rotary Patterns

These patterns are seen in faster gaits such as a lope or gallop. In a transverse gait, the order of footfalls is left front followed by right front and then left hind followed by right hind. It may also be initiated by the right front in the order right-front left-front right-hind left-hind. Compare this to a rotary gait in which the order is right-left-left-right or left-right-right-left. A transverse galloping pattern with the animal moving left to right, then, is shown on Diagram 4.5A. There will be a gap between each of four-print patterns (4X) and the next.

114

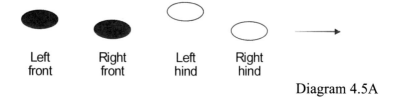

Left	Right	Left	Right
front	front	hind	hind

Diagram 4.5A

A rotary galloping pattern, on the other hand, is shown in Diagram 4.5B.

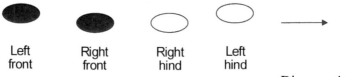

Left	Right	Right	Left
front	front	hind	hind

Diagram 4.5B

In both patterns, the two front prints — on the left in the diagrams above — register first, followed by the two hind prints.

In both cases, the front feet reach out in front of the speeding animal and strike the ground. As the animal's body moves forward over these planted feet they, in a relative sense, pass backwards under its body. Meanwhile the hind feet leave the ground behind the animal and move forward, passing on either side of the retreating front legs so that for a moment the two sets of legs, viewed from the side, are criss-crossed under the animal's body. (To imagine this, think of the common cartoon drawing of a dog chasing a car.) Finally the hind feet plant forward of the prints made by the front feet. The order in which the front feet and then the hind feet strike the ground determines whether the pattern left behind is transverse of rotary. While the diagrams above illustrate galloping gaits, the same order of footfalls can be used for loping and bounding.

Because of their shape, sometimes these patterns are called T-patterns and C-patterns respectively. They can have many variations and the choice that the animal makes between the two seems to have as much to do with the ground in front of it as any supposed increase of speed of one over the other. A transverse pattern, especially one where the hind prints appear almost even with one another, allows the animal to shift its direction right or left to account for the maneuvers of a dodgy prey animal or to avoid an obstruction in the predator's straight-ahead path. A rotary pattern, however, is more common where straight ahead speed is needed, as in escaping or in running over ground that is level and without obstruction. Examples of both these gait patterns by a red fox are illustrated in Figure 4.3.

Here the fox used a rotary pattern at the bottom of the frame, changing to a transverse gallop for the remaining. In this case the transverse patterns are as stretched out as the rotary, with each print equally separated from the others in each pattern. Therefore, both patterns probably represent equal speed. The animal may mix and match these patterns as it chooses. The uniform surface in the photo suggests that the switch probably reflects random choice.

In rotary lopes, occasionally the pattern that results looks like the trail shown in Diagram 4.6.

Front   DR   Hind

Diagram 4.6

This is not the trail of a three-legged animal! The middle track in each pattern is actually a direct registration of the right hind over the right front print with the hind concealing the front one beneath. This is a slow loping gait occasionally seen in weasels and porcupines. A couple of winters ago while tracking with Joe Choiniere in the woods behind his place in Hubbardston, Massachusetts, we found a lot of porcupine trails like this, representing a slow loping gait. This was probably as fast as the porky could move. Since the woods were also full of the trails of red foxes hunting, the porcupines may have found it prudent to make speed from tree to tree!

Direct Registration and the Walking Gait

To follow the trail of a red fox walking the perimeter of a pond covered with an inch of snow is to follow beauty in frozen motion: the sparkle of the morning light on the snow, the graceful arcs of the line of tracks as the animal rimmed the pond, stopping briefly to investigate a spot on the shore before starting a new arc to the next spot that caught its attention. Each foot was placed so precisely, 15 inches apart with each mark in the snow lying directly ahead of the one before like stitches by a sewing machine. This is a spare and efficient beauty.

116

Figure 4.3. The galloping trail of a red fox. The animal was moving away from the camera. The closest 4X pattern is rotary, the rest transverse.

The fox learned in its first winter to walk in this manner, not to create beauty for the eye of the human follower, but to conserve a tiny bit of energy with each step by pre-packing and perhaps pre-warming the snow under the front foot for the benefit of the following hind.

The first question is, "How did the fox, as it walked along, make one mark in the snow with two feet? That is, how did the animal "direct register?" Unlike any of the faster gaits, this one is easy for the human to simulate by getting down on all fours. Earth-bound as we are, we cannot simulate the third, vertical dimension of fast gaits, but luckily a walking gait has, practically speaking, only two dimensions. First, move your left knee, simulating a quadruped's left hind foot, forward. Now in order for it to replace the left hand on the ground, that is, in order to direct register, the left hand has to move. Just as the knee approaches the hand, lever the hand forward so that its heel rises and allows the knee to slip underneath. At the last instant before applying weight to the knee, lift the fingertips clear and the step is completed. To finish the gait cycle, move the left hand forward and place it out ahead. Now your "footprints" are in the shape of a parallelogram, with all four on the ground at the same time. The next move is with the right knee, which moves forward and replaces the right hand, which in turn moves forward and plants out in front, reversing the parallelogram.

Note that throughout the gait cycle, three "feet" are planted at all times, with only one foot moving at a time. This preserves tripod stability at slow speeds and qualifies the gait as a walk. (If any two feet, one front and one hind, are out of contact with ground at the same time, the gait becomes a trot of one sort or another.) If you perform this little exercise in snow or sand, you will see that the trail you leave behind is quite wide and rather clumsy-looking compared to the elegant trail of the fox. But it is a human's natural crawling motion, the one a baby uses crossing a rug. Our facility with it is perhaps more evidence (beside my aching lower back) of a quadrupedal human past.

Although I have already suggested a reason or two for the animal having this gait in its repertoire, I should elaborate a bit since the habit appears in the behavior of so many animals like foxes, coyotes and bobcats (figures 4.4, 4.5, and 4.6). In snow, or any other soft, deep footing, there is an obvious advantage to the animal in pre-packing the snow with the front feet for the support of the hind. Soft snow

Figure 4.4. A direct register walking trail of an eastern coyote. It may be more than one!

Figure 4.5. The direct register walking trail of a bobcat. The double registration of the 4th track from the bottom and the slightly wider straddle than red fox or coyote suggest the more flexible legs of tree climber.

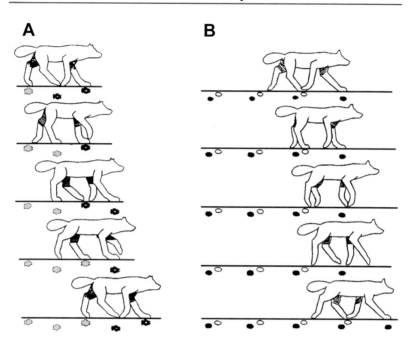

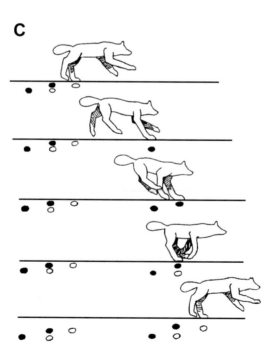

is such an energy-waster in winter when energy is so dear that predators that hunt on their feet use every stratagem available to them to save it. A coyote, covering miles in every night, can conserve a tiny amount by this pre-packing. He can also do it by restricting his movements to packed deer trails. And he can do it, as well, by stepping directly into the walking tracks of the coyote ahead of him. Often I have followed a line of perfect coyote direct registrations in the snow, thinking that I was following a single animal, only to have the following animal's trail diverge momentarily to lift a leg on a stump before falling back in line with its companion. Occasionally a whole "family pack" of this species will proceed in this manner, but to look at the trail you'd swear there was only one animal.

The tamping down of snow by the front foot in the direct-register walk may also pre-warm the track for the hind foot and for the feet of others that may be following immediately in a line. Once again this may be a tiny amount of energy, but repeated several thousand times a night by a species that hunts on the move, it can add up to a significant saving in energy.

Although direct registration makes sense in the winter, why does the adult coyote on the sandy margin of the reservoir in the summer leave the same pattern in its walk? Part of the reason may simply be the physiology of habit. Move the same way thousands of times in one medium and you are likely to perform the same motion in other mediums, as well, where such efficiency is less critical. In fact, the easiest way a tracker can distinguish an adult coyote or fox from a juvenile is by the precision of its foot placements. The youngsters have yet to

◀ **Figure 4.6. (A)** Walking: pattern of "alternating DR," that is, a zigzag trail of alternating direct registrations. In similar patterns the hind feet may understep, registering behind the front, or overstep, registering ahead of the front, or indirect register laterally. **(B)** Trotting (displaced): pattern of "repeating slant 2X SR," that is, patterns of two prints arranged on a slant that repeats from pattern to pattern rather than alternating. Note that in this pattern, common among canids, all the front prints are retarded and lined up on one side of the trail while all the hind prints are advanced and lined up on the other. **(C)** Lope-bound: pattern of "transverse 1-2-1," that is, a pattern that begins with a single front print followed by a pairing of front and hind arranged even or on a slant and finished with a single hind print. The order of placement here is "transverse," that is, in the order of right-left-right-left.

121

experience winter and so have yet to perfect the habit of direct registering, as they will during the following winter or not survive it.

The use of direct registering by a bobcat at seasons other than winter may have practical value. Bobcats, unlike coyotes or foxes, are more stealth hunters. As the bobcat stalks through dense white pine regrowth, for instance, it reaches out gingerly with one front foot and visually chooses its placement to avoid any twigs or other noisemakers. But as its hind foot moves forward, the bobcat cannot see it or its placement, since the cat's body has rocked forward as well, blocking its view beneath and behind it. By replacing the front foot during a stalk with the hind on that side, the animal is sure that putting it down and applying weight to it will make no more noise than did the front. In effect the foot placement has been pre-tested against the possibility of a snapping twig.

Even with "coincidence hunters" like foxes and coyotes, however, direct registration can have practical advantages for movement over broken terrain. Once the predator detects lodgings for its front feet in the lower half of its field of view, it then can put the hind in the same spots without having to pay any more attention to them. Instead it can keep its head up and eyes scanning its surroundings, alert to for potential prey that it may spot before being spotted itself.

One might ask why domestic dogs don't direct register as neatly as their wild cousins. The answer is the same as for all the other sloppy characteristics that distinguish the trail and behavior of a domestic dog. His survival doesn't depend on perfect foot placements but rather it depends on pleasing his master, who will feed him at the end of the day and provide him with warm shelter. Over the centuries dogs have been bred away from their wolfish origins toward a variety of characteristics: guarding, racing, retrieving, pointing and so forth. None that I know of were ever bred for efficiency of motion. Only life in the wild requires that characteristic.

Verticality

There is no one-to-one relationship between patterns and the gaits they record. Some patterns, for instance, can be made by more than one gait: a slant 2X pattern can represent a high bounding gait or a flat displaced trot. Practically speaking, patterns are two-dimensional while gaits are three-dimensional, having not only length and width but also height. This height or "verticality" of a gait is an invisible factor and accounts for much of the confusion in relating patterns to the movement that produced them.

Verticality is the vertical component of the force vectors in a stride. That is, in any gait cycle, part of the animal's energy is devoted to horizontal movement and part to vertical movement. If you attached a light to an animal's shoulder or hip and then obliged the animal to move across one's field of view, the light on its body would describe a series of arcs that I call "arcs of stride." If you diagram a few of these arcs, the vertical vector would be a vertical axis from the ground to the high point of the curve and the horizontal vector would be the horizontal axis along the ground.

A Displaced trot   B Weasel bound

Diagram 4.7

On a flat, firm surface the most efficient gait would be one that maximized the horizontal vector and minimized the vertical (Diagram 4.7A) since energy used to push the animal upwards is wasted in terms of the ground covered. A coyote is a long-distance hunter. It should not be surprising, then, that on a surface free of obstruction such as a lightly traveled dirt road, the gaits of this animal and others that hunt by their feet are ones with very low arc of stride such as the displaced trot that occasionally slows to a walk for sniffing ground scents or for approaching something interesting. The short-legged weasel in soft snow, on the other hand, cannot reasonably use these gaits since their flatness would cause the animal to plow a lot of snow. Instead it must choose a gait with a higher arc of stride, introducing more verticality to clear the snow surface (Diagram 4.7B). The gait invariably chosen is a bound, one common version of which leaves a pattern that can be termed a "repeating slant 2X DR," that is, a pattern of two tracks, each a direct registration representing the impression of a front and hind foot, arranged on slants that repeat rather than alternate. If you draw lines in the snow across the leading edge of

123

these tracks, the slanting lines will be parallel. Diagram 4.8 shows what this pattern in the snow looks like:

Diagram 4.8

On a smooth, solid surface such as a snow crust, on the other hand, a weasel has no reason to introduce energy-wasting verticality. So it stretches and flattens its gait into a lope, the pattern of which looks something like Diagram 4.9.

Diagram 4.9

An interesting exercise that helps to understand the differences and transitions between a DR bound pattern and a lope is to make a pair of transparent overlays. On one draw with a grease pen two black ovals, representing the front prints, arranged in a diagonal bound pattern. Place the second overlay over the first and draw the two blank ovals (the hind prints as shown in the illustration above) directly over the front ones, the result of which creates a DR bound pattern. Now slide the upper transparency forward, first a little to show a double-register bound pattern and a bit farther to show a slow transverse lope, then farther yet to show a faster lope. Now invert one of the overlays and play with it to see a 4X double-registering-bound pattern, which, when slid forward, becomes a rotary lope pattern.

The lope pattern shown above is often referred to as a 1-2-1 pattern and has many variations that depend largely on the speed of the animal. (The farther to the right that you shift the upper overlay, the faster the gait.) It is really a similar physical motion to the weasel-bound but is stretched out so that, instead of the hind feet direct registering in the prints of the front and then pushing upward, more energy is devoted to pushing the animal forward.

It should be noted at this point that the patterns described above are "aligned" gaits, with the hind foot on each side passing directly onto or over the impression of the front on that side. However, the loping gait has a displaced version, as well, most often used by members of the weasel clan. Its pattern may look like Diagram 4.10.

Diagram 4.10

Here the animal has twisted its spine diagonally to the direction of travel so that the hind feet pass by the front feet, before the latter leave the surface, rather than over the vacated front prints. The order of footfall is left-front, right-front, left hind, right-hind. Note that the temporal order differs from the spatial order.

This displaced version has the advantage of lower arc of stride than the aligned one, requiring less energy per cycle. However, the trade-off is that it has a slower maximum speed since the forward motion of the hind is limited by the presence of the front foot of the same side. This is a major disadvantage for a predator that hunts on the go. However, this slower displaced gait may be necessary when moving over breakable crust. In fact I once spent half a day tracking two fishers who taught me a lesson about this. Fishers, large members of the weasel family, are dimorphic with the female substantially smaller and lighter than the male. Both were hunting over fresh snow, the surface of which had been stiffened by some mist at the end of the last storm. The female was able to use the faster aligned pattern throughout. The male, hunting the same range as the female, tried occasionally to speed up to the female's velocity by using her aligned lope. But a faster gait implies a higher arc of stride and, every time he did so, the increased impact as he came down on his front feet caused them to break through and he would stumble, forcing him to revert to the slower but flatter displaced lope.

Here are two more terms that will be encountered from time to time in our discussion of track patterns and gaits.

Resistance: "Resistance" is a term I use for the degree to which the terrain or ground conditions inhibit efficient horizontal movement. The flat, hard road or firm crusts mentioned above are surfaces with low resistance. Habitual road travelers like coyotes will exploit these flat, even surfaces by using an efficient displaced trot with the least verticality of any of the running gaits. On the other hand, deep, soft snow on a forest floor covered with blowdown and windthrow, or dense growth at the edge of a beaver pond, are both high-resistance terrain. An animal's reaction to such resistance varies. A bobcat moving through dense cover is more likely to use an aligned gait, which gives it less frontal area so that it can insinuate itself between and around obstacles without creating disturbance. If it wishes to cover ground in this terrain, that is, if it is traveling more than hunting, it may choose an aligned trot that has the same narrow frontal profile as a walk but adds verticality that allows the animal to bounce over obstacles in the dense growth at its feet. Given the tendency of bobcats to frequent denser growth than either foxes or coyotes, their preference for aligned gaits should not be surprising. Bears, on the other hand, are hardly stealth hunters. They may use their weight to crash through brush and blowdown or may squeeze under or clamber over fallen branches and tree trunks. Each tactic is a moment to moment response to movement through high resistance terrain. The coyote, the bobcat and the bear in one way or another either introduce verticality or minimize it according to their traveling and hunting needs.

Context: In analyzing a pattern on the ground to determine what gait was used to make it, one might keep in mind that track patterns should not be viewed in a vacuum; one should look at the context and anticipate the vertical vector as an alternative for overcoming resistance. This may not be easy to do. In trying to distinguish a direct-register walk from a DR trot, for instance, one has to look closely for other information since both leave the same pattern. First, look at the leading edge of each track and examine the plume of debris. Is it a short, laterally narrow mound or does it stretch out forward of the track? In the case of a red fox, a short plume suggests a walk since these animals are very dainty steppers that lift their feet free of the snow before placing them down again. A longer plume, on the other hand, suggesting more energy and speed, points to the aligned DR trot. Look as well at the conditions. Is there a reason for a DR trot? Foxes are only likely to use this gait when there is considerable track depth or other resistance. In snow the animal is sinking down with each foot placement and wishes to vault over the intervening snow, a motion with a high arc of stride that otherwise would be energy wasteful. Finally, one might look for evidence of impact. This is usually snow or sand rimming the track or splash marks in slushy snow radiating

126

outward from the track. Such impact evidence suggests a high arc of stride typical of higher energy gaits such an aligned trot.

High-energy Patterns and Gaits

All possible gaits for quadrupeds can be divided into two groups: low-energy and high-energy. Low energy gaits are those with little verticality, that is, with a low arc of stride. These include walking and displaced trotting. High-energy gaits require varying amounts of energy to be used in pushing the animal up as well as forward. Thus they have more verticality, that is, a higher arc of stride. Their energy use ranges from moderate in the case of an aligned trot to very costly in the case of a bound or gallop. The horizontal speed of the animal only roughly reflects the amount of energy expended. In the higher energy gaits the height of the vertical vector has also to be considered and this, in turn, is dictated by the softness of the surface. A lope is both slower than a gallop and requires less energy. A bound on the other hand can be quite slow horizontally but requires a great deal of energy because its arc of stride is so high. As a result, this last gait is used mostly on surfaces with a lot of resistance.

Let's begin by discussing how quadrupedal and bipedal gaits are similar and how they differ. From there we will look at a couple of high energy gaits and the patterns that they leave behind.

Bipedality and Quadrupedality

Since most of us are familiar with how we ourselves move, it may be helpful in explaining quadrupedality to examine our own bidpedality as an introduction. Running by humans is a series of leaps off of one foot at a time. As we come down from each arcing, forward leap, we advance one foot in order to stop ourselves from crashing to the ground. So during this phase of the stride the extended leg has a propping function and, at the jogging gait that most of us use, a braking function as well. As our body passes forward over the planted foot, however, the function changes. Now, with the foot to the rear of our center of gravity, we are able to apply some force to it in order to push ourselves forward and up into the next leap. Once this foot, now well behind our center, is forced to leave the ground and move forward (it being attached to our forward-moving body), a moment of suspension is created where both our feet are off the ground. This

127

moment of suspension lasts until the other foot, advanced forward of our center, comes down from its leap and strikes the ground to complete the stride cycle.

Quadrupeds do the same thing when they run, but they do it with four feet rather than two, divorcing the propping and braking function, on the one hand, from the propulsive function. In an easy lope, the front feet prop and the hind feet propel.

The difference between a quadruped's easy running lope and a gallop is the difference between jogging and sprinting by a human. As we jog along at an easy pace, the braking function of the first part of our stride is pronounced; we introduce it so that we can stay within our cardiovascular limits, allowing us to continue longer than we could if we eliminated the braking and sped up to a fast run. This braking action is one of the reasons jogging is relatively slow. In a sprint, however, the braking function is reduced, with most of the energy of the gait delivered to the propulsion forward and up. The sprinting gait also flattens out, with energy from the vertical vector taken away and introduced into the horizontal vector. Not only is the arc of stride lower than in a jog, but the body position in a sprint is slightly lower, allowing each foot in contact with the ground to remain in contact slightly longer. Although many of us can jog a mile, few have the cardio capacity to sprint a mile. Sprinting is wasteful of energy partly because of the rapidity of leg turnover and partly due to the lower body position, which requires deeper knee bend as the body passes over the foot. It also allows less restoration: during the propping phase of each stride in a jog, the body is allowed an instant to rest before propulsion is required. During these instants the running muscles can at least partly replenish themselves from stored energy and eliminate by blood-flow the waste build-up as well. These instants of restoration add up; the lack of these instants in sprinting or galloping rapidly exhausts the fittest of our kind and of most wild animals as well.

Lope versus Gallop

The popular notion of the term "lope" is an easy run, and the usual meaning attached to "gallop" is a high-speed, high-energy sprint. While some tracking books assign esoteric meanings to the words or, in at least one case, see no difference at all between them, I will use the popularly understood meanings here.

128

Galloping is a high-energy and inefficient gait used by most animals either in the closing moments of a pursuit to a kill or, more often, as an escape gait. The pattern it leaves in snow or sand is similar to that of a loping gait, the mechanics of which it superficially resembles. However, in attempting to interpret a running animal's trail it is important to try to distinguish between the two from the evidence they leave behind. In so doing we are not only imaginatively recreating the event but we are also enabling some speculation about motive, since these two similar gaits are used for very different purposes.

Just as with the human jogger, a loping quadruped must substract from propulsion, devoting part of its stride cycle to preventing itself from collapsing to the ground. However, having four legs it can separate the functions and use only the front legs for propping. These legs are kept fairly straight, without the deep bend that would put them in position to deliver power to the stride, and this straight, propping posture is less stressful. Furthermore, as the gait cycle is slower and the body position higher than in a gallop, the lope is a more efficient way of covering ground over a firm surface. Being a neutral gait, in a sense, it can easily transition with minor modification to other gaits, more energetic or less so, as changes of surface, slope or situation require.

In a gallop, on the other hand, motive force in propelling the animal forward is applied not only by the hind legs but also by the front. The only native wild mammal that I know of in North America that can sustain a gallop for any distance is the pronghorn of the open regions of the American West. For this animal, galloping is a means of escape from traditional predators such as wolves but more often from the human predator with which it has coevolved for centuries.

Both a quadruped's lope and its gallop are running gaits with a rolling, rockinghorse motion; the hind end of the animal is down when the front is up, and vice versa. Lights positioned at the animal's hip and shoulder would rise and fall in a pair of out-of-phase sine curves as the animal moves past.

Both gaits leave either transverse or rotary 4-print patterns. Distinguishing between the two gaits from their patterns can be tricky, especially when the animal blends the slower gait with the faster, as animals will. It may help to describe the similarities and differences in the two gaits in detail to get a start on distinguishing them by their patterns.

The Loping Gait

In a lope, almost all of the force in the stride is generated by the hind legs. But these legs are not asked to propel the animal forward forcefully. Instead, at the moment when the legs are criss-crossed (when viewed from the side) under the animal and the hind legs are poised to propel forward forcefully, the animal "pulls its punch," so to speak, and introduces a sort of half-hearted push. As a result, the front feet, extended forward, contact the ground prematurely, first one and then the other. As the animal's trunk moves forward, the two front feet prop the front end of the animal without imparting force to its motion. As the trunk of the animal moves farther forward over the two planted front feet, the hind legs are moving forward even faster, passing on either side of the front so that for a moment the front and hind legs once again are criss-crossed. Just when the front feet are forced by the forward motion of the animal to leave the ground, the hind feet plant, first one and then the other, in position to push the animal forward into the next stride cycle. As the animal is powered in this way into the next shallow arc, the front feet, advancing out in front once again, contact the ground just before the hind feet leave the ground. This is an important instant, since it spells the difference between a lope and higher energy gaits where the front feet are still suspended in the air at the moment the hind feet leave the ground. Technically, a lope, then, is described as having little or no "extended suspension" where all four feet would be in the air, two extended forward and two backward, at the same time. It also has little or no "gathered suspension." When the front and hind legs are criss-crossed under the animal's body in a lope, either one set or the other is in contact with the ground or there is only an instant when all are suspended (Figure 4.7).

The Galloping Gait

In a gallop, the animal is trying to get as much forward speed as possible. To do so, it modifies the neutral loping gait in a number of ways. First of all the animal wants to get as much forward propulsion out of all its legs as it can. Since air time is wasted time and verticality is wasted upward motion, the animal lowers it trunk closer to the ground. In this position it reduces the vertical radius of the ellipses its front and hind feet describe during the gait cycle and increases their horizontal

Figure 4.7. The elegant trail of a loping red fox on the firm snow of a frozen pond. The separation between 4X patterns is after the first four prints from the right. As the fox moved from right to left, the two larger front prints in each pattern landed first followed by the two smaller hind, which are advanced. Note that the middle two prints in each pattern are closer together. This is a fast lope. If the middle two prints were reversed in order, it would be slower lope.

length. This means that both its front feet first and then its hind feet are in contact with the ground for a longer portion of the stride than with the higher position of the loping gait. To visualize this, see the animal reaching forward with its front feet for the ground. The lower arc of stride allows it to make contact a little sooner and the lower body position allows it to make that contact a little farther out than with a higher arc and body position. In effect, the animal reaches out and grabs the ground with its front feet, pushing it under its trunk. Once again the hind feet, moving forward at this moment, criss-cross the front in a moment of "gathered suspension" with all four feet simultaneously in the air and then make contact with the ground well ahead of the front prints. This contact is farther ahead of the front prints than in a lope due to the higher speed of the animal. Once in contact with the ground, the hind feet begin actively pushing the animal forward. Thus, in a gallop, there is an almost continuous application of horizontal force alternating between front and hind legs.

I first observed this phenomenon as a boy watching a cat chasing a gray squirrel. The squirrel did not use the usual high bound normal to these animals. Instead it held its body low and greatly increased the speed of its stride cycle. The transverse pattern of the squirrel's gait brought its hind feet down nearly side-by-side, a temporal and spatial simultaneity that allowed the animal to ziz-zag left and right as it raced forward, utterly confounding the cat.

The galloping gait is highly inefficient because the rapid turnover in the stride cycle taxes the animal's cardiovascular system at a higher rate than it can sustain. Also, the lower body position is more tiring than the higher one of a lope. Prove this to yourself by standing up and bending your knees slightly. A normally fit person can hold this position for quite a while. But if you bend your knees more to lower your body, you will see how much more quickly your leg muscles fatigue. This is why most animals employ galloping only briefly either in the closing moments of a capture or in an escape. Another reason for using this gait sparingly is that it requires strength in the front shoulders or chest. Few quadrupeds have much musculature in this area. Deer, for instance, have very scrawny front shoulders, more adapted to absorbing the shock of landing on the front legs than in imparting power to them. Compare this part of their anatomy to that of the pronghorn, cited above, with its more muscular chest. Or compare it to the chest of a race horse, an animal which, although it is bred specifically for galloping, can sustain this gait for only a little more than a mile.

Loping and Galloping Patterns

Sometimes the difference between a lope and gallop in a pattern of four prints can be deduced from their distribution while sometimes in can be inferred from the conditions and terrain in which the pattern in imbedded. For instance, does the pattern extend into the distance for some length even though cover might be found on either side? If it does, consider that it might be a lope, since most animals normally use galloping only briefly. Are the front nails of the animal greatly extended? This suggests a desire to "grab" the ground in a high-energy gait — a gallop. Other details may tell the difference as well. The first depends on the surface in which the patterns are found. If a surface such as snow, soft earth or sand allows displacement of material under the animal's foot, then distribution of debris can often be used to tell the difference among various gaits with similar patterns. The most significant of these in a gallop is the presence of either a mound of pushed-up substrate or even debris *behind* the front prints (Figure 4.8). Since the animal is actively pushing backward with the front feet, there should be some evidence in this area. In a lope, on the other hand, the front feet merely land and prop the animal's trunk while the hind feet provide the motive power. As a result, there should be little disturbance behind the front prints. Some care is needed here since on some surfaces the impact of the front feet descending from their arc of stride may cause debris to be thrown. However, that debris will be all around the prints of the front feet with no more of it behind than beside or in front. Since surfaces vary infinitely, some judgment must be used here and the resulting evidence added together in order to distinguish these two gaits. Like many identification and interpretation problems in tracking, any determinations are more likely to be made on the sum of the evidence rather than on one particular detail or another.

As far as print distribution is concerned, if the two central prints in the pattern overlap or even if they are closer together than the spaces between the other prints, then the pattern is probably a lope of one shade or another. This overlap or shortening is evidence of lack of gathered suspension: the animal's hind feet are landing at the same moment or only slightly after the front feet lift out. In a true gallop, the hind feet register well ahead of the front prints both spatially and, after the front feet lift out, temporally as well. Also in a gallop, the straddle and pattern width are apt to narrow from that of a lope. Of course one needs a loping pattern by the same animal for comparison, but in general a gallop is suggested if four prints are distributed over a long pattern where the two middle prints are not closer to one another than to the others and with a narrow pattern width. The patterns for loping and galloping are similar — both present either a transverse or rotary 4X pattern. Here are several examples of both gaits, although there are many variations. The black ovals represent front feet.

Loping Patterns

In the first pattern of Diagram 4.11 look at the two central prints and notice that the right front print is advanced ahead of the left hind, indicating a slow transverse lope. In the second pattern, the left hind

Figure 4.8. This galloping pattern of an eastern coyote is so fast that it is hard to tell whether it is rotary or transverse. Note the debris behind the front prints at the bottom of the frame, indicating force applied by the front legs.

Diagram 4.11

print of these two is advanced, showing a slightly faster transverse lope. In the third, the pattern is rotary, with the right hind overlapping right front in a double registration. This is a "3X" pattern, that is, only three tracks show in the pattern. Sometimes the middle hind print lands perfectly over the print of the front in a direct registration, giving the impression of the 3-legged porcupine we saw earlier!

Galloping Patterns

The first pattern in Diagram 4.12 shows a transverse gallop and the second a rotary gallop. The straddle in both examples is somewhat exaggerated for clarity; in cases with animals that have narrow chests and hips, like foxes, these patterns may describe nearly straight lines, making it difficult even to distinguish transverse from rotary.

Diagram 4.12

Additional Notes

In the field, while following an animal that blends one gait into another, the differences in the patterns left by lopes and gallops as well as other gaits may blend as well and so may not be as apparent as in the examples used above for illustration. You may soon find yourself using terms like "lope-gallop" or "lope-bound" to describe them. This is good; it means you are visualizing the event after the fact and putting yourself in position to correctly interpret the behavior of the animal when it was hidden from human witness.

In some loping trails the speed of the animal may be so slow that the space between each 4X pattern and the next will be only slightly greater than between individual prints within the patterns. A slow loping skunk may even leave a trail with no defining inter-pattern space. Such a trail may look like that shown in Diagram 4.13.

Diagram 4.13A

This appears disorganized and patternless, as if the skunk were drunk on fermented apples perhaps, but close examination and identification of front and hind prints will show the boundaries between the lope patterns and bring order to the seeming randomness as shown in Diagram 4.13B.

Diagram 4.13B

Foxes and coyotes do this a lot, as well, often as a series of rotary 3X patterns with no appreciable separation. Once again, distinguishing front from hind prints and places where the middle two tracks register directly or indirectly may be needed to determine the separation between patterns, as shown in Diagram 4.14.

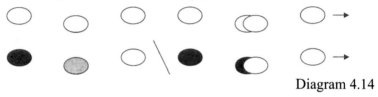

Diagram 4.14

Note that the straddle here, slightly exaggerated for demonstration, shows that this is a rotary lope rather than a walk or aligned DR trot. In an actual red fox trail of this sort, the tracks will form a nearly straight line, making distinguishing the gaits more difficult. However, the step length will be greater than the normal 15-inch walking step length for this species.

In the slower "coincident" gaits (coincident in the sense that the shoulders and hips of the animal rise and fall together) of walking and trotting, successive step lengths of either the front or hind feet will be fairly uniform. A walking red fox, as I indicated, will usually show a repetition of 15-inch steps with little more than an inch of variation while an eastern coyote will normally show a 23-inch step length with a variation of only a couple of inches. In the rolling gaits of loping, galloping and bounding, however, there is normally an alternation of step lengths — short-long-short-long. See the above illustrations for this alternation. This is another way of distinguishing between these two classes of gaits where the spacing of tracks is confusingly uniform.

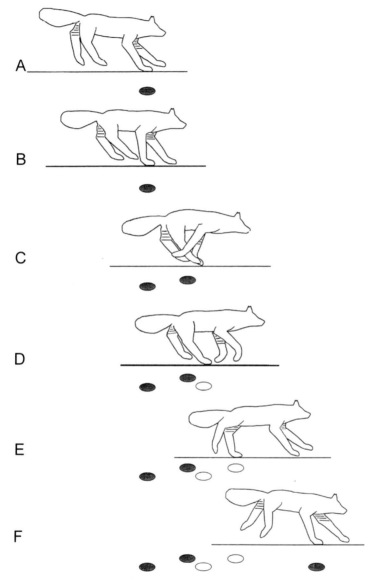

Figure 4.9. Canid transverse loping profiles. A lope is an easy run with a higher body position than a gallop. The front legs are merely props that do not impart force to the gait. There is little or no extended suspension since the first front foot contacts the ground just as the hind lifts out, as in profiles E and F. The high body position is less strenuous than in a gallop, and so the animal is able to sustain this gait for a long period.

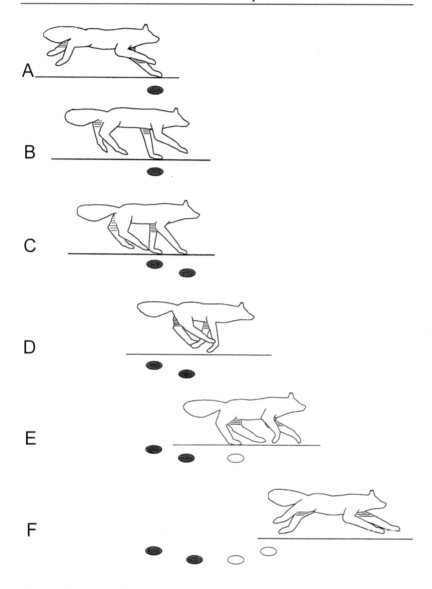

Figure 4.10. Canid rotary gallop profiles. A gallop differs from a lope in several ways. Instead of the front legs being used simply as props, they help impart force to the gait. The body is held closer to the ground so that the feet are in contact with it longer. There are two discreet moments of suspension. Profile D is in "gathered suspension," and profile F is in "extended suspension." All larger long-legged quadrupeds, such as dogs, cats and deer, can use this gait for at least brief periods.

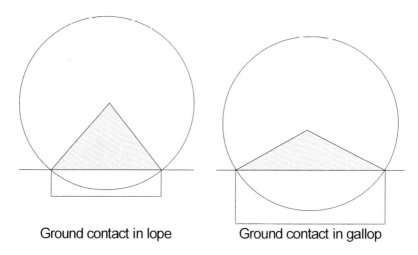

Ground contact in lope   Ground contact in gallop

Figure 4.11. Ground contact duration of lope versus gallop. Because a lope maintains a higher body-position than a gallop, it is a less strenuous gait. However, it is also less powerful because the length of the stride when the foot is in contact with the ground and able to apply backward force is less. In the diagrams the center of the circle is the animal's shoulder or hip joint, and the pie-shaped arc is the distance when its foot is in contact with the ground. Only when a foot is on the ground can it apply force to the gait. The price to be paid for more power is the muscle fatigue caused by the deep flexion required in the gallop. Since this gait is tiring to sustain, loping is used for long distance travel while galloping is reserved for escape or for a short-range attack at the end of a stalk.

Bounding

The third high-energy gait used by wild mammals is the bound. It differs from loping and galloping in having a higher arc of stride than either of the others. In a bound, the animal sets itself on hind feet that are arranged either perpendicular to the direction of travel or on a diagonal and launches itself upward as much as forward. At the descending end of the arc, the greater impact of this gait is absorbed by the muscles, tendons and ligaments of the front legs. As the hind feet follow the front downward, they either register in the vacated front prints or at least in the same vicinity, creating a much tighter pattern than occurs with loping or galloping. Which of these occurs depends on the species and on the surface. Mustelids often direct register their hind into their front in a piece of magic that will be discussed in detail in the next section while other animals as diverse as squirrels and deer typically register the hind feet forward of the front. Some examples of the patterns that may result are shown here in Diagram 4.15.

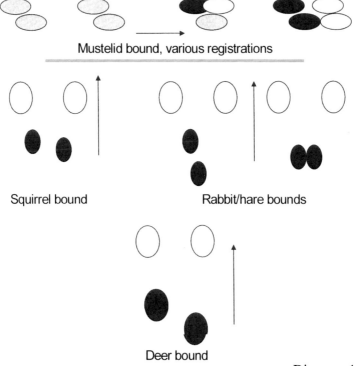

Diagram 4.15

There are many other variations on the theme, but all of the above patterns and the gradations among them have the same central characteristic: the animal sets its hind feet to give it a simultaneous two-legged boost into a high arc of stride. The diagonal arrangement of the mustelid bound represents a special case and will be dealt with separately. For the others, the power comes from the hind legs, the front legs being used, as in a lope, merely to prop the animal's front end.

With a high arc of stride, obviously this gait is wasteful of energy. However, it often is used by the above animals and many others in high-resistance terrain. Weasels use it to vault out of deep soft snow where plowing through would be even more wasteful. Deer use it to vault over obstacles in the woods such as fallen limbs. Squirrel and lagomorphs use it habitually to cover ground in situations where other mammals with different anatomies might lope or gallop. In this last case, the patterns involved can stretch out into a high speed run close to the gallop of other animals but in most species with the hind prints arranged nearly side-by-side. I have, however, found jackrabbits in the flat, open ground in the Southwest running in a pattern identical to a rotary gallop, with one hind print well advanced ahead of the other.

The three high-energy gaits, then, are the lope, the gallop and the bound. All three in their gross mechanics resemble one another: hind legs propel the animal upward and forward; the front feet come down diagonally to absorb the impact of descent; the hind legs move forward and criss-cross with front legs that at this point are slanted backward by the forward motion of the animal's trunk. The hind feet then plant and power the animal forward again. The main differences among them appear in the dimension that is invisible in the patterns: the verticality or vertical vector as expressed in the arc of stride. The gallop has the lowest arc of stride with the shortest vertical vector and the longest horizontal one; the lope has a somewhat higher vertical vector and shorter horizontal one; the bound has the highest arc of stride of the three and the shortest horizontal vector. The three gaits are used for different purposes: the gallop mainly to attack or escape, the lope to cover ground over long distances rapidly but without exhaustion and the bound to vault over obstacles in terrain with high resistance.

The Mustelid Bound

As indicated in the above section, the bounding employed by members of the weasel family is a special case. Rather than two hind feet being placed perpendicular to the direction of travel, the common mustelid bounding pattern is a 2X DR on a *repeating slant*, that is, sets of two tracks each composed of a hind print directly registering over a front and arranged on a diagonal relative to the direction of travel and with the direction of the slant repeating from pattern to pattern. With increasing ground speed from left to right a string of such patterns looks like those in Diagram 4.16 and Figure 4.12.

Diagram 4.16

Occasionally a third hind print will appear as a double registration, usually in front of the retarded front track, or two such registrations will show just ahead of the front prints on both sides as variations on the 2X theme. This seems to happen more often among the smaller weasels.

The mechanics of the "rolling" or "rocking horse" gait that leaves this evidence are complex. Since it is a bound, the arc of stride is quite high. As the front end of the animal descends one front foot lands (the left in the above example) and cushions the impact. The horizontal motion involved in the gait shifts the mass of the animal forward so that, an instant later, the other (right front) foot lands diagonally in front of the left and the animal's weight begins to transfer to it from the left front foot. This is where it gets complicated. One or the other or both of the following things happen as the hind feet descend toward the original two prints. First, the left front foot begins to articulate forward onto its tip in much the same way that the front vacates a print for the hind foot in the direct-register walking gait discussed earlier. This at least partly allows the descending left hind foot to slip in under it and replace it. However, I suspect that with this particular bounding gait something else is going on as well. The impact of the front of the animal in the descent, is absorbed as potential energy in the muscles and tendons of the animal's front feet, front legs and chest, putting its front quarters in a deep posture, ready to propel the animal's front end upward and forward an instant before the hind feet land and provide the main propulsion to the whole body. Thus, unlike in a lope, the front legs add at least some power to the cycle of motion. The effect visually is a very floppy gait that looks as if the two ends of the animal are functioning independently, one coming down while the other is headed up. Especially in the smaller weasels, it's like watching a rag being jerked up and down on the end of a stick.

In a loping gait, by way of contrast, the front feet come down and are then passed by the hind feet, which plant out ahead and power the front of the animal forward into the next stride cycle. In the mustelid bound, the front end doesn't wait for the hind to plant but

Figure 4.12. Mink in a 2X DR bounding gait. This mink's trail shows perfect direct-registering bounding patterns in a couple of inches of slushy snow.

vaults upward and forward under its own power, at least partly independent of the animal's hind legs.

It is interesting to follow a weasel's trail as it moves from soft snow under evergreens out into a sun field where the snow is firmer. What starts out as a slow — in terms of ground speed at least, if not the animal's motion — 2X pattern of tracks that are nearly even, begins to speed up to a more diagonal arrangement as the animal begins to steal energy from the vertical vector and transfer it to the horizontal for more ground speed. As the surface firms even more, the animal takes advantage of this by emphasizing the horizontal component even

143

more and the hind feet begin to overstep the front, eventually moving well forward into the familiar transverse lope pattern. Such a progression is shown here in Diagram 4.17.

Diagram 4.17

So, we have been able not only to identify the mammal and interpret its gait from the patterns it leaves behind, but we are also able to connect the trail to its background and interpret the animal's motive in choosing one gait over another. This is another instance of "eco-tracking;" we have climbed inside the animal a little bit and seen, or rather felt, the world from a perspective different from our own, even though we are physically outside the animal's body and mind.

You may have noticed that the patterns on the left above look very similar to the patterns of the displaced trot so often used by canids. Technically the fox or coyote's pattern can be termed a "repeating slant 2X **SR**" as opposed to the weasel's "repeating slant 2X **DR**." In the displaced trot, each track is a "print," that is, the impression of a single foot, with the larger front retarded and the smaller hind advanced. In the mustelid bound, however, each impression is a "track" composed of the prints of one front foot and one hind. With only a little experience a tracker can tell the difference at a glance based on the surface in which the trail is imbedded: canids only leave this pattern on a firm, even surface while weasels only leave it on soft surfaces. Furthermore, the difference in the print size between front and hind of a canid should be apparent on most surfaces.

It should be noted that Elbroch in his excellent reference book, *Mammal Tracks and Sign*, presents the mustelid bound as a "2X lope." I believe the discrepancy between his term and mine is semantic. Because the hind feet do not land and take off together, the mustelid bound differs from the classic bound of squirrel, rabbit or deer and has some characteristics of a lope in the initial propping action of the front legs. However, it's high arc of stride and close grouping of prints make it resemble other bounds more than the straight ahead lope. For these reasons I will continue to include it in the bound category. Here, once again, we are trying in a sense to digitize an analog motion. As I have suggested before, it is our civilized habit to quantify and

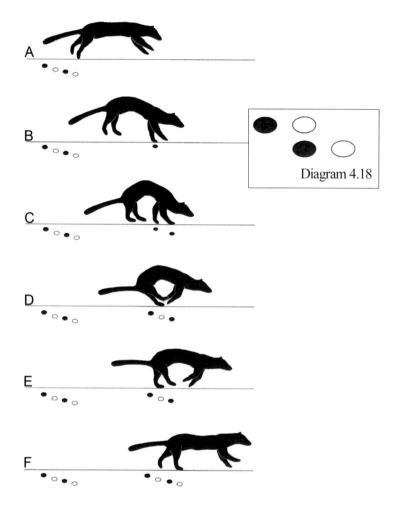

Figure 4.13. Weasel lope profiles. On level, firm surfaces weasels usually lope rather than bound. This gait has a lower arc of stride and thus is more efficient since less energy is devoted to vertical motion. The gait shown is a displaced lope, with the hind end of the animal shifted to the right to facilitate passage of hind feet past front. For more speed the weasel may opt for an aligned lope where the front foot on each side leaves before the hind lands. The pattern that results is a diamond-shape (Diagram 4.18) with the animal moving from left to right. This gait involves a little more verticality and thus greater impact on the surface as the animal descends from each arc. On delicate, breakable crust this can be a problem, with the animal breaking through with its front feet. In such a case the animal is likely to revert to the slower gait.

145

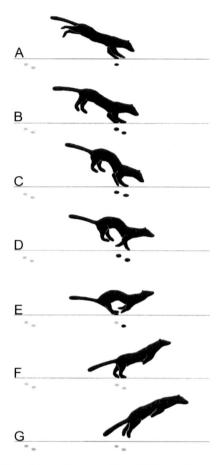

Figure 4.14. Weasel bounding profiles. This gait is used in soft conditions where the animal is sinking into the snow more than about a third of its leg length. The front feet pre-pack the snow for the benefit of the hind, which direct register over the front tracks (indicated by graying of the track ovals). This firmed snow provides a stable platform from which the animal can vault powerfully upwards and forwards. On a firm surface this gait would waste energy in the vertical direction. But on soft surface the high arc of stride allows the animal to vault over the resistance of snow between its landing spots. The pattern that results is a "2X DR with repeating slant," that is, sets of two direct registrations arranged in a series with repeating rather than alternating slants. Canids also leave repeating slant-2 patterns when they are displaced trotting. However, each track is a single print rather then a direct registration. In the case of weasels the tracks will be the same size; with canids the retarded front print will be larger than the advanced hind.

then manipulate the groupings that result. The wild animal, of course, labors under no such restriction, and in terms of our visualization of its motion, the discrepancy is immaterial.

"The Raccoon Catastrophe"

When Newtonian physics failed to predict a certain behavior of light waves, the resulting graph was termed the "ultra-violet catastrophe" because it demonstrated in one electrifying instant that Newtonian physics, the physics that we all apply to the world around us and that works well enough in our daily lives, is wrong. That moment gave rise to quantum physics, which today purports to explain the universe as it really is.

I can't resist using a paraphrase of that astounding moment to describe the normal gait of a raccoon (Figure 4.15) as demonstrated in some videotape I recorded years ago on the main Gate 30 beaver pond in Quabbin Reservation. While admittedly vastly overblown as a title, I justify it because, having worked out what I believed were rock-solid principles of animal motion with my western mind, I discovered that the raccoon moved in a way that defied and gave the lie to my neat categories. The gait that I observed I call the Raccoon Half-trot. (When the civilized mind can do nothing else to explain what it sees, it fractionates!) Here is its description.

Raccoon Half-trot

Technically, a "trot" is a gait where two feet of the animal are on the ground at the same time. Those two feet may be on diagonally opposite sides of the animal as in the aligned trot of a bobcat or the displaced trot of a canid. Or, more rarely, they can be on the same side of the animal, with the right front foot and the right hind planted simultaneously. This is the gait of a "trotter" as the horses that pull sulkies in a race are called. It is a very flat, and therefore efficient, gait, but one that requires an even surface since its lack of verticality prevents the animal from bouncing over obstacles. This gait is very rare in Nature. I may have seen a moose do it years ago on soft mud by way of limiting the penetration of its feet (this gait has a very low arc of stride and the animal's weight is split between two feet at a time), but this was well before I became interested in animal gaits and so I have no data or accurate memory of it. Other than that questionable

Figure 4.15. Raccoon walking trail. Here the raccoon's larger hind print is retarded in each pattern. If the animal speeds up, the hind will move forward, parallel with the front. At some point as it advances ahead of the front, the gait transitions from a walk to a trot.

instance, I have never seen a wild animal perform it. My video, however, shows a raccoon almost doing it.

In a normal quadrupedal walk the hind foot on the right side moves forward and just before it touches the back of the right front foot, the front foot lifts out and moves forward, vacating a landing place in which the animal can perform a direct registration. Since, except for an instant when the feet trade places, three feet at the time are planted, we call this a quadrupedal walk (or a "tetrapodal" walk if you prefer the Greek). In a raccoon half-trot, however, the sequence is modified slightly. As the right hind foot moves forward and when it is about half-way to the front foot, the front foot departs, not waiting, as in a walk, for the hind foot nearly to touch it. Thus the two feet are not moving forward in perfect coordination as in an equine trot, but rather the front advances somewhat before the hind foot arrives. And rather than landing directly in the print of the vacated front foot, the hind passes over that location and advances far forward to land in the vicinity of the front foot of the opposite side. This, of course, requires more than an instant when both hind and front foot are off the ground. What is this gait then? Two feet are off the ground for part of the cycle, suggesting a trot, but the mechanics are very nearly that of the normal diagonal walk. Indeed, my short video clip seems to show no suspension that would be expected in an aligned trot, that is, the animal's trunk stays level rather than rising and falling. Instead the animal's entire weight shifts onto the front and hind of the same side while the feet on the other side move. Seen from front or rear, the animal's body rocks precariously right and left as it balances on one side at a time. Our neat categories break down with this gait — a minor catastrophe for the tracker intent on plugging it into a box, but of no concern whatever to the raccoon, which has no idea that a human is trying to write in a book about it.

The pattern this gait leaves shows the hind foot registering somewhere in the vicinity of the front foot of the opposite side, either retarded in a slow version, arranged evenly in a faster version, or advanced in the fastest. Diagram 4.19 shows a continuum of slow to fast from left to right with the black ovals representing front prints.

Diagram 4.19

In the first three patterns, the hind, represented by a white oval, registers retarded rearward of the black front. In the middle two patterns the speed has increased and the hind foot has caught up, registering parallel to the front. In the right two patterns the speed and the overstep have increased even more so that the hind foot now advances ahead of the front. These are the normal patterns made by a walking/trotting raccoon. The huge overstep is permitted because of the exaggerated curve of this species' spine, which locates the hind leg much closer to the front than the body length of the animal would suggest. Which of the above print arrangements results from this gait seems to be a matter of how much energy the raccoon chooses to introduce into it although at some point in the advancement of the hind print a moment of suspension must occur, transitioning the gait into a true trot with a vertical vector. Just where that moment occurs is difficult to judge. Possibly some evidence of increased impact may show around the prints made in a trot, but with such a small animal that evidence is apt to be vestigial at best.

It is possible that other mammals use this gait, as well. Certainly the patterns remind one of the "wiggle walk" that I will describe below (page 201). Also, I can rationalize no other explanation for the patterns that an opossum sometimes leaves behind, which I have elsewhere termed a trot. Unfortunately, I have no videotape of either animal performing this gait to prove or disprove its use.

Chapter V

Putting It Together

Now that we have examined in detail the inner and outer parts of track and sign interpretation, it is time to combine the two and apply them to real-world tracking problems. Before we do this, however, we need to look at a little background with which to prepare ourselves. This chapter opens with a discussion of predation, a process central to our interpretive efforts since every animal spends a great deal of its life searching for and eating something else. Together, knowledge of eco-tracking and an understanding of this continual food-acquiring process provide us with the ability to reason in a more sophisticated way when we try to analyze the Nature that we see before us. I have come to understand over my own brief lifetime that few phenomena in Nature yield to simple explanation and that one should look with suspicion on any analysis that draws a straight line between cause and effect.

These preparations and cautions are followed by the first opportunity you will have in this book to apply what you have learned to eight real-world tracking problems from the pages of my field notebooks. All of them look at field evidence, translate that information into the animal in motion, and require you to come to some conclusions about its behavior. In these incidents, I do the work with you riding along on my shoulder. Later (pages 429-448), you will get eighteen chances to do this on your own.

How Predation Works

As trackers, we should want to know more about the animal whose sign we encounter in the field than simply its identity. By interpreting the track, trail or sign beyond identification we can imaginatively recreate the animal's experience, gaining insight into its hidden life. This once again is "eco-tracking." And this information can be expected to be candid. If we attempt to see the animal directly, our presence more often than not affects the animal's behavior — it senses our presence and flees. Few animals "know" that they leave tracks behind to be interpreted by humans. Bears sometimes dissemble, but whether this is knowledge in the human sense or is simply mindless instinct instilled by selection is an open question. At least we can safely assume that the animal's sign is seldom a deliberate lie intended to deceive a following animal like ourselves. This sign, including tracks and trails, can tell interesting stories about an animal's behavior as it responds to its immediate habitat, answering the questions: What was the animal doing? Why was it here?

To prepare the way for a deeper investigation of animal behavior as evidenced by their sign, we should know how basic predation works.

A few years ago I heard a gentleman announce at a meeting of trackers that once fishers arrive in an area, they wipe out the local porcupines. On another occasion in the Federated Forest I encountered an old hunter who claimed that fishers were the reason for the decline in snowshoe hare numbers in the area. This is the incident elaborated in "The Inner Game: Cause and Effect" (pages 54-55). A third incident occurred up at Pinkham Notch where a coyote sympathizer claimed that the wolves recently reintroduced into Yellowstone were "wiping out" her favorite mammals.

Porcupines and Predation

Not only do predators sort themselves out into a hierarchy of dominance but prey animals do as well. The fittest and strongest dominate the less fit. In the case of porcupines there are only so many places in a forest in which to hide one's vulnerable face from bobcats, fishers, coyotes and foxes. The dominant porcupines appropriate places such as rock crevices or hollow logs for themselves, driving off the subordinate members of their species (Figure 5.1). These non-dominant animals, then, represent a vulnerable surplus. It is these creatures on which the available predators concentrate. If you have ten porcupines but only five suitable crevices, the sub-dominant animals are left to shift for themselves in an unfriendly world. Any available predators that have some skill at killing porcupines will then concentrate on this surplus. These hunters may include not only fishers but also coyotes, bobcats and even foxes.

Nature abhors three things. The first, as we were taught in school, is a vacuum. However, another that we weren't taught is equilibrium,

Figure 5.1. A dominant porcupine gets the best places to hide its vulnerable face from predators.

and a third, the one that obtains here, is a perfect defense (or conversely, a perfect offense). If porcupines were equipped with hard quills all over their bodies with which to punish any attacking predator, no porcupines would be killed. These invulnerable animals would then breed others with characteristics like themselves and overpopulate, destroying the available food supply. All then would die of starvation or of the diseases facilitated by starvation. Instead of macro-predation by larger predators that would keep their numbers down they would be catastrophically depopulated by micro-predators — bacteria and viruses that invade weakened animals. To avoid this catastrophe, evolution has inserted a flaw in every defense. In the case of the porcupine, all its hard quills, the ones that can do the damage, are on the rear half of the animal, leaving its face vulnerable. This provides an attack point for predators. Only the fittest porcupines, then, the ones that can outmuscle or intimidate the weaker of their own species and appropriate places in which to hide their vulnerable end, survive to procreate genetically similar offspring with the same strengths and weaknesses as the parent animals.

Once the vulnerable surplus of prey animals has been taken off, the predators ignore the remaining fitter animals as too hard to catch. Instead, they switch to some other prey species that happens to be providing a vulnerable surplus at the moment, such as voles, rabbits or deer. Thus it is an error to insist that fishers wipe out porcupines in an area. Instead of eliminating their food supply, they simply farm it, leaving the dominant and dangerous prey animals to procreate the next generation of the predator's food supply.

Fishers and Snowshoe Hares

As for fishers eradicating snowshoe hares, drops in the populations of this lagomorph are cyclical, as I pointed out above (pages 54-55), caused not by fishers but rather by the development every 7-10 years of anti-browsing toxins in the hares' staple browse plants (Figure 5.2). These plants are in a struggle among themselves for sunlight and nutrients. They must try to grow faster than their neighbors to dominate the supply of each or they will succumb. It is as if every now and then the plants get tired of being browsed back by hares and, rather than devoting all of their energy to competition among themselves, begin to devote some of it to developing substances in their

154

Figure 5.2. Heavy browse such as this by snowshoe hares stimulates anti-browsing toxins that weaken the hares and make them vulnerable to predation.

twigs and buds that are distasteful or even toxic to the animal browsers. This costs them growth, but at some point they "decide" (apparently a hormonal decision) that their survival is jeopardized more by the browsing than by the local competition. The result for the hares is that they weaken from starvation and become vulnerable to local predators such as bobcats, coyotes, foxes, hawks, owls and, of course, fishers. Thus their fur begins to show up disproportionately in the scat and pellets of these predators, leading the old hunter to assume that the predator was to blame. This is the error of proximate cause. The faulty syllogism goes this way: fishers kill hares and hare populations are in decline, therefore, fishers are the reason for the decline. This sort of error occurs repeatedly in the "lore;" the proximate agent must be the causal agent. The notion that there may be an ultimate cause several steps back rarely gets recognized as does the possibility that both the supposed cause and its effect may both actually be effects caused by some third, unseen factor.

Rather than being a catastrophe for the species that is affected by the decline, in this case the hares, these cyclical depressions are actually critical for the species' survival. The low end of the sine curve that describes the populations of all wild animals in their natural state is the testing period for the genetic fitness of individual animals. Only the fit survive and go on to procreate offspring with genetic characteristics appropriate to the habitat. Since habitats in Nature are always shifting, the periodic testing of the species for fitness is critical to its adaptation to continually changing circumstances. Equilibrium, as I indicated above, is the third thing that Nature abhors. If a species is not allowed to swing through periods of plenty and privation, its evolutionary fitness is short-circuited and it becomes a domestic animal rather than a wild one. In this regard one might consider the dubious wisdom of game management policies designed to provide a stable population of huntable animals from year to year for the satisfaction of hunters whose license fees pay the managers' salaries.

Wolves and Coyotes

The coyote sympathizer who presented the program at Pinkham Notch claimed that the wolves reintroduced into Yellowstone a decade ago had "wiped out" the coyote population. Later in the presentation she allowed herself to admit that there were in fact some coyotes left in the Lamar Valley where the wolves had been released. That this new dynamic might represent an appropriate relationship between hunter and scavenger seemed to escape her. In fact, the wolves did bring down the coyote population to what many would say is a more appropriate level. However, coyotes are smart and the ones that were left soon realized that they not only could survive with the wolves but even prosper with them. The surviving coyotes figured out that they could live in the seams between the various wolf packs because the wolves themselves avoided these boundaries for fear of aggressive interaction with neighboring packs. And because the wolves could bring down elk, prey animals much too large for the coyotes to handle themselves, an entirely new source of scavenge suddenly became available. The wolves were imported to their traditional range to reintroduce cyclicity to the elk herd, improving its genetic fitness, and the coyotes benefited by having access to elk carcasses after the wolves had fed, hazardous access to be sure, but access nonetheless.

A Predator-prey Relationship

Here is a tracking problem appropriate to our discussion of predation. The photos show the trail (Figure 5.3A) and a pair of tracks (Figure 5.3B) from the trail that was found in December on thin early snow in a grassy, abandoned orchard in central Massachusetts. The trail wandered about on a low hill about 300 meters from a large beaver impoundment. The step length of these direct registrations was about 4 inches and the trail width about 3 inches. The close-up shows detail from another segment of the same trail, here with single registrations of front and hind.

• What is the identity of the animal?
• What may have caused it to travel to this location?

Quite an eco-tracking story can be surmised from this scant information. First, we must identify the animal that made the track and trail. The animal was walking, its large hind feet having obliterated most of the smaller front prints in the photo. In a trail of this sort we should always first eliminate raccoon, the great fooler. We can drop it from consideration due to several details. Not only are the prints somewhat smaller than those of the average raccoon, but the print width of the front foot, which shows on the right in the detail photo, is much smaller than the hind; raccoon front and hind are of nearly equal width.

157

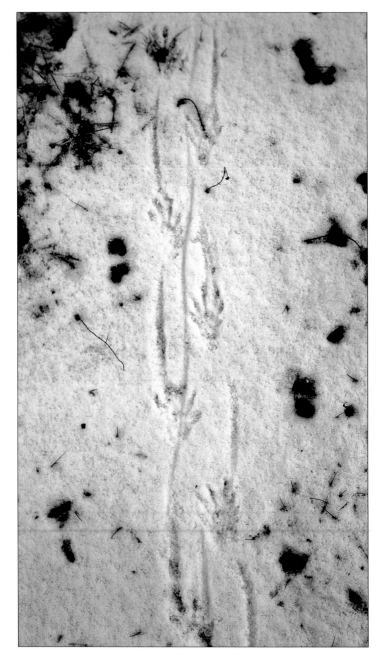

Figure 5.3A. A mystery trail.

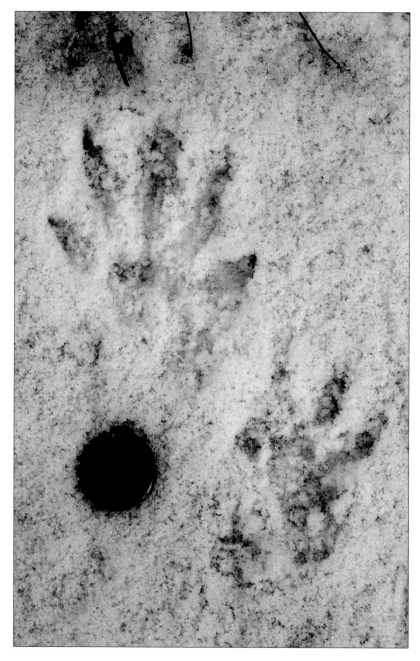

Figure 5.3B. A close-up.

Secondly, although raccoons can diagonal walk in the way shown in the trail photo, on firm surfaces they almost always greatly overstep, leaving alternating slanted patterns of two prints (pages 147-150). Thirdly, the tapered toes are quite different from the club-like toes that a raccoon print shows when its toes are extended. Finally, the secondary pad, although somewhat obscured by the soft snow crystals, fails to show the smooth leading edge that we expect from raccoon, appearing instead rather lumpy and irregular.

Looking carefully at the toe prints in the detail photo, we can see an "echo" outline, most visible in the middle toe impression of the larger hind foot. This along with all the other evidence points to a muskrat walking through the orchard. We might find this hard to believe as we look around at the habitat. Muskrats, after all, are water animals. As part of our training we have learned to step back and look at habitat in order to consider what animals should be there. So what would an aquatic muskrat be doing in an orchard 300 meters from water?

In his book *Muskrats and Marsh Management*, biologist Paul Errington detailed the life cycle of muskrats and their predation by the small semi-aquatic weasel, the mink. As with the porcupine, muskrats sort themselves into dominant and inferior. Their principal predator, the mink, can swim under water, protected from the cold and wet by its luxuriant fur, and catch muskrats in their domes and dens, or at least in those that were inadequately built or sited. The dominant muskrats get the best hiding places in which to protect themselves and the inferior animals get the rest. Just as the fisher went after the more poorly protected porcupines, the mink goes after the vulnerable surplus of muskrats, leaving alone the ones that are harder to catch. As the winter progresses, food increasingly becomes scarce, with plant-dependent species as well as their predators entering the annual testing period. In the course of the winter the strong do better, commandeering the most productive feeding sites for themselves while the weak are pushed to the margins and go into decline. Desperation for food leads the weaker animals, denied sustenance under the ice of the pond, to leave their protective habitat and embark on seemingly inexplicable "muskrat rambles" on adjacent dry land. These desperate excursions may or may not result in food but certainly do expose these animals to all the predators that lurk in the woods and meadows, predators from which the animal would normally be safer in its watery home in the frozen pond.

Anthropomorphism

The coyote sympathizer at Pinkham Notch disputed the need to reintroduce wolves into Yellowstone, an introduction intended to provide effective predation on a burgeoning elk herd, for the good of the herd itself. Instead of the wolves picking on the weak, old, young and untested, the wolves, she insisted, were selecting their prey at random. If this were true, much of what I have just described would be demonstrably false. Therefore, an examination of the evidence for her claim is needed, and this discussion of predation is an appropriate place to do it.

Her evidence was a video of a wolf pack pursuing a running pack of elk cows and yearling calves. In the video, all the elk are running together at the same pace and appear to be healthy. The wolves, she claimed, pursued these running elk for a mile or so and then randomly focused on one of the animals, which they attacked, dragged down and began feeding on before the elk was even dead. This footage did much to upset the audience and turned them emotionally against the wolves, exactly the effect that the presenter was trying to produce.

Leaving aside the brutality of real predation, whatever the predator or prey, let's consider the anthropomorphic fallacy buried in the interpretation of this incident. Modern humans are a sight-dependent species. Our binocular vision is very good at distinguishing depth, detail and color. The acuity of our other senses has declined with our increasing remoteness from the requirements of a primitive life. The range and breadth of our hearing pales compared to that of a dog, and our sense of smell is the most deficient of all. We marvel at the ability of a fox to locate a cached bird under two feet of crusted snow in a featureless field with no other clue to its location than what the fox's nose tells it. Because our other senses have dulled and we have become sight-dependent, it is easy for us to assume the same of wild animals when we see them in action. The presenter assumed that the wolves, in running the elk herd, were "looking" for signs of weakness rather than apprehending them with other senses. This is what we might do, but it may not be what the wolves in the video were doing. Instead, it is possible, in fact likely, that running the herd was an effort to make the animals sweat and breathe hard. As the wolves run along, beside and behind, they smell information in the scents given off by the fleeing animals that certainly communicate all sorts of information about the internal condition of these animals, including their state of health. Gradually winnowing down the group to fewer and fewer animals allows the wolves to attribute the scent of vulnerability to a particular animal, which they then separate, attack and kill. While the death of the individual is gruesome to the human eye, it may benefit the elk herd as a whole. If the elk had an invisible condition that would affect the animal in the future, better to weed it out of the herd now rather than let it reproduce, with potentially the same defect in its offspring. It is even possible that elk sweat and breath are designed by the prey species itself to contain this information for the benefit of the wolf in order to rid the herd of weaker animals so that they can neither use up forage nor produce genetically inferior offspring. Nature is beautiful in its economy and complexity, but it is not always pretty.

A certain amount of humility about jumping to easy conclusions is appropriate in observing natural processes. It is good to remind ourselves as trackers reconstructing an event after the fact that few things in Nature are simple. What may seem like obvious cause and effect patterns to the human mind and the human senses usually turn out to be far more complicated, with remote causes hidden from immediate view. In trying to "see" the event for ourselves, moreover, we should not only view it imaginatively from the viewpoint, both literal and figurative, of the animal, but also consider the functions of its other senses from that viewpoint.

Unlearning a Few Rules

"Rules are made to be broken."   — Anonymous

"One spends the first several years of one's experience learning the rules, and the next several unlearning them."
     — The Author

"Measurements are the tracker's training wheels."
     — Kevin Harding

Tracking sometimes seems to me like a Russian grammar. There are so many exceptions to the rules one uses to learn the language that it seems useless to learn them in the first place. However, at the very least, principles provide a starting point from which to depart. Not to learn the principles first would require the aspiring tracker to learn every scenario that he might encounter one by one — a task that would take more than a lifetime. The trick is to understand that there are almost always exceptions so that one's mind is aided but not enslaved by the rule. One must learn to embed each phenomenon in its context, that is, in the conditions in which it was created. This requires some experience, for which we are told correctly there is no substitute, as well as some flexibility of mind. As this is a somewhat advanced tracking guide, that is, a guide for those who have already learned the rules, we can jump right in to the "unlearning" phase with a few examples.

As we have seen earlier, one must resist the assumption that a line of tracks was made in the same circumstances in which it is viewed. We may come upon the tracks in early morning in open sunlight on heavy crust, but the tracks may have been made at dusk after a warm day where temperatures had softened the surface into fragile spring "collapse-snow," a common condition in late winter and early spring. These two conditions can, as we will see, produce very different trails of the same animal.

A contributing rigidity of mind, appropriately enough in a discussion of rules, involves the ruler, that is, involves measurements, which are often either too rigidly applied or the parameters of which expand with experience beyond the bounds of usefulness. One learns early on, for instance, that a "typical" step length for an eastern coyote is 22-23 inches with a straddle only slightly wider than the width of the feet themselves — 3 inches or so. What does one make, then, of a coyote leaving a step length of 18 inches with a straddle of up to 6 inches? We also learn that a major difference between a coyote's footprint and that of a domestic dog is that the dog will show widely splayed toes with prominent nails while the coyote will show tightly grouped toes and lightly registering nails, especially the medial and lateral nails. What then, if the animal that left the 18-inch step length with the 6-inch straddle also showed a track with splayed toes and prominent nails. A domestic dog, right?

While measurements can be very useful as part of the larger identification process, they cannot be applied arbitrarily. Although I have just insisted that one must learn selectively to ignore some rules, it is important to know which ones to ignore and when. Here's one that you must, dare I say, *never* ignore: Embed the track or trail in its circumstances. I learned this "over-rule" years ago at Gate 30 in Quabbin in the snow conditions I described above: a foot and a half of wet spring "collapse snow," the kind of snow that is partly corned with the seasonal thaw and that supports at one moment while collapsing the next. Here was the trail of a mammal leaving nearly perfect direct registrations, heading down a side path over unbroken snow. The appearance of the trail showed a lot of zigzag out to the 6-inch straddle and 18-inch step length mentioned above. I puzzled over the trail for a while, but it was not until I put myself in the animal's place and used my imagination to feel myself moving the animal's body that I understood the answer. It was shortening its step, much as I would on breakable crust, in order to lessen the impact of each step and so to avoid a collapse. I myself have done much the same on breakable crust to avoid the unpleasant, energy-wasting sensation of a sudden break-through. Furthermore, as we saw with the red fox earlier, the animal widened its straddle so that in the event of a break-through, it would fall to the medial side, that is, toward the centerline of its body. Such a fall can be countered by the feet opposite the break-through. Were the coyote to use its usual, more efficient narrow straddle, it would have a 50 percent chance of falling laterally where there would be no opposite feet to do the catching. Should that happen, it would fall over and have to expend energy floundering in the snow to regain its feet. A dog that will be fed by its master can afford such profligacy. In winter, a coyote, which hunts with its feet and is on the tightest of energy budgets, cannot.

In addition to the atypical appearance of this trail, the same context contributed to the distortion of the animal's footprints. In firm, cold conditions, a coyote will keep its toes together for mutual warmth and withdraw its nails into the fur between and above its toes. In those circumstances, as the rule indicates, it will look very unlike most dogs whose foot morphology more closely resembles that of their immediate progenitor, the wolf. The wolf's toes tend to splay in most conditions and the nails register prominently. However, in an effort to distribute its weight over as much area as possible on a surface that may give way at each step, the coyote may spread its toes and arch its nails forward into a fair imitation of a dog's or wolf's print.

Something had told me this was not the dog that the rules said it should be. Perhaps it was the perfection of the direct registrations and

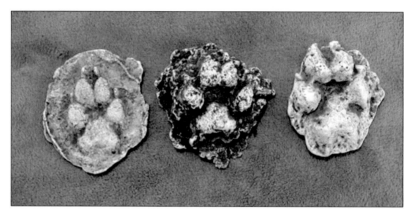

Figure 5.4. Canid casts. Which casts are of eastern coyote and which domestic dog? Although the middle cast with splayed toes may look more like the domestic dog print on the right than the coyote on the left, it is in fact a coyote spreading its toes for support on soft mud. Context is important.

the lack of an accompanying trail of the dog's master. Moving over this surface was hard work. Would Rover persist at this pointlessly without the urging of its master? After tracking the animal for a while I decided that despite the rules this was not a domestic dog, after all. Only a wild animal whose survival depended on moving over this undependable surface would continue walking carefully and precisely deeper into the forest.

Rezendes tells a story about this. Years ago a big study of coyote behavior at Quabbin Reservation required him quickly to train up a group of volunteers to go out in the winter and record the trails of these animals. At the end of the first mid-winter day, which featured dry, soft snow, the trackers came back with reports of coyotes at this location and that location. Fine. A month later, at the approach of spring, the group assembled again, but his time they were sent out to track over the sort of damp, corned "collapse-snow" mentioned above. They came back at the end of the day amazed. "The coyotes have all disappeared," they said, "but now there are domestic dog trails all over the woods!" See Figure 5.4 for comparison of dog and coyote tracks.

Other predators also widen their straddle in similar conditions. The books, including my own tracking guide, will tell you that red foxes usually provide a walking 15-inch step length and a narrower straddle than even a coyote. However, in those same collapse-snow conditions of later winter and spring, these foxes too will widen their straddle. In indistinct trails where padding is obscure this can lead to identification problems versus bobcat trails, which are often distinguished from wild canid trails by the comparative wideness of their straddle. While bobcats can stretch their step amazingly, their usual step length while walking on a firm, even surface is around 15 inches, the same measurement as is usual for a walking red fox. Track shape in these circumstances can be of little help — as red foxes, like coyotes, spread their toes for support, their normal hexagonal print, longer than wide, becomes wider than long. However, the leading edge of a bobcat's print normally shows a smooth arc while that of a red fox in these circumstances almost always shows angles caused by the indistinct impressions of extended claws.

Gray foxes can be even more confusing. With front toes spread for support, this fox's front print can look round like a cat's. Furthermore, its straddle is generally a little wider than that of a red fox. While its normal walking step length is shorter than either red fox or bobcat, it can lengthen it considerably, once again just like a cat, at some point transitioning to an aligned direct-register trot. Even though the gray fox has a generally smaller print size than most bobcats, there are small or female bobcats that can provide some overlap with a male gray. Here you will have to follow the animal for additional information. If it varies its step into a single-registration gait, you may note the different shape of front and hind — a combination of round front and narrow hind is often a clue to this species.

The morals of the story are:

- Do not apply rules arbitrarily.
- Embed the evidence in its circumstances. Imagine, if you can, that you are moving inside the animal's quadrupedal body and with its motivation.
- Look at the sum of the evidence rather than pinning your determinations on one or two clues.

Interpreting Some Trails

I have studied at one period or another in my life a half-dozen languages, everything from classical Greek to Russian. The code-breaking aspect of language- learning is the part that has always fascinated me. In a sense, interpreting the trail of wild animal is like interpreting a foreign language. The tracks represent the words, each section of trail a sentence, and enough of them in sequence a paragraph. Altogether they represent an entry in a diary of a wild animal's daily experiences. One can then follow along after the fact and "read" the diary, speculating on the animal's inner impulses as it "wrote" the paragraphs. In this way the tracker can gain inside knowledge of the animal he is studying. In seeing what the animal saw, reading its reactions in its "words" and "sentences," one can unite the animal with its habitat, see it in its context and gain enlightenment about that hidden world of life that surrounds us, upon which we ultimately depend, but about which we are otherwise largely ignorant.

A few of the better tracking books available in bookstores these days can be helpful as "dictionaries" in teaching us the vocabulary of this foreign language. I include among them my own *Trackards for North American Mammals* and *The Companion Guide to Trackards for North American Mammals*. The present book, on the other hand, is about the syntax of those words strung together, in effect a "grammar" toward that end. Words by themselves contain little meaning. It is their grammatical relationships that communicate the sense of the words, and the context that provides us with the perspective that will lead to understanding. So it is with tracks and trails.

Reading the Diary in an Animal's Trail

As our experience as interpreters of this foreign language accumulates, our reading becomes more expert. A typical progression of this sort may be sequenced as follows.

The beginning tracker may see a passage written in the snow and be able to understand no more of it than this: ***An animal passed here.***

By studying the better manuals or attending tracking workshops, he may soon begin to understand a little more of the message. In these sources he was taught to note the efficiency with which a wild predator moves over the winter landscape. The appearance of that efficiency, expressed in a series of perfect direct registrations, allows him to translate a bit more: ***A wild predator passed here.***

Further study helps him to notice the direct registrations are 22 inches apart and in an almost straight line. The manuals, then, help him to understand the identity of the predator:

167

An eastern coyote passed here.

And a growing familiarity with pattern diagrams helps him to understand the gait the animal was using: *An eastern coyote walked past here.*

Noting the smaller size of the footprints from the illustrations in his "dictionary" he decides on the sex of the animal whose diary he is learning to read. *A female eastern coyote walked past here.*

This interpretation is confirmed by the January date and a spot of blood next to a scat he finds along its trail: *A female eastern coyote in mid-winter estrus walked past here.*

But what was this animal doing? Was it merely traveling to a scavenge site, or searching for a mate, or escaping detection? A look around and a reading of several signs at once allows him to translate the answer. He notices that the coyote is "root humping," that is, aiming its trail directly at the trunks of trees and passing over the root hump of each. He sees as well that there are many trapezoidal bounding patterns of a small animal in the thin snow at the base of the trees where these animals had been foraging. *A female eastern coyote in estrus walked past here, hunting squirrels.*

Given enough time and experience, the increasingly expert tracker might be able to read ever more subtle information into the "words" and expressions in the diary. He might be able to decide, based on track depth, deformation and the crystal growth in the compacted snow of each track, how long ago and at what time of day or night the incident took place: *A female eastern coyote in estrus walked past here early this morning, hunting squirrels.*

And so on, seeing, relating, and interpreting evidence that the inexperienced eye might be incapable even of detecting. At some point, he passes his personal test and joins the long chain, very nearly irretrievably broken, of expert readers of animal sign. Just as the student of languages with training and experience eventually becomes a competent translator, he or she becomes a competent tracker.

A Fox and a Hill

Here is the winter trail of a red fox at Mount Misery in Lincoln, Massachusetts (Diagram 5.1). The snow had crusted heavily and then an inch or so of soft snow had fallen on top. The fields at Mount Misery are often visited by foxes to hunt for voles in the fields. This trail led away from the fields into the woods where it ascended a small hill with a short terrace half-way up. As the fox left the field, the trail consisted of pairings of two tracks on a repeating diagonal. Each pairing consisted of a front print retarded (black oval) and a hind advanced. The step length was about 17 inches and increasing. As the fox approached the hill, the gait pattern changed to sets of 4 single registrations with the order from rear to front of fhfh. At the base of the hill, both hind were advanced as in front-front-hind-hind (the middle of the second sequence illustrated in Diagram 5.1). As the trail ascended the slope, the order of the four prints once again became front-hind-front-hind. About half-way up the slope toward the small terrace the gait changed back to patterns of two — front-hind, as it had in the field, this time with shortening pattern lengths. This pattern persisted to the terrace where it changed to the four-print pattern of front-hind-front-hind, which remained that way for three or so cycles until the slope once again ascended. A short distance up this slope, the pattern changed again to patterns of two once again (end of the third sequence) and then as the fox neared the top of the hill the pattern changed to a narrow line of alternating direct registrations: front/hind, front/hind (gray ovals) which continued over the crest of the hill.

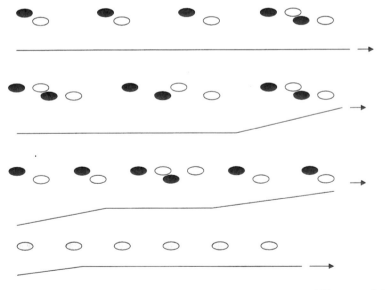

Diagram 5.1

169

That's a lot of pace changes for a fox! The alternation of three or more different patterns for this animal over a hundred yards or so might seem inexplicable unless one takes the terrain into account. Initially, as it left the field, the fox was in a displaced trot, its default traveling gait. Once it saw the hill approaching, it speeded up, first lengthening its trot pattern and then switching into a transverse lope which gained speed at the base of the hill (at maximum speed both hind feet are advanced ahead of the front: ffhh, in what almost qualifies as a lope-gallop). The fox was trying to gain some momentum on the flat to help it up the slope. As it climbed the first part of the hill, it allowed its momentum to expend itself, slowing first back to the slower lope (each hind print registering to the rear of the front on its side: fhfh) and then, as it neared the terrace, back into the displaced trot (fh fh fh . . .) with a shortening pattern length reflecting its loss of speed.

Once on the terrace, it tried to do the same thing as it had at the bottom of the first slope, that is, speed up to develop some momentum for the latter half of the hill. Because the terrace was narrow, it was unable to attain the full lope-gallop it had used at the base of the hill, only managing a standard lope (fhfh). As a result it carried less momentum into the second slope, falling back quickly, first into a trot with declining speed, and then, when its forward momentum was completely lost, into a direct-register walk-off at the top of the hill.

Any predator that travel-hunts during the winter must be very careful about how it uses its energy. In this case the fox decided to speed up on the flat where it cost it the least to do so in order to carry expendable momentum into the hill where the animal could allow it gradually to dissipate. On the terrace, it tried to get a little speed going again but carried less into the upper slope and so slowed all the way down to a walk as it neared the top. This is undoubtedly the most efficient way to climb the slope in these conditions and shows a keen awareness by the fox of its energy budget. The carefulness of its behavior suggests that this was an adult that had survived previous winters and had learned how to do it.

A Galloping Fox

For all native North American animals other than pronghorns, galloping is used as a brief attack or escape gait. Few animals have the chest musculature or the cardiovascular capacity, it seems, to

Figure 5.5. Red fox rotary gallop. This strenuous gait is used for attack or, in this case, for escape.

sustain this gait for very long. That's why I wondered at the patterns I found on the surface of a frozen pond at Mount Misery (Figure 5.5). The area consists of agricultural fields of corn and hay adjacent to woods. A narrow pond about a hundred yards long with a heavy margin of brush on either side occupied part of the border between field and forest. The presence of a red fox in such a habitat was no surprise, but what this particular fox had done was. In a skim of snow over ice for the length of the pond was a trail composed of patterns like those shown here (Diagram 5.2).

Diagram 5.2

The pattern lengths increased from 22 to 29 inches as the trail progressed down the pond, indicating that the animal was speeding up. Having studied the chapter on patterns and gaits, you may recognize this as the pattern of a rotary lope-gallop and recall that this is a very energetic gait.

Here indeed was a puzzle. The trail was very fresh, so at first I thought the animal might have been fleeing from me. However, the fact that the trail went down the length of the pond rather than quickly dodging into the brush along its edge suggested otherwise. Furthermore, this certainly wasn't an attack. Typically, a fox maneuvers carefully toward its prey pouncing or using a brief rush over a short distance to capture. This animal was running for a hundred yards. I noted nail marks, testifying to the urgency of the run, and some slippage, as well, always to one side. In terms of energy conservation, none of this made any sense.

At the end of the pond the trail disappeared into dense brush near the outlet. I took some measurements along the trail and stored the information, hoping for some enlightened intuition to strike on the way home or in the evening when I had the leisure to go over the evidence. Deciding not to try to force through the brush at the outlet and not trusting the ice there either, I diverged up into the fields and around through the woods, distracted by other animal trails. Eventually, I worked my way around through the woods and came down to the little step-across brook that flowed from the pond. In its frozen bed I once again picked up the trail of the fox, now walking with its usual direct-register trail appearance. I followed for perhaps 50 yards and then suddenly came upon the answer to the riddle. There on the ice of the brookbed, surrounded by fox tracks, was the bloody leg of a dismembered white chicken. Of course! The farmer who worked the fields had a chicken coop in back of his house. This was an escape, after all, but not to the nearest cover. Better to get far away, beyond the range of the farmer's dog or shotgun. And anyway, a fox with dinner in its mouth doesn't need to be careful about wasting energy.

Subsequently, a careful examination of the trail on the pond showed occasional linear marks alongside the front prints of some of the patterns where the downward arc of the fox's head at the end of its extended suspension caused the legs of the chicken to flop and touch the skim of snow. The lateral slips on the ice that I noted before were caused presumably by a laterally unbalanced load.

A Bear at the Trailhead

It's not often in a tracker's experience that he gets to see the animal in motion and then locates the track patterns that the animal left behind. But such was the case one day as I pulled off the Kancamagus Highway into the Sawyer River Trail parking area. From my left I saw a black shape run across the lot for the space of perhaps twenty yards and into the brushy border — a running bear, I quickly realized. But as it disappeared in a few seconds, I was unable to testify to much else and was left to reconstruct what had just happened from memory. As with many short-lived phenomena, my memory was not of what had actually happened, but only of my memory of what had happened — a memory of a memory, so to speak. Many experiments have shown that such recollections are highly unreliable. However, in the coarse granite detritus that the Forest Service uses in their trail-head parking lots, I was able to distinguish the fleeing animal's track patterns, which provided me with much more detailed and reliable information about what had just occurred. While what I saw took only a few seconds, I could ponder the track record at length and reconstruct in my mind's eye much more about the event than the witness my physical presence had provided. Here, from left to right, are the patterns I discovered (Diagram 5.3).

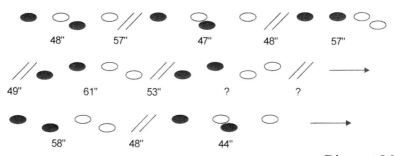

Diagram 5.3

The double slants represent the greatly compressed inter-pattern spaces, and the question marks are for illegible journal entries in pencil that have faded over time. Apparently the bear detected me just before the sequence begins. First, it took two lopes of 48 and 47 inches respectively, with 57 inches in between. Note that in each of these patterns the first hind footfall is located to the rear of the second front footfall indicating a relatively slow ground-speed. Being a bulky animal, it took these two lopes to overcome inertia and get into full stride. After that, for the next four patterns, the hind prints are both well ahead of the front, indicating that that animal took four long full-gallops with an increased pattern length of 57, 61, (?) and 58. In the last sequence, just before the bear entered the brushy border of the lot, it slowed dramatically to a lope once again. Here, the pattern length is down to only 44 inches. Perhaps at that moment it felt that safety was near and it could afford to slow down, or perhaps it slowed to search out an entry point into the cover.

The inter-pattern spaces are worth noting for the bio-mechanical difference between a lope and a gallop. Measurements for these spaces in the lope-gallop-lope sequence — 57, 48, 49, 53, ?, 48 — vary less than the pattern lengths. The pattern lengths for the gallops — 57, 61, ?, 58 — increase substantially over the 44-47 inches of the loping strides. Why don't the inter-pattern spaces vary proportionally with the pattern lengths?

As explained in the section on high-speed gaits (page 127), the front legs in a lope are used more for props than for propulsion. In a lope, the motive power is provided mainly by the hind legs, with more arc of stride in between patterns. This explains the large inter-pattern distance between the two initial lopes. In a gallop, on the other hand, air time is lost time. To maximize speed the animal must maximize contact time with the ground, gripping it and pushing it behind him through as much of the stride cycle as possible. The resulting flatter and briefer arc between the gallop patterns explains the decreased length of the inter-pattern distance. For the same reason, the animal lowers its center of gravity during the gallop in order to reduce the radius from shoulder to foot so that each foot contacts the ground longer than it would with the higher body position of a lope. This lower body position and almost constant application of force to the ground would be very inefficient and tiring in the long run, but the bear knew it only had to get to the cover at the side of the lot for safety.

Once again, the track record allowed me to get inside the animal and feel its experience rather than settling for the perspective of the outsider looking in at a sudden and brief encounter easily distorted by memory.

First Snowfall

Here was an oddity! I was tracking the Saugus River near Camp Nihan the day after the first December snowfall. In the two inches of soft powder I found a series of amazing canid patterns left by a pair of animals (Diagram 5.4).

Diagram 5.4

At first, I didn't know what to make of them. The prints appeared to be coyote by shape and measurement as well as gestalt. However, I have seen certain dog breeds that present a pretty good coyote print. It must be a domestic dog, I thought. No wild animal would be so wasteful with its energy in the winter. For a while, I followed these patterns — ragged racing prints combined with the sorts of patterns illustrated in Diagram 5.4 — from the woods out into Cedar Glen. Nowhere was there a sign of the dogs' owner. I looked at the tracks again — inconspicuous nail marks, especially the medial and lateral, tight grouping of toes, squaring off of the corners of the secondary pads. Everything about the track morphology, at least, seemed to say coyote. I puzzled some more. Insanity can occur in wild animals as well as civilized ones, I thought. But simultaneously, in a pair of animals? Then one after the other I noticed two things. First, the print width on both animals was a little undersized. In Massachusetts the normal front track width for an eastern coyote is about 2¼ inches. The front width of the pair of animals I was tracking was just slightly over 2 inches. Then, looking even more closely at the prints, I noticed a few crystals of snow in the toe impressions.

I sat on a stump and spooned some hot soup out of a thermos. Jays in a pine grove mocked my ignorance. And then, inspiration! The tracks had been left by undersized coyotes during a light snowfall (the snow crystals within the prints), the first of the year. These were grown pups from this year's litter that had never experienced snow before. And they didn't like it. It disoriented them to surroundings that had become familiar, and worse, it stung their toes with its coldness. These animals were trying to levitate! Or perhaps they were jumping to snap at the falling flakes as I had seen young domestic dogs do. Or both.

There are two parts to any tracking experience. First is absorbing the scene. The second is thinking about what you have absorbed. The first requires close attention. As Rezendes used to say, "Look, and now look closer!" The second requires reflection and intense mental activity. Sometimes inspiration comes on the way home from a hard day's absorption, or in the evening, or the next day, after some time has elapsed and the subject revisited in a different frame of mind. Most learning, it seems to me, involves relating what you see before you to what you already know. Unconsciously rummaging through a lifetime of experience may take some time. You may be lucky and stumble on the solution at once as I did with these young coyotes. Or it may occur hours, days, years later. "Intuition is the unconscious application of everything you already know to the matter at hand." And that application may take some time.

Two Fishers

Above (page 125), I briefly described tracking a male and female fisher over a breakable snow crust. I elaborate on this incident here as it presents some important insights. It was a January day many years ago in north Quabbin when I was tracking with Paul Rezendes. The shadows from the low early winter light provided excellent relief to tracks such as those of a loping female fisher whose trail we came upon just inside the reservation gate. The snow surface was fresh with about a foot of newfallen snow that had included some mist late in the storm, just enough to stiffen the upper layer. This surface was firm enough to allow the female to cruise around her range in a loping gait with the pattern shown in Diagram 5.5.

Diagram 5.5

You will recognize this by now as a medium speed aligned lope. We followed this female on her rounds deeper into the woods until we came upon the trail of another fisher that intersected that of the female at several points. Where the female had shown a track width of about 2¼ inches, this new trail showed a front track width almost an inch wider — a male (Diagram 5.6). This animal, too, was using a loping gait but its pattern looked somewhat different.

Diagram 5.6

This is a slower-speed, displaced lope, its speed indicated by the fact the hind print of the middle two prints in each phrase lags behind the front rather than being placed even with it as in the female's patterns. These two loping patterns can tell us a lot about the relationship of female to male fishers.

This species shows a distinct gender dimorphism, that is, males are a third to half again larger than females. This may seem like it should be a hunting advantage for males, but that assumption would be wrong. Smaller size, as it turns out, is a distinct advantage for the female. First of all, she can hunt into smaller holes than the male, places where the abundant small prey like squirrels den and hide. Secondly, climbing up a tree after squirrels costs her less energy since she is lighter. As an added benefit of smaller size and lighter weight, while hunting in tree holes she can simultaneously reconnoiter them as possible denning sites for birthing and raising kits. Typically, New England forests are managed for timber; trees are not allowed to grow old enough to develop natural cavities from limb-fall before they are harvested. As a result, cavity-denning species like fishers need to rely on the excavations of woodpeckers and, in the case of a fisher, large woodpeckers, of which in the Northeast there is only one — the pileated. A smaller size allows her to exploit their abandoned nesting and roosting cavities for her own purposes. Not only can she use these as birthing dens, she can also rest in them, above any disturbance below, and take advantage of the good insulation of sapless winter wood. In such a managed forest a male fisher must make do with damp, cold ground-dens.

But there is one more advantage to small size — speed. Fishers, like foxes and coyotes, are travel hunters, putting in many miles in a night revisiting known hunting habitats where they have had success before as well as revisiting old caches in case they fail to make a fresh kill. The fisher that can get around to such sites the fastest has a hunting and survival advantage. In three seasons of the year the advantage should lie with the more powerful male. But the fourth season, winter, is the critical one for all animals. This is the season of privation, the testing period for genetic and physical fitness. And, on snow, the female in most conditions has the advantage, for she can stay higher on the surface than can the heavier male.

Given the short legs of members of the weasel family, an adaptation for hunting in holes, travel is often a matter of loping or bounding, reserving the walk for stalking or when fatigued. In the rocking-horse motion of a lope or bound, vertical force is wasted force. The deeper an animal sinks into snow, the more energy it must use to vault itself upward to get clear of the snow between it and its next landing spot. As we have seen above (page 123), this upward vector of force lengthens at the expense of the horizontal vector. And this not only requires the expenditure of energy in a useless direction (the predator's desire is to maximize the horizontal vector to get around to its hunting sites), but also slows the animal down. This latter effect is shown in the two simultaneous trails we encountered on that winter day.

Careful comparison of the patterns reveals subtle differences. Both are lope patterns, but the pattern length of the male was longer than that of the female as a function of the greater length of the male's body. However, a surprise at first was that the inter-pattern spacing was slightly less with the male than with the female. He was moving more slowly, and the respective print arrangements within the 4X patterns show why. Looking carefully at the female's four-print pattern, you see that the hind prints are slightly more advanced in her loping patterns than are the male's. That means that she was moving faster in the horizontal direction than he was. Furthermore, note that the male had to resort to a displaced lope, angling his spine diagonally to the direction of travel. This allowed him to get his hind feet past his front while minimizing his vertical vector. An increase in the vertical force for the heavier male would have driven him through the thin crust either on take-off or on landing. If he had tried to align his gait without greatly increasing his vertical motion, he would have tripped over his own front feet. The lighter female, on the other hand, could afford more spring in her gait, spending enough time off the ground in gathered suspension that she was able to keep her spine and feet aligned with the direction of travel, that is, her front foot on one side was leaving the surface before her hind foot on that side proceeded forward over its vacated print.

All of this shows that the lighter female was able to take better advantage of the drizzle-stiffened surface of the snow than could the male. Indeed, in following the male, we occasionally saw where his pattern changed to one similar to the female's. He was trying to speed up. However, these patterns would only last a few gait-cycles before the first front footfall would break through the crust and the male would stumble and flounder. The displaced pattern that he had been using and to which he returned after each break-through has, as with any displaced pattern, a fairly flat arc of stride, so that the first front footfall in each pattern comes in at a shallow angle and with little downward force. This was really the only way he could get around on such a fragile surface. Speeding up to the female's pattern required aligning his body, springing higher into the air to avoid tripping himself and thus coming down heavily at a steeper angle. After each stumble, he would settle back into his slower displaced pattern, conceding the hunting advantage to the female.

A Mystifying Trail in the Snow

One winter many years ago in the Federated Forest campground I came across the walking trail of an odd animal. The depth of each track was about 4 inches and the pattern was the presumed direct-registering diagonal walk pattern used by many predators that are abroad in the snow. The track width was about 2¼ inches and the length about 2½ inches making the impressions longer than wide. Add a step length of 23 inches with a very narrow straddle and the trail sense indicated an eastern coyote. However, the shape of each track mystified me. Two nail marks, close together, marked the front of each track but the rest of it was lumpy and amorphous. The tracking conditions were pretty good, with well-defined impressions, and the trail was fresh, so that neither conditions nor aging under a February sun were enough to distort the tracks to any great extent. Intrigued, I decided to follow this odd trail to a different surface somewhere along its length that might solve the mystery.

From the point where I had picked up the trail in the primitive campground at the state forest, it went one way up toward Soapstone Ridge, the other way down toward the Gate 36 North beaver pond. I decided to foretrack the latter as the easier course. Down the dirt jeep road it went, across sunfields and the shade of a red pine grove. Still, the tracks showed the same lumpy outline. Had I discovered a new mammal? At the bottom of the hill, the trail led out onto the wind-packed surface of the pond. Snowshoeing behind and off to the side, I followed out onto the pond avoiding, as did the trail, the weak area near the beaver lodge. Half way down the pond, the mystery was solved. The trail suddenly became two trails! A coyote's continued straight down the pond while the trail of a bobcat came in from the woodline on the left and joined it, but heading in the opposite direction. The bobcat had used the pre-packed coyote trail as an easy way to move around its hunting range. On the left of Diagram 5.7 is the mystifying track, a combination of the two tracks on the right.

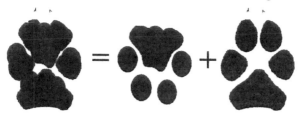

Diagram 5.7

Although the normal walking step length of a bobcat is about 15 inches compared to the 23-inch step length of a coyote, bobcats have long legs and a flexible anatomy, able to stretch their step lengths to accommodate the trail of any animal whose route they may encounter in the woods. Since this incident, which occurred early in my tracking experience, I have seen these cats not only shorten their stride to that of a porcupine, skipping every other track, but also stretch it all the way out to the walking step of a white-tailed deer at 25 inches! Seeking winter efficiency as they move around their range, and possibly like housecats, disliking loose snow on their feet, they habitually use the pre-packed trail of any animal they come across, even stepping perfectly into their own tracks from a previous evening. Down at Gate 35 in Quabbin I came upon this very act, with the cat walking in the opposite direction from its original trail. But this time I was mentally prepared by my previous experience in the Federated Forest and correctly analyzed the evidence.

These incidents show, once again, how following an animal's trail can open a window into its candid behavior, insight denied to direct observations of the animal itself. Tracking and interpreting can, over time, allow you to accumulate a sense of a wild animal's personality. Here, imagining the bobcat in the act of going out of its way to travel in the pre-packed trail of another animal shows both its resourcefulness and its opportunism. As the tracker's catalogue of these kinds of encounters increases, he or she gets to experience more and more the animal's sense of its world.

A Purposeful Bear

On November 19, about a decade ago, I was tracking in north Quabbin when I happened upon the trail of a black bear in early snow. Its track width of 4½-5 inches indicated a large animal. The trail crossed a utility road in the park heading into dense laurel and other brush. The trail across the road looked like that in Diagram 5.8 and had the following dimensions: pattern length, 70 inches; pattern width, 10 inches; inter-pattern distance, 47 inches; track depth, 3 inches.

Diagram 5.8

All the above information presented a "boar," or male bear, moving as fast as it could in a transverse lope/gallop. Since the trail leading up to the crossing was a conventional walk with direct registrations, the change of gait seemed to be an effort to get across the road (and escape observation?) as quickly as possible. Because it was late in the day and I was running low on energy, I decided to backtrack the animal through the easier terrain on the other side of the road.

For the first hundred yards or so the animal had continued to walk as it approached the road. Carefulness, I thought. The road was a danger area and so the animal preferred to cross unobserved. However, other bear trails I was familiar with crossed roads with no apparent change in gait. What was going on? I wondered in a somewhat fatigue-fogged brain. After the initial hundred yards of back-tracking, the gait changed a little. The step length, which had been between 20-23 inches, lengthened to 31-33 inches. Still the pattern was an alternating diagonal pattern of direct registrations. I reasoned that this meant the animal had approached the road walking, but farther back it had been trotting in an aligned direct-register pattern. The fact that in the longer DR pattern the straddle was very narrow so that the tracks

181

appeared in a nearly straight line supported this theory. With the road still some distance ahead of it, the bear felt safer, I decided.

Along this line of tracks I discovered a urine stain running through several stride cycles. The bear was peeing on the move; at one point urine in the track suggested that it had peed on its own foot. A bit farther back and deeper in the woods, the bear had gathered itself and jumped a brook. Beyond that, the gait had been a combination of walking and trotting.

Eventually, I approached the foot of a hill where the trail (Diagram 5.9), with a pattern length of around 50 inches, was coming down toward me.

Diagram 5.9

This was a lope-bounding gait involving some energy expenditure, moderated by the descending slope. As I continued to back-track up to the ridge of the hill, the trail came to a junction with another bear trail in a walking gait coming down the ridge from the right. To my left the trail came *up* the hill toward the junction also in a direct-register walk. At first it looked as if two animals had somehow become one!

A closer look showed that the one on my left was actually a two-way trail, with the ascending one laid directly over a descending one obscuring the underlying prints! I followed this line of tracks down the ridge to its base where the trail crossed a small seep-brook. Just on the other side the trail disappeared. I sat on a log and pondered, reversing the evidence I had found so that my back-tracking became fore-tracking in my imagination.

The bear had walked down the slope of the ridge to the base of the hill, stepped across the small brook, turned, and then walked carefully back up the slope in its own prints. Half-way up the ridge it had leaped to the right off the walking the trail and lope-bounded in the 3X pattern shown in Diagram 5.9 down to the bottom of the slope, then off through the woods, trotting and loping toward the road where I first picked up the trail.

At this point, I decided to follow the animal back to the road and investigate the other side. Along the way I noticed that the bear had aimed its trail directly at the gaps in two stone walls. Obviously this animal knew the area well, suggesting that it was not still on its fall shuffle far from its home range in search of late season beechnuts and acorns. This animal was "home," but not until I followed its sprint across the road and into the dense brush beyond did I realize how "home" it was. In the laurel scrub, the bear began to walk again, heading straight into the thickest tangles it could find. In the southern Appalachians they call these thickets "laurel hells" and I was soon reduced to crawling on hands and knees to get through them. On the other side of the thickets, things got worse as I was forced to enter alder clumps over a seepy swamp where the surface crust threatened to collapse under me at any moment. Finally, after 3 hours total and with the light fading, I gave up my pursuit. The bear had won. It was almost as if it had set out to frustrate me. Through my increasingly fatigued mind I suddenly realized how right I was. It was November. This was a bear heading into den for the winter and trying to make sure that nothing and no one could follow him there. Hence, the sub-terfuge of walking back in its own tracks, and the leap off this trail, and the care with which it approached the road, and the speed with which it crossed it, and finally the impenetrable thickets he dove into on the other side.

Bears are rarely hunted in this part of Massachusetts and never within Quabbin reservation. However, this bear had ancestral memories of being attacked by humans during its winter torpor. This I watched in a documentary film once that showed it as the common method of bear hunting among the Cree in Canada. It seems a clear evolutionary message had been encoded in the genes of my animal — be especially careful in disguising your trail when heading into winter den.

But was the bear smart enough to be aware consciously that a human could recognize and follow its tracks, or was this dissembling simply blind instinct buried in its genes? I recalled reading the results of a study many years ago on wolf-coyote interaction in western Canada. One of the findings was that when coyote trails struck those of wolves, the coyotes significantly more often than not followed, that is fore-tracked, the wolves rather than back-tracking them. Coyotes are both victimized by wolves, which will kill them as food competitors whenever they get the chance, but also are dependent on wolves to bring down large ungulate prey that the coyote cannot manage. Once the wolves feed on an elk, the coyotes (as well as magpies, eagles, jays, foxes, wolverines and any other scavenger in the neighborhood) can then feed on the leftovers. Thus it is in the interests of a coyote, encountering the trail of wolves, not to move away in the direction of safety but rather to follow the trail and so be first among scavengers at an abandoned kill. Whatever the risks, the coyote's best choice is to fore-track the wolves before everything is eaten. But how does the coyote know which way the wolves were headed when they passed? They can't see or hear the wolves themselves, which presumably are long gone. The strength of the scent of the wolves should be equal on either side of the trail junction, so the sense of smell doesn't help. How, then, does the coyote know? Although it was not the point of the original study, I suspected from the evidence that the coyote must have been able to read the tracks of the wolves as we would and perceive the orientation of their feet. The coyotes were trackers! And presumably, if they could perceive track orientation of a wolf, they could also do so for the tracks of any other animal that might represent either potential threat or potential food.

I have a prejudice that bears are at least as intelligent as coyotes, and so I have preferred to believe that the bear that November day knew that it was leaving behind evidence of its own passage in the impressions it made in the snow. This was not the sort of instinctive behavior that induces a bittern to inappropriately assume a vertical cattail pose in the midst of green grass. This had the feel of intelligence — the bear "knew" that it could be followed and made every effort to conceal or otherwise obstruct its trail.

It is, of course, possible that this was just instinctive behavior, after all, bred into the bear over generations. The bears that didn't behave deceptively as simply random behavior when heading into den got found and killed and therefore failed to procreate the next generation with a genetic bias for this kind of behavior. Personally, I prefer my smart bear hypothesis, but for no reason other than a personal affinity with bears.

Chapter VI

Species Notes

There are many very good sources of information about the life histories of North American mammals, several of which are listed in the bibliography for this book. Such has not been the case for information about tracking these animals, however, as until recently there has been little accurate guidance in this specialized skill. I have devoted this chapter of *The Next Step* to clips that illustrate one or another aspect of this art. These include anecdotes from tracking adventures I have participated in over the past twenty-seven years as well as odds and ends of information that do not appear either in my own *Trackards for North American Mammals* and *The Companion Guide to Trackards* or in other reliable guides currently available. These notes are not an attempt to replace the information available in these publications but rather to add to it. As wildlife tracking is a newly rediscovered and rapidly developing art, the reader has the opportunity to add new information as soon as he or she gets past the novice stage and can reliably attribute the evidence that is found both to the correct species and to the corresponding behavior of individuals of that species. Few other fields allow one to occupy a position on the cutting edge of study without accumulating a lengthy resume of degrees, appointments, grants awarded and academic hurdles negotiated.

The chapter begins with a discussion of "sign," a collective term for any evidence that individuals of any given species might leave behind.

"Sign"

The term *sign* properly refers to any information that a wild animal leaves behind, including tracks, tooth marks, scat, fur, bones, etc. However, it is often used as well to mean all of the above except tracks and trails, and that is how I am using the term in this book. Thus, I have put it in quotation marks in the title to indicate that here we are dealing only with evidence other than tracks and trails. There is an abundance of different sorts of this sign, only a few of which are mentioned here by way of introduction. Other sign will be dealt with in the notes for the individual species.

Scat and Urine

In seasons when there is no snow on the ground, animal droppings, or "scat" (from the Greek *scata*), is easily the most common sign of the presence and behavior of wild animals. Not only can it tell you which species was present but also what it was eating, thus connecting the animal with its habitat.

The more reliable identification guides will provide you with one or more representations of the scat of a particular species as an aide in identification. As this is an advanced tracking guide, I expect that the reader will already have some experience with scat identification or will at least have one of the reliable guides at hand. (As a general observation, I have found that the less accurate the images of tracks in a guide are, the less accurate the scat renditions will be as well.) The comments in this section are intended to elaborate the subject and provide some caveats.

The scat of any species can vary greatly, according to food, season and the size of individual animals. Furthermore, the scat of one species may masquerade as that of another. Let's take that last first. Many animals, especially predators, use their scat as well as urine to mark their presence or passage. This may be a self-assertive expression of territorial ownership or it may be an effort to provide details about oneself to a prospective mate. Although we can't really tell how much information is in the scat and urine of a wild animal, the interest that other animals show in it suggests that sniffing a deposit is, as is often said, like reading the local newspaper for them. Because of its importance, then, some species will counter-mark any such deposits they find in their hunting range. This may be one coyote placing evidence of its own over that of another coyote or leaving them side by side to show that a particular area is owned by a mated pair. Predators of different species may counter-mark, as well. For instance a passing coyote may leave its scat at an otter roll where it detects the fishy odor of otter scat. Foxes and other predators also may leave acrid urine deposits near or even over the scat of other predators, leading the careless tracker to identify the underlying scat as fox. In my experience counter-marking is most often the reaction of one predator to the presence or passage of another, especially one of its own species.

Raccoons often leave their scat at a particular spot over and over again. Due to the varied diet of this omnivore, individual deposits of their scat at such a scat station may differ from each other in appearance. This, in turn, might suggest counter-marking by more than one species. Recognizing such a site as containing all raccoon scat, whatever their appearance, may be critical for the tracker due to the dangers of fractionating the droppings of this particular species (page 387). The point is that the tracker needs to be careful in attributing found scat to a particular species since appearances can be deceiving.

It is easy to get so absorbed in the details of the appearance of scat that we forget about the context (pages 55-56). The manner and location in which a scat is deposited may be more helpful than the appearance of the scat itself since the scat of all similar-sized animals using the same food source will look similar. A fur scat deposited in a prepared scrape located in an area of dense undergrowth can reliably be attributed to whatever felid of appropriate size ranges in the area. A similar scat but with a scrape beside it and located at the intersection of a trail system in more open circumstances can be attributed to a canid of appropriate size. A collection of scat with different appearances and composition at the base of a tree next to a stream or pond is likely to be from a raccoon. Bean-sized droppings on a boulder a few feet from shore in a pond are most likely from a muskrat. Fish-scale scat about the width of a cigarette and located on a muskrat mound, on the other hand, is likely from a mink. This is the sort of contextual information that any good identification guide should supply, and you should pay attention to it since this information may supply the confirming detail in the array of evidence that goes into any successful identification.

The scat of members of the same species can vary greatly with changes in seasonally available food. This is especially true of omnivores and scavengers. For instance, bear scat can be a runny pile of cherry pits or a tightly wound tube of fibers, to cite two of many variations. The only similarity among them is that bear scat is usually very large. However, a bear cub or yearling can leave a dropping that can look a lot like the scat of a coyote that happens to be eating the same

187

thing. I have often found scat composed of apple skins that could be attributed to either animal, based just on its amount, size and appearance. Once again, some attention to context often tells the difference since bears leave their scat more randomly than do coyotes, the deposits of which are usually more purposefully placed. When raccoons are feeding on late summer berries and other fruit, they, as well, can leave a surprisingly large deposit that I have seen mistakenly identified as bear scat. Because of its diarrheic effect such droppings can be found anywhere, so context may not be very helpful in identifying it. Look for any tubular shapes in the amorphous mass; their approximate diameter will often distinguish these two species.

Scat size will also vary among different members of the same species. Many predators are dimorphic, with males being larger than females. Correspondingly, their scat will vary as well. As with tracks, trails and other sign, identification of scat often results from considering the sum of the evidence rather than one particular characteristic. Add up and evaluate the information. The result may not always be certain, but you are more likely to be right by doing so.

The randomness or purposefulness of a scat location often can be helpful in identification. Herd animals and other herbivores eat low quality food and thus must ingest a lot of it to sustain themselves. As a result, they void scat frequently and randomly. Bears seem to void both scat and urine at random as well, substituting other calling cards such as marking trees to advertise their presence or passage. Since most predators sit at the narrowing apex of the food hierarchy, on the other hand, they must be thinly distributed over a large area in order to survive. There are exceptions, of course. African lions must hunt in prides in order to bring down large and fleet animals. In North America, wolves also hunt in packs of related animals, also because they hunt prey larger than themselves. In the Northeast, only the eastern coyote hunts in this fashion, and then only when it is attempting predation or scavenging on deer or moose. In these cases of pack-hunting, scat may be deposited more randomly than with the lone hunter protecting a lean territory and advertising for a distant mate.

Not only is the size, shape and context of scat valuable in identifying a species, but urine deposits are helpful as well. Many have characteristic odors that identify them immediately. The sweet odor of castoreum on a mound of debris next to a pond tells you that the mound was scraped together by a beaver rather than by an otter, which species also pulls together vegetation upon which to urinate. The odor of fish oil on such a mound indicates the latter species. Red fox urine has a strong skunk-like odor while coyote urine mixes that with the odor of burnt hair and is much less pungent. Both deer and moose urine in the winter, when they are feeding on buds of shrubs and saplings, has the pleasant herbal odor that I refer to as "woodwine." Porcupine activity in winter smells like turpentine. Gray fox and fisher urine has the odor of chemically-treated paperboard. Location of the scent mark, its context once again, may also help. Bobcats often choose the stubs of dead trees to mark with their ammonia scented urine while red foxes in winter habitually mark any top of fir or spruce that sticks up above the snow by about a foot. Once again, a good identification guide should provide this kind of information.

Skulls and Bones

In some states it is illegal to collect animal skulls or bones without a license. The reason is that there is a short but spectacular list of invariably fatal diseases that one can catch by careless handling of bones that have not been thoroughly cleaned and disinfected. When a wild animal dies, cleaning the skeleton of its soft parts begins with invasion by bacteria as well as by embryos from the eggs of flies and various beetles, all of which feed on the fatty tissue. Eventually, given enough time, these agents will thoroughly clean the bones. But until this cleaning is complete, as long as there is any fat at all left on the skeleton or in the marrow of its bones, it may harbor interesting diseases like bubonic plague, rabies and anthrax. For the most part, despite having a license, I usually leave bones alone, contenting myself with photographing them and looking for gnaw marks on their surface for evidence of other animals in the vicinity.

Bones, of course, have a great deal of calcium in them and as such are a source of this important mineral for other animals, both to support their own bones as well as to form the bones of fetuses. Careful inspection of bone remains will often show the incisor marks of rodents (Figure 6.1). These marks may overlap so badly that measuring them for identity is impossible. Look for random marks outside the major area of concentrated gnawing for single test gnaws and estimate from them the size of the feeder.

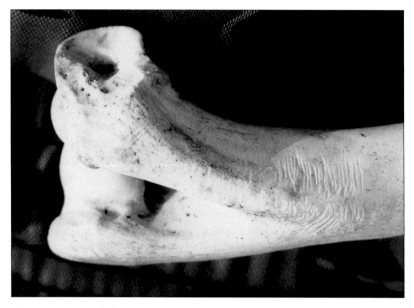

Figure 6.1. Examining a bone carefully may show evidence of gnawing. Here a small mammal with two central incisors, such as a mouse or vole, recycled calcium for its own benefit.

Dogs have large carnassial teeth and robust molars powered by strong facial muscles. With these teeth they can crack the bones of very large prey animals like deer. Such breakage eliminates the smaller cats as the predator/scavenger since they lack the robust teeth of a canid. To remove flesh from bone, they instead use their raspy tongues to lick off the meat, leaving clean rib bones as evidence of their identity. Bones removed from a disintegrating carcass and hidden in the snow at some distance from the skeleton is evidence of an effort by a non-dominant group-scavenger to avoid submitting food to the feeding hierarchy within its pack. In the Northeast, such evidence points to the coyote.

Fur

After scat, perhaps the most commonly found animal sign in non-snow seasons is fur. It may be part of other sign, as in the contents of a scat. It may be at a kill site where an animal was dismembered and the inedible parts, such as the tail, were left behind. It may be stuck to the sap of a bear's mark tree. It may have been scraped off on a branch or rock as the animal passed by or it may be found snagged by barbed wire as an animal passed underneath. I had only just read that elk refuse to go under barbed wire when I found a commotion of elk tracks in Arizona directly below a wire fence and a clot of fur on one of the lower barbs (Figure 6.2). (The careful and observant tracker can find himself in a position to correct traditional assumptions that have seeped into even respectable sources.)

By itself, the fur of a wild animal is a fascinating study that deserves a book of its own, but as of this date I know of no such work. Trolling through my many years of experience I have come up with the following observations, which are far from exhaustive but may still be helpful to the journeyman tracker.

Many animals molt fur seasonally, pushing out one coloration with new fur more suitable for the approaching change of seasons. Many weasels molt from reddish brown in summer to white in the winter. Snowshoe hares do the same, losing their mottled brown camouflage and replacing it with white. For some other animals the

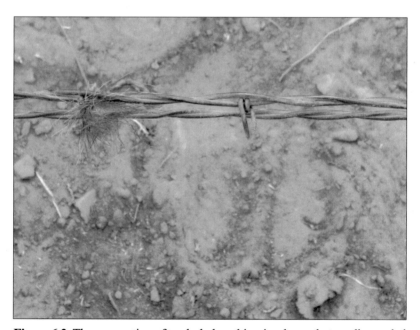

Figure 6.2. The commotion of tracks below this wire shows that an elk crawled underneath, snagging its fur on a barb.

change is more subtle. The reddish brown of a white-tailed deer in summer is replaced over most of its body with two-toned fur as the winter approaches. White fur is hollow and that hollow central core of the fiber makes it a better insulator; colored fur contains melanin granules in its core that make it more conductive. A deer in winter has it both ways. Each fiber of the fur, to which it has molted in expectation of cold, is white for most of its length. However, it has a gray tip. Thus with the fur laying flat against its hide, the animal looks mostly gray. This color allows it to blend with the gray tree trunks of the northern hardwood forest in winter as well as to absorb sunlight better than if its fur were white to the tip.

Deer fur has different consistencies on various parts of its body. Tail fur tends to be thick and kinky while body fur is finer and straighter. Contrary to accounts that I have seen in print, all white fur will bend at a sharp angle like a soda straw. This is because it is hollow. (I have seen this characteristic attributed to deer alone as a way to distinguish deer sign from that of some other white-furred mammal.) Clumps of white deer fur are often found on the ground after spring thaw. This is not necessarily evidence of winter mortality since deer shed this insulating fur in the spring in preparation for the warmer months. Examination with a loupe may be needed to see whether the fur has been mouthed and cut by a predator, has a normal root nodule at is end indicating it was shed in molt or has traces of blood on this nodule suggesting that it was pulled out during a struggle with a predator.

Fur shows up in predator scat all the time. Elsewhere in this book I related often finding fox scat composed totally of cottontail fur as the predator returned to feed on a cached carcass repeatedly until it was consuming nothing but fur, the silky feel of which identified its source, even after having been run through the fox's digestive system. Below (page 303) I mention often finding scat in spring composed of yellowish fur. Eventually I figured out that this was the coyote itself biting off itchy molt fur from its flanks and swallowing it, perhaps for oral gratification, given its lack of nutrition. A friend found a lay site during the spring where the bed was completely surrounded by a ring of this molt fur, pulled off by the animal as it rested. The presence of a few large feathers in the surrounding brush had led him to think that he had found a raptor plucking site on some prey animal. If anything the opposite might have been true: the coyote may have been the predator and the bird the prey. However, it is more likely that he was looking at the temporal compression of two independent events that occurred at the same location.

Red foxes retain the reddish guard hairs on their pelt throughout the year, but in the fall they begin growing a thick layer of white wooly underfur. Seen in the winter, the animal looks quite robust, but this is illusory. If you observe a red fox in summer when it has lost the un-needed insulation of its underfur, it will look emaciated. A red fox's dense winter undercoat of white, hollow fur allows it to lie out at sub-freezing temperatures with its large brush wrapped over its nose and feet like a sleeping bag and feel no effect from the cold. The fur of the more southerly gray fox, on the other hand, does not seem to be as effective. In the winter I more often find sign of this animal on south-ern exposures where it can get maximum benefit from the low sun.

Red squirrels also molt, but so inconspicuously that their winter pelt looks little different than their summer. However, since their win-ter coat is not much more insulative than their summer one, these animals spend a lot of time under the snow. Once the snow depth is greater than about 8 inches, they begin to tunnel between nest and cone cache, protected from both cold and predation. In the spring you can find these freshly exposed tunnels in thaw snow where constant coming and going has worn them down to about 4 inches across, com-pared to about 2 inches for voles. When we see red squirrels above the snow in the winter, they are usually racing for cover. An exception is created by any concentrated and reliable food source such as a bird feeder where these normally solitary and aggressive little animals barely tolerate the presence of others of their species while they forage on the ground for fallen seed, the fat of which, added to their energy budget, offsets the effects of the cold.

The red squirrel's bushy tail is interesting. It might appear at first to be nothing more than an adornment, but I began to notice one win-ter that I could predict the temperature outside my home on a winter morning by how these squirrels held their tails while they fed. On very cold mornings they held it curled closely over their backs while on warmer mornings they tended to hold it at a variety of angles. It oc-curred to me that their tails had a blood supply and that by curling that tail over their backs they were recapturing in its furry breadth some of the heat radiating from their bodies, thus recirculating it back through their body. These tails frequently show up at kill sites since predators such as foxes and raptors often nip them off before departing with the rest. After all why carry the weight of anything that is not nutritious? The ruddy color of these tails, with a darker band near the ends of the fibers, will identify them as this species.

While we are on the subject of the usefulness of tails I should relate a find on the spoil of a red fox's den one early winter. A vixen had apparently been investigating the site as a prospective birthing den for later in the season. Her body outlines could be found in the shallow snow at the entrance as well as on the root hump above the renovated woodchuck tunnel. On the spoil I found a number of cottontails — not the rabbit, just the tails! These lagomorphs had been enjoying a cyclical abundance around then, and the vixen apparently was concentrating on them exclusively, distributing the nipped-off tails around her lay sites like so many discarded cotton balls.

As a final gesture to the importance of tails we should consider the black tips on the tails of two of the three North American weasels. The general wisdom is that these black tips on otherwise totally white bodies of the winter animals are intended to confuse the aim of an owl swooping down on the otherwise invisible animal as it scampers over the snow. With only the black tip of its tail seized, the animal has the main part of its body free to wriggle loose. A defect with this theory is that the least weasel, appropriately to its name the smallest North American mustelid, has no black tip on its tail. If it is valuable for the two larger cousins, why isn't it valuable for the littlest weasel? One could argue that the black tip on the tail is the obligatory defect in an otherwise perfect defense (page 154). But this doesn't work either since the same question arises — why no defect in the least weasel's defense? I admit I have no answer for these questions, only a dissatisfaction with current theories I've read.

Fur color is an interesting topic. Why, we might wonder, do different members of the same mammal family have differently colored fur? Otters, fishers, minks and all three of the smaller animals we refer to as weasels are members of the same family. In winter, however, the larger members stay more or less dark brown while the smaller members often turn white. White fur, as we have noted, is hollow and therefore a better insulator than dark fur. Part of the answer may have to do with body size. The weasel shape is a poor one for cold

194

climates since its length, compared to its width, results in an unfavor-able surface-to-mass ration. A dark brown moose is so large that its biggest problem even in winter is how to get rid of heat rather than how to retain it. An otter is large and thick enough that its dark, oily, waterproof fur is no penalty in the winter, especially since it allows the animal to dive into water where the temperature is never lower than 32 degrees Fahrenheit. This is true of the smaller mink as well. How-ever, a mink has so little body mass that it must use behavior to assist in heat retention. For the coldest part of the winter it rests in under-ground chambers such as muskrat tunnels or abandoned beaver lodges where it curls up into a ball to reduce surface exposure to the cold. Even at that, this nervous little animal must feed a raging metabolic furnace to keep its body temperature up. Since any momentary failure in the supply of prey means doom, mortality in the smaller weasels is high.

Fishers are dark, as well, but profoundly non-aquatic. The name of this species is amusing since these animals don't fish and won't enter water, preferring long detours in their hunting route to avoid even getting their feet wet. In all the years I have collected mud prints of various animals, fishers seem to be the most careful about even stepping in mud, more so even than the fastidious bobcat. Like other weasels, fishers use chambers above and under the snow — such as hollow snow-covered logs and tree cavities — for refuge. It may partly be for this reason that I find their trails more frequently in the older-growth woodlands of metropolitan parks than in the relentlessly man-aged even-aged forests farther out in the countryside. They also use their dark fur for solar absorption, picking sunny, windless spots such as a squirrel or raptor nest in dense evergreen forests for resting dur-ing sunny winter days.

Winter is not necessarily white. In the North Temperate Zone where snow accumulates and lasts through the winter, the forest is a mix of white snow, darker shadows created by the low winter sun and the dark trunks of trees as well as the shade of those trees. Therefore there is little advantage in terms of camouflage for a predator at these latitudes to be white. Certainly white fur is a better insulator, but with natural insulation around, such as tree cavities and the snow itself, it is not critical for these animals to have white, insulative fur. A deer, how-ever, may profit by having white underfur because it can't ball itself up in a cavity somewhere. Better, then, for a fisher to have dark, absorptive fur that blends well with tree trunks around which it hunts for squirrels and hares. Furthermore, while resting in a raptor nest in a deep, windless conifer forest, the fisher can use that same dark fur not only to absorb short-wave radiation directly from the sun but also absorb long-wave radiation from the warmed evergreen boughs around it.

What about the smaller weasels, then? Don't they pay a penalty at north temperate latitudes for molting their fur to white from end to end of the fibers? Wouldn't it be better for them, for whom conservation of body heat is a constant challenge, to have dark tips on their white fur for solar absorption? Perhaps their hunting tactics in the winter hold the key. Their small body mass and long, thin shape are designed to allow them to pursue small rodents into their tunnels. There they not only find their livelihood but also are protected from the elements above the snowpack. It is for this reason that many more weasel sightings occur in months when there is no concealing snow. Presumably their fur is designed to take more advantage of the insulative value of snow than of sunning opportunities above it. Furthermore, when they do emerge onto the surface, they pay no camouflage penalty against a partly white background. In this regard, there seems to be equal value in being either white or brown, at least in temperate regions where there is a lot of vegetation.

In the Arctic, by contrast, the sun is so low that the solar absorption of dark fur is less efficient since at mid-winter there is little or no sun for most of the day. Although there are shadows from time to time behind protruding ice floes, the dominant color of winter at interior high latitudes is a relentless white. There the arctic fox can afford for its fur to remain totally white both for its better insulation as well as for camouflage. The polar bear, as well, remains more or less white, better to resemble an ice floe at a seal's breathing hole. However, in the dead of winter when there is no sun above the Arctic Circle, even this bear goes into den for a few months below the insulating snow. When it emerges, the sun is low but provides at least a little warmth, which one might think would be reflected away from this mostly white animal. But here the equation of heat retention and absorption is a little different. The dense fur, white from end to end, is hollow for warmth and so dense that there is little air movement through it to contribute to convective heat loss. Furthermore, the skin of a polar bear is black. Look at photo of a polar bear and you can see this dark skin through the shorter hairs around its muzzle. If you were a Lilliputian standing among the roots of a polar bear's pelt, you would be standing on a black floor surrounded by a light-reflecting and -refracting forest of white where the dark skin was absorbing the ambient light and there was no wind. Add to these fur advantages the enormous heat-retentive bulk of a polar bear and its small ears (to reduce radiational heat loss) and you can see why polar bears in zoos prefer to spend so much of their time in cold water.

Marchand, in his book *Life in the Cold*, speculates about why arctic foxes along the Alaskan coast often turn a bluish gray in the winter rather than white. It is the Arctic, after all, where, with the loss of the warmth of winter sun, white fur is more useful than gray. The likely answer is that fur color for many animals is a compromise between camouflage and insulation. Its adaptive capacity makes a decision, in a sense, as to which is more likely to result in the animal's own survival as well as successful procreation and survival of offspring. Is it a greater advantage for a particular species in a particular habitat to be better at heat retention or better at hunting? The inland arctic fox has it both ways. Its white fur is among the most insulative in the world. And since the habitat over which it hunts is white for the length of the arctic winter, it pays no penalty for this. But coastal foxes hunt different things in a different environment. These coastal editions of the species spend a lot of time scavenging in the tidal wrack line for dead fish or anything else that is edible. They also hunt arctic hares for which purpose camouflage is useful, since the protection these hares employ is to rise on their hind legs and scan the flat beaches and volcanic ash fields around them. Here, white would not be the best choice of color for their predator since the predominant habitat tone is a gray, foggy seascape devoid of snow in the tidal zone. Furthermore, the proximity of the sea, which is a heat sink warmed by ocean currents, moderates the temperatures compared to those on the white snowfields farther inland, obviating the need for perfect insulation. The compromise that evolutionary adaptation has made here is toward better camouflage, that is, to favor the so called "blue fox" over a white one.

Kill Sites and Pluck Sites

In the course of your tracking experience you will often come upon sites composed mainly of a pile of feathers. A couple of distinctions

Figure 6.3. These blue jay feathers were plucked by a bird of prey, either a Cooper's hawk or a sharp-shinned hawk.

need to be made at the outset. First, is it a kill site or a pluck site? That is, did the attack on the bird occur there or somewhere else, the carcass having been transported and plucked enroute to den or secure feeding site. The first thing to look for is blood. When an animal expires, its heart stops pumping. So if there is no blood at the pile of feathers, it's reasonably certain that you are looking at a pluck site or perhaps a feeding site.

The next thing to determine is whether the predation was by a bird of prey or a mammal. Sometimes a hawk or owl will expel a pellet with the indigestible remains of a previous meal at the new feeding site. If it is a single, discreet pellet without odor it is a pellet and the hunter (or at least the scavenger) was a bird of prey. The condition of the feathers may also reveal this. Lacking teeth, the hawk or owl will pull the larger wing and tail feathers out of a bird carcass (Figure 6.3), while a mammalian predator may saw them off with the teeth on the side of its mouth (Figure 6.4). Also a mammal will often mark a feeding site with a scat or urine. Carnivore

Figure 6.4. A red fox removed some of the feathers from a kill to lighten the load on the way to its den. Note the cut vanes.

scat has a distinct odor while pellets have little or none. Urine marking may be obvious, as with red fox, or subtle, as with the gray fox.

Of course, nothing is ever that easy. You also need to be alert to the possibility that other animals may have scavenged the site after the fact, attracted by the odor of carrion. Did the fox leave a scat because it had fed there, or is this simply an assertion of territorial ownership to warn off an interloper who had actually performed the kill and subsequent plucking or feeding?

Species Notes: Black Bear

Black Bear Gaits

Bears use a variety of gaits to get around. High energy lopes and gallops as well as bounding patterns were discussed above (pages 127-141). Here are two more that I've encountered over the years.

Walking/Trotting

In deep, soft, early snow, bears use the same direct-register walk that other animals use in similar circumstances. At other seasons, where little efficiency is to be gained from direct registering, they vary this to understep and overstep aligned walking and trotting. The patterns look like this, with a slow understep walk transitioning to a faster overstep walk/trot (Diagram 6.1).

Diagram 6.1

The point where walking transitions to aligned trotting is often difficult to tell, but as the step length increases, eventually a walking gait in which three feet at a time are in contact with the ground must change to a trot where only two feet at a time are in contact. I sometimes try to judge the transition by the amount of toe-in I see in each print, assuming that at slower speed there will be more, whereas as the speed increases there is a tendency for the gait to straighten out. The same can be true of straddle: as speed increases, straddle usually decreases. Another clue that can be used sometimes, depending on the firmness of the surface, is to look for an increase of "pressure effect" around the tracks. In an aligned trot such as the one above commonly used by bears and bobcats, there is substantially more vertical vector of force than in a walk. That is, there is a higher arc of stride. As a result, the landing feet should strike the ground with greater impact. In the loose sand or granite detritus that is common in the beds of dry northern streams that are used by many animals, including bears, to get around their range, this impact can be read as mounding around each print. In other mediums you may find splash rays or thrown debris. The point at which these pressure effects become more pronounced or less so often marks a change to a gait with a higher or lower arc of stride.

Wiggle Walking

One spring on Popple Mountain I found this interesting pattern (Diagram 6.2).

Diagram 6.2

The surface was fairly firm spring snow, but not firm enough to support the impact of anything faster than a walk. The pattern resembled the familiar one left by walking and trotting raccoons where the hind foot advances and lands not in or near the impression of the front foot on that side but instead continues to advance and lands all the way forward to the vicinity of the front foot on the other side! Raccoons can do this because of the uniquely radical arch of their spine that positions their hind legs closer to their front than in other animals of similar size. But a bear has no such radical arch. If I imagine myself inside a bear's body on all fours, the only way I can manage this pattern is by rotating my pelvis radically in the horizontal plane, in effect to twist my rear end as I advance each hind foot. This radical overstep walk I have thus termed a "wiggle walk." The only mammals in my experience in the Northeast that can perform this evolution are raccoons, bears and bobcats, all with the flexible anatomies required of tree-climbers. I did once see an animal actually perform this gait in person, so to speak. At Gate 35 in Quabbin one cold January morning I spotted a bobcat walking down the plowed road a hundred yards ahead, demonstrating this exaggerated wiggle before it sensed my presence, leaped a snowbank and disappeared into the woods.

In the case of the bear on Popple Mountain (Figure 6.5), the poor snow density would not have supported any appreciable verticality in the gait. The additional fact that the track depth was only about an inch leads me to believe that this was a low-impact walk, not a trot, and that this walk probably looked a lot like that of the cat I followed at Gate 35 early that winter day.

Scat and Feeding Sign

Black bears are seasonal feeders whose diet changes according to what is available at the moment. In the spring, just out of hibernation

Figure 6.5. Black bear wiggle walking on spring snow.

bears concentrate on newly emerging plants such as skunk cabbage, a plant that generates its own heat to thaw the snow above it. Although skunk cabbage is an arum and so contains oxalate crystals that act like ground glass in the mouth and stomach of potential foragers, black bears apparently generate an enzyme in their saliva that renders this anti-browsing defense less effective. As a result, especially during the early spring before the plants develop a lot of this crystal, you are likely to come across bear scat composed of the tightly wound fibers of this plant.

As summer approaches, many plants that were food during the spring harden up and either become woody or otherwise indigestible. Bears are also "top-end feeders," that is, the passage of food through the digestive system is so rapid that only the readily available food value of anything they put in their mouth is actually transformed into energy. The rest is excreted as an only a partially digested mass with both visual and olfactory clues to what the bear was eating.

Because of their rapid and inefficient digestion, plants that harden up as they grow provide little sustenance, so that by the time summer arrives, bears begin starving. During the summer months one of their main occupations is searching for ants and grubs in decaying wood or ground nests. A good place to look for this evidence is in an old lumber cut where decaying stumps and scrap logs have been disintegrating for a few years. In such locations, sunlight reaches the ground and promotes the presence of anthills, which bears find and dig out. Here you may also find places where old logs have been rolled to reveal creatures such as ants and grubs under the log (Figure 6.6). The bear approaches such logs and scrapes its claws over the top until they catch on a crevice, giving it purchase to roll the log out of its bed. The resulting claw marks often can be seen on the log (Figure 6.7). The original depression where the log had lain may also be visible.

Bears can destroy old logs without rolling them, as well. When a rotting log falls apart naturally, the pieces of bark and chunks of wood fall off and lay parallel to and close to the trunk. However, when the log is torn apart by an animal, the slabs of bark and wood lay at an angle to the log (Figure 6.8). Since these logs are often visited by both bears and fishers, the animal that did the work can be judged by the amount of violence in the destruction. Bears are a lot stronger than fishers and have longer legs. Therefore, bear work on these logs may

Figure 6.6. A bear has rolled this log out of place searching for ants, grubs or anything else living underneath.

Figure 6.7. A closer look shows the clawmarks where the bear pawed the log to roll it over.

Figure 6.8. The angle of the torn pieces shows this to be bear work. Clawmarks were found here as well.

show slabs of debris at some distance from the bole. A close look at the surface may show claw marks, as well, where the bear scraped its claws across the trunk until they caught on a seam, allowing the animal to pull off a section. Even if these scrapes are not evident, look for compressions along the edges of the torn wood segments where the bear applied power to its claws to pull them off. Although the spread of a male fisher's claws approaches that of a bear, a combination of this spread and the level of violence should help to tell one from the other. It should be noted that skunks also tear apart rotting logs for ants and grubs. Since they are small animals, one can expect even less scatter of debris than with the work of a fisher.

One other caution needs to be considered in attributing torn-apart logs, however. If the fallen tree was formerly standing deadwood, as opposed to a log felled by a tree-cutter, the impact of the fall itself may cause the trunk to disintegrate and scatter pieces of dead wood widely around the site. I found one of these recently where the trunk had been torn at by various feeders enough to weaken it so that it broke off about 6 feet up the trunk and fell butt first to the ground. The initial impact dug the butt into the soil and then the rest of the tree fell heavily, partially shattering and casting wood pieces in all directions. The signs of high impact were everywhere, not only the initial dig into the ground, but vegetation torn off of surrounding trees as the heavy bole came down.

Both in early spring as the surface ground thaws and in early summer when food is scarce, bears search for worms, salamanders, centipedes, grubs, ants and other insects under rocks as well. I recall finding a small ledgy clearing on Mount Wachusett one spring where a bear had systematically dislodged and overturned every slab stone in the field in such a search. I have read of bears at this season even sucking up blooms of springtails (or "snow fleas" as they are often called) as they gather thickly in spring pools.

In ant digs, one can often find a footprint in the spoil, the last one made just as the bear stopped digging. This is also true of other animals from foxes to otters that dig or scrape the ground. Rock walls can contain potential food, from chipmunks to ants. These are easy to attribute as the strength needed to displace large boulders is possessed by only one genus of wild animal in North America (Figure 6.9).

Figure 6.9. The strength needed to move boulders in this stone wall to dislodge chipmunks shows the digger to have been a bear.

Because there is so little to eat in the summer, the other occupation of bears in this period is sex. Since they spend the rest of the year either eating in preparation for hibernation or actually hibernating, mid-summer is the only time when there is leisure for mating. However, the timing of gestation and birth could pose a problem. Bears give birth to their cubs during their winter sleep and active gestation beforehand only takes at most a couple of months. Without a biological mechanism called delayed implantation, mating would have to occur in late fall or early winter, a most inconvenient time when bears are either topping off their winter fat or heading into den. What happens is that mating takes place at mid-summer when there is little else for a bear to do, and then the blastocysts or fertilized embryos remain suspended in the female's uterus until November. The bear may produce a half-dozen or more of these embryos. How many will actually develop into growing fetuses and how many will instead be reabsorbed depends apparently on the weight of the female as she enters hibernation. If her feeding range is very productive she may develop more than two; if it is poor and she is underweight, she may develop only one or even remain barren for that winter. Two winters ago someone in my town photographed a female with four cubs! Bears had only recently moved into our area and clearly found plenty to eat. Unfortunately, some of what they discovered was in the form of bird feeders and grain bins in horse barns. This has given them something of an unsavory reputation locally, as has their unfortunate habit of invading campgrounds in the summer. Bears are smart, with a long memory for where they have found food. The mid-summer starvation period coincides with the peak of the camping season, and bears have figured out what a rectangular box at a campsite is likely to contain. Once they are gratified in this way, campground visits will be come part of their regular summer circuit.

In late summer, as the berry crop ripens, bears forget about sex and campgrounds, and begin instead to search out prime berry patches. A good place to find bears and their sign at this season is in old log yards in the woods. These staging areas, usually about a quarter acre in size, rapidly grow back to blackberry and other rubus brambles after a cut. Here you may find scat (Figure 6.10) and wide corridors in the canes stomped down by bears as they lumber about searching for ripe berries. Due to the rapid movement of food mass through

Figure 6.10. Various black bear scats. (A) Late fall apple scat. (B) Spring scat composed of beechnuts, refrigerated in the snow-pack all winter. (C) Summer scat of undetermined content, found in dry streambed. (D) Blackberry scat from a cub or yearling. Note the undigested berry and the slight taper of the segment on the right that is often a feature of smoother bear scat. (E) Blueberry scat.

their gut, the scat will look and smell like the food source. As a berry ripens, the waxy cuticle on its surface thins and the sugar becomes concentrated, ready for ingestion. But inadvertently the bear will also consume some berries that are still green. These berries with their thick cuticle may move through the gut and appear in the scat whole, undigested and easily recognizable (Figure 6.10D).

Unfortunately, corn also ripens in late summer. Bears have a well-developed sweet tooth and will head for the farmer's cornfields at this time. To escape harassment by the farmer, the bears in my area have figured out a way of feeding relatively invisibly. They may enter the field surreptitiously, easing between the cornstalks until they reach the middle and then gradually hollow out the cornfield from the inside, more or less like a worm in a perfectly presentable apple. Seen from the air, these hollows look much like the celebrated crop circles that are sometimes found in the middle of mid-western farm fields. Once again, the resulting scat will look like its source, yellow and grainy with many unripe kernels evident on its surface. Its size should distinguish it from the scat of raccoons, which also frequent corn fields in August. However, it may take some care to tell raccoon scat from those of bear cubs, led to the field by their mother. Although raccoons often deposit their droppings in collections at scat stations, bears drop theirs randomly.

As the berry season fades, the nut crop ripens, drawing bears away from fruit and berries to nut trees. When the forests of the north were clearcut for softwoods to feed the paper industry, the axes spared beech trees growing among the spruce and fir since apparently they were of little use for anything more than flavoring Budweiser. After the softwoods were cut out, the beech were left behind to inherit the earth, spreading by rootstock and producing nuts that grew into little beech trees that in turn grew up to dominate the new forest. In late autumn after leaf fall, whole mountainsides in northern New England take on a pinkish hue caused by the mingling of gray bark of beech and the reddish-orange of mountain ash berries. These beech stands become attractants to bears that are putting the finishing touches on their fat deposits before heading into den for the winter. Like any nut, a beech nut is a little globule of fat. The actual embryo inside the husk is only about the size of a pinhead. The rest is sustenance for the embryo to live and grow on until it is large enough to put out a leaf and capture the energy of the sun directly. And of course, the nut's energy can also be used by the bear to sustain its own life.

In the fall two kinds of bear sign are common. The first is the claw marks that appear on the trunk of smooth-barked trees that the bear climbed to get at the nuts before they fell (Figure 6.11). In beech groves on remote mountainsides, tree after tree shows this kind of

Figure 6.11. Clawmarks (see arrow) are a common sight in groves of beech trees where the bears climbed and descended after nuts.

Figure 6.12. Not a giant bear! As the tree ages, both the clawmarks and their distribution spread on the bark.

abuse. Although these marks are dramatic and often-photographed evidence for this species (Figure 6.12), they are often misinterpreted. Sometimes people try to judge the size of the bear by the spread of the claw marks. However, this spread actually doesn't correlate well with the bear that left the marks. As I pointed out above (pages 49-50), as a beech tree ages, its bark stretches, expanding the spread of the prints and the impression of the size of the bear. Nevertheless, the claw marks can still be useful if one considers the width of the individual marks. When they are first created, each is only a 2-3 millimeters wide. If the marks on the tree that you are inspecting are each perhaps a whole centimeter wide, then they are very old. You can still extrapolate from them to a rough idea of the size of the bear since each mark spreads at about the same rate as the entire set of 4 or 5 marks (4 because the small medial claw mark often fails to register).

Once the animal is up in the beech tree, it may build a "nest" (Figure 6.13). In order to reach nuts far out on the upper branches, it breaks them off, pulls each branch in and stands on it to reach the morsels without having to climb out on a limb. As branch after branch is added, the mass takes on the appearance of a nest. However, the

Figure 6.13. A bear "nest" in a beech stand.

bears don't sleep in these structures; they are simply a by-product of feeding.

As the nut crop is gradually consumed, the curve of consumption declines. Simultaneously, as the weather turns colder the curve of energy expenditure rises since the animal needs to devote more and more calories to heat in order to maintain both bodily warmth and the athletic process of food-searching. Once the declining curve of energy availability crosses the rising curve of energy expense, the bear heads for hibernation. It is not believed, as is often imagined, that bears enter hibernation in response to hints from declining photoperiod. Instead, lack of food is what seems to do it. Indeed, I once tracked a bear up Firescrew on Mount Cardigan in New Hampshire on New Years Day in over 2 feet of snow. What could it possibly expect to find to eat up in the boreal zone of a mountain in that season and in those conditions? The puzzle was solved on the summit. Mountain ash grows as a small tree on northern hillsides, but as with other growth, the higher you go on a mountain the shorter the growth becomes until on the top of Firescrew, mixed among the stunted spruce and fir, the mountain ashes were merely shrubs with the berries at a height convenient for browsing. Although their berries are normally rather tannic and unpalatable, once they freeze, more sugar develops in them and they become moderately sweet and pleasant to eat. With its sweet-tooth, long memory and keen sense of smell, this bear both sensed a good berry year for this species and remembered where they were most available. Clambering through the summit drifts, it feasted on the shrubs ringing the ledgy clearing. Clearly the curve of energy availability, for this bear at least, was still rising.

During hibernation, the animal's bodily processes slow down to minimize loss and the animal dozes away the worst of the winter months. In early April the males come out of hibernation and begin searching for food. April is the killer month; animals that were hungry through the winter are starving by April and on their last legs. Nothing green is yet growing to revive them, and many succumb to this period of genetic testing. It seems cruel to the civilized mind, but this is the end of the annual period where a wild animal's fitness is tried and weak genes are weeded out of the breeding pool. It might seem odd, then, that a bear would emerge from hibernation during this period of minimal energy availability. Evidence I found many years ago showed me one reason.

While hiking on White Ledge on the eastern slopes of the White Mountains late one spring many years ago, I found in the trail a fairly fresh scat composed of beechnuts! Wait a minute, I thought, beechnuts are produced in the fall. Why would a fresh bear scat in the spring consist solely of their husks? I answered my own question the following December in Lucy Brook on the back side of Mount Attitash while participating in a Christmas bird count. On the banks of the stream I discovered the tracks of a bear in about a foot of snow. In several places the bear, working its way upstream, stopped to dig in the snow. Eventually it left a scat, a beechnut scat. It occurred to me that nut fall in the mountains stretched into the first snows. A bear prowling the surface could use its phenomenal sense of smell to find individual beechnuts suspended and refrigerated in this snow. Then I remembered the scat on White Ledge and realized that a bear just out of hibernation could do the same thing on spring snow, excreting a

Figure 6.14. This may look like shotgun vandalism, but actually a bear mouthed this sign and tore off a chunk or two.

fresh beechnut scat even though it was six or seven months since nut-fall. In a subsequent winter up on Maple Mountain I tracked a large male, just out of hibernation, that was using its nose to find what it could under the half-foot of remaining snow. What it found was a rusty old gas can from a log yard, a few moldy moose bones upon which it stood for a while and then, lumbering up the hillside to a beech grove, tore up patches of snow for a total of perhaps half an acre in search of those same refrigerated nuts.

Other Bear Sign

"Bear Signs"

Bears also leave behind evidence of their passage other than tracks, trails, scat and feeding sign. One of these is the habit of ravaging signboards that they come across in their range. A common sight in bear country is a trail sign meant to direct human hikers that has been chewed along its edges. When I first encountered one of these years ago up in the Passaconaway intervale, I didn't immediately associate the damage with bear work (Figure 6.14). Instead I passed

Figure 6.15. My favorite. A bear apparently felt gnawing on this particular trail sign was appropriate.

over it as the vandalism of hunters using the boards to site in their shotguns. Later it occurred to me that New Hampshire hunters use high-powered rifles, not shotguns, at least for deer, bear or moose. Returning to the sign, I realized that what I had taken for pellet holes were actually the marks of a bear's canine teeth. After that epiphany I began to find these ravaged signs everywhere I looked. My favorite was one I located in the Secret Mountain Wilderness in central Arizona. A trail had been named by the Forest Service the "Bear Sign Trail" (Figure 6.15), and a board had been erected on a post to show the way. The local bears found the board and gnawed its edges as if to illustrate!

According to Ben Killam, a well-known bear biologist from New Hampshire, this vandalism results from the bear's curiosity. The orphaned bears that he raised investigated anything novel that they came across in their range, usually by mouthing it as human babies do in order to learn their environment. From the many damaged signs that I've examined, it seems to me that something else is in play as well. The Forest Service makes its signs about two inches thick out of locally available softwoods and then stains them. The presence of the sign provokes the bear's curiosity, to be sure, and the scent of its paint stain provokes testing for palatability, but most of these signs show shims driven into the ground at their base to steady them. It seems the bears were in the habit of not only mouthing the signs, but they also sank their teeth into them and then rocked back and forth, loosening them in the ground. The boards, it seems, are just the right thickness and bite-ability for some pleasurable tooth and jaw exercise.

Mark Trees

Another kind of sign that bears leave behind is the notorious mark tree (Figure 6.16). A bear selects out the branchless trunk of a resinous tree along its hunting route and begins by clawing down on it, creating a sappy scrape. Then it turns and rubs its back against the damage. A bear can reach most parts of its body with its claws or teeth for grooming, but for a good back scratch it must use some borrowed tool, in this case the roughened bark. While it is scratching it may turn its head to the scrape and dig out a chunk of wood with a diagonal bite of its canines. The procedure is so ingrained that it is probably instinctive; the bear is not only getting a good back-scratch but also marking its range or its route for the benefit of other bears that may stray into its area.

This profile, down-scraping and with two slanting tooth marks in the middle, is easy to distinguish from a moose gnaw, which it may superficially resemble. But moose mostly choose sweet-barked trees like maple and mountain ash while bears choose sappy ones such as red pine or balsam fir. Moose, lacking upper incisors, gnaw upward on the bark while bears scrape downward with their claws. Checking the little tails at the end of the vertical marks will show most of them at the bottom of a bear mark and at the top of a moose gnaw. Finally, a close inspection of the sappy bark in the wound will often reveal a few strands of fur stuck to the resin. You would expect these to be dark brown, like the bear that left them, but they soon bleach out to a curly blond appearance.

Bears gnaw other objects beside signboards. Telephone poles are also clawed, gnawed and bitten as well as forestry construction such as stream bridges on lumber roads. I suspect the odor of creosote as well, perhaps as the odor and taste of paint stain makes them attractive.

Figure 6.16. A balsam fir marked by a bear. Note the diagonal gouges of the animal's canine teeth.

It may be tempting to judge the size of the bear by the height of the damage from the ground. However, Killam says that his cubs used to jump up on trees in order to claw down. I suspect this may have been an effort to make themselves seem bigger and more important than they really were. A better way to judge the size of the animal is to look at the width of the canine marks in the diagonal bites: big bears have big jaws.

On a number of occasions I have also seen bears, just out of hibernation in the spring, shred the outer bark of spruce trees, apparently to extract the inner pulpy cambium for food. And once Joe Choiniere showed me a large, old white oak in an Audubon Sanctuary in Barre, Massachusetts, where a bear had debarked this tree up to the height of a man. Once again this seemed to be feeding sign, as a bear just out of hibernation was looking for a little protein in its diet in the form of beetles and grubs that it might find under the loosened bark of an old tree.

Species Notes: Beaver

Beavers get a lot of bad press: they are a major vector for the giardia cyst, they flood low-lying property and roads and, of course, they cut down trees. We have been so successfully conditioned by timber interests into believing that a perfectly formed tree, unviolated by surrounding organisms, is the highest virtue in Nature that many people regard beavers as wanton vandals. Of course, Nature is perfectly indifferent to this evaluation. In cutting down trees, beavers allow sunlight to reach the surface and then by using the wood to impound water (Figure 6.17), these mammals create two of the major ingredients for life on Earth. As any walk through the woods to a beaver pond will reveal, these impoundments are far more active in wildlife numbers and diversity than the surrounding forest. While most of the energy of a typical managed forest is in the canopy and supports only those animals that can climb or fly up into it, the energy in and around a beaver pond is distributed at all levels (pages 89-91). As we near the pond, the canopy of the trees that supports tanagers and vireos declines to a brushy edge that hides the nests of yellow warblers and the tangles that provide cover for yellowthroats, rabbits, mice and bog lemmings. The emergent grasses and sedges feed wood ducks and muskrats. Farther out, lily pads feed beavers and muskrats. Moose

Figure 6.17. Beavers are safer on water than on land. They build dams to raise water levels so that they can bring the edge of the water closer to food, reducing their exposure to danger.

wade out into the pond and use their long snout to reach down into the muck and root out aquatic plants. In fact, the increase in the number of beaver ponds in New England in the last couple of decades is one of the reasons for the rapid rise of moose populations in the area. The aquatic plants of beaver ponds provide the sodium that a moose needs for life. Sunlight and water also spawn insect hatches that, in turn, feed fish and birds like tree swallows, kingbirds, phoebes and cedar waxwings that fly over the surface to catch them. The fish themselves are food for kingfishers, mergansers and otters. Herons and bitterns (Figure 6.18) feed on the frogs hiding in the pond's edge growth, and red-shouldered hawks add to this by searching the margin for these frogs as well as snakes.

Because of this diversity, beaver ponds are also a boon to the tracker. There he can find the trails of bobcats and foxes that prefer the wind-packed snow on the ponds to the soft snow in the forest interior for easy transit around their hunting range. Coyotes hold convocations and howl to the moon from that same frozen surface. Bears, which need a lot of water in summer, drink from its edges. Besides

Figure 6.18. In impounding water beavers create habitat for many species of wildlife like this American bittern that uses the cattails and sedges along the pond margin to hunt and hide.

using the ponds for transit, bobcats also hunt the dense margin for its many hidden creatures and use dead snags sticking out of the ice as scratch and scenting posts. Deer walk out to the beaver lodge to browse the leaves of the host's cache that protrude above the ice. Otters and mink regularly visit, gaining entrance under the ice through bank holes and weakness in the northern margins caused by overhanging evergreens. The list goes on and on and can be regularly accessed by any tracker interested in the concentrated winter behavior of animals.

General Sign

Most people who spend any time in Nature these days are familiar with the obvious signs of beaver presence: the dam and the freestanding lodge (Figure 6.19). The dam is constructed in order to raise the water level of the pond, allowing the beavers to reach desirable growth around its edges. In the water, beavers are formidable; fast swimmers with powerful jaws, they easily intimidate or escape any predator. On land, however, their heavy weight, which in water has neutral buoyancy, makes them ponderous and vulnerable. Once the fringe of desirable trees, such as aspens, closest to the edge of a new pond are culled out, the beavers, by raising the water level with their dam, shorten their trips to the next rank of trees farther back in the woods, and so minimize their exposure to danger.

The other obvious sign is the lodge. In the classic beaver lodge, the animals find a shallow spot away from the shore where they build up a pile of mud and sticks until the mound clears the surface enough to excavate a dry chamber within. This lodge is their winter defense as well as a safe summer nursery for their kits. In the summer, any predator wishing to get at the babies must first swim out to the lodge, dive and then enter a narrow tunnel up into the chamber. Beavers usually leave a babysitter with the kits, and an otter entering such a

Figure 6.19. That this is an active lodge can be seen by the recent mudding and adding of vegetation as well as the cache of branches in the water to the left.

tunnel is heading for trouble; any animal that has jaws strong enough and teeth sharp enough to cut down trees is an animal nothing should want to face head on! Beavers often build more than one lodge on a pond, using the first until it becomes so overrun by vermin that life in it is miserable, at which point they will move to the second (Figure 6.20). After that one too is infested by fleas, the beavers rehabilitate the first and move back, alternating the use of the two or more lodges over the years. To tell which one is in use at the moment, look for fresh mudding, fresh branches in the construction and a winter cache of branches stuck into the mud, or sometimes left as a free-floating mat, next to the lodge.

The other use for the lodge is winter defense. Once the pond freezes over, potential predators can trot out over its surface to the lodge without having to swim. However, there they will find it impregnable, mudded in the late summer and fall to freeze into a solid fortress in the winter. Inside, the beavers sleep, exit the lodge when they feel hungry, swim underwater to the winter cache of tender, water-softened branches of their cache that are frozen into the ice, nip off a section and haul it back into the lodge to be consumed, all without

Figure 6.20. When the original lodge on a pond becomes infested with parasites, the beavers may build another, alternating between the two according to the current level of infestation.

once exposing themselves to predators or the winter elements above the ice. Another food source is the starchy tubers of lily pads in the mud at the bottom of the pond. These too can be dug out and carried back to the lodge without the beaver having to emerge from under the ice.

As I have suggested before, Nature abhors three things: a vacuum, equilibrium and a perfect defense. If the beaver's defense had no flaws, beavers would overpopulate and overwhelm their food supply, deforest the land and end up starving and diseased from malnutrition. Humans may do this, but other animals do not. Any natural defense must have a defect, and the defect built into the beaver's defense is in its mouth. The jaw and tooth arrangement of a beaver is that common to any vegetation-feeding rodent: two sharp incisors at the front top and bottom cut off the food and pass it back across a toothless gap to a long set of grinding molars that chew the woody food and mix it with digestive enzymes before it is swallowed. The problem arises in the front incisors. Any instrument for cutting abrasive material such as wood must be sharpened periodically just as a chisel must. If the beaver had teeth of the sort you and I have, they would soon be worn down to nothing and the animal would starve. The physiology of the beaver gets around this by having the sort of front incisors that continue to grow throughout the life of the animal. That works fine in the summer when the beaver is cutting down trees; as the teeth grow, they are worn down by the cutting. But in the winter the beaver is restricted to the soft starchy tubers of lily pads it finds at the bottom of the pond and to the cambium under the water-softened bark of young branches in its cache, neither of which is very abrasive. During the winter, then, the front upper and lower pairs of incisors continue to grow longer. What is the beaver to do? I have read that beavers solve the problem by scraping the lower and upper incisors together to sharpen them. This theory seems to have been advanced to debunk the old idea that beavers needed abrasive wood to do the job. And it may be true that they use this method to sharpen their incisors, but in my experience this is not enough. They also need to wear down the continuously growing teeth. The evidence for my opinion lies along the margin of any beaver pond I have visited at elevations below or south of the boreal zone. Walk around one of these ponds and you will encounter a tree with very hard wood, in my area usually a red or white oak, that has been gnawed at intermittently and clearly without the intention on the beaver's part of cutting it down (Figure 6.21). Look closely and you will see that the gnawing went well past the cambium layer, the only part of the tree that the beaver actually eats, and into the underlying wood. So this gnawing can't be for food, in which case only the cambium would be gnawed, nor for construction material, where the whole tree would be cut down. Look again and you may detect patches of the gnawed wood of different shades, indicating that the gnawing was done serially over a long period of time, over years perhaps. I am at a loss to explain this activity as anything other than obligatory grinding down of their central incisors, despite the revisionist theory. This is important because the need to gnaw something very hard obliges the beaver to leave its safe refuge and go ashore during the winter, a defect in its defense that exposes it to the very predation that the rest of its strategy is designed to avoid.

Not all terrain adapts itself to the classic arrangement of a beaver pond — a dam and a free-standing lodge. Some ponds will lack a spot in their middle shallow enough for the foundation of a lodge. In this case the beaver may settle for second best. Since the margin of any pond has shallows somewhere along its perimeter, a beaver may build its lodge adjacent to the shore, even leaning the structure against it. Of course, this location makes it more vulnerable during the summer than if the lodge were far from the shore, but apparently the beaver must accept this risk in order to use the pond. Recognizing the vulnerability of this situation, however, the beaver may gnaw out a chamber not in the stick pile but rather in the earth of the bank, using the mudded wood pile as a defense. At other times, the beaver may not even cover the entrance in this way, leaving the underwater access unprotected (Figure 6.22). But while the water is high enough, at least, this practice has the advantage of eliminating a visual clue to the existence of the tunnel. I suspect that such tunnels are used more for resting and feeding than for the protection of kits since at mid-summer or during droughts, when the entrance is exposed, the beavers can simply stop using it.

Under some circumstances the beaver may construct neither lodge nor dam, even for raising young. The Swift River, located in the Passaconaway intervale of northern New Hampshire's White Mountains, is well named. Any lodge built in its flow or any attempted dam would be swept away in the spring floods. Here the beavers use the

Figure 6.21. The different colors of the gnawing on this red oak show that it was done over a number of years.

Figure 6.22. Entrance to a beaver's bank den exposed by low water. A tunnel several yards long leads upward to a dry chamber above the high-water level.

river as a linear habitat moving up and down its length to cull out any desired food tree within easy reach of its bank. For a lodge, they use a bank tunnel with an entrance concealed underwater so that, to the casual observer, there may seem to be no beavers there at all. However, attention to the bank will show the tell-tale diagonal cuts on streamside brush and saplings that betray the presence of this animal.

The food sources that a pond provides, then, consist of the cambium of trees, branches in the winter cache, pond-lily corms (Figure 6.23) and, lastly, the lily pads themselves. The last of these, the lily pads, the beaver eats by curling them up from both sides and feeding the tubes through its teeth and down its throat like a chipping machine.

How long a beaver pond has been in existence can be told roughly by the growth around it. Because beavers prefer hardwoods for food and construction, they gradually eliminate this class of trees from the margin of a pond. In addition, they may girdle any hemlocks they find along the shore, but often without cutting them down. Tom Wessels suggests that they do this because they want to kill them in order to create sunspace for feed trees they prefer. While I can draw an evolutionary connecting line through this, it is tenuous. A friend has

Figure 6.23. In the winter under the ice beavers can dig out the starchy tubers of lily pads from the pond bottom to supplement their diet of branch bark.

suggested, only half in jest, that since they regularly eat much woody, indigestible material, they need the tannin in hemlock to settle their stomach! In any event, the one tree that beavers usually leave alone is the pine. In my area, a pond surrounded by white pines from which most of the hardwoods have been culled is an old pond indeed. Nevertheless, I have seen these ponds go on and on successfully over the years, their denizens apparently subsisting on lily pad parts and the woody plants that grow up in the sun along the pond edge. Despite their preferences, however, beavers do consume white pine sometimes. I have videotape I shot years ago of a beaver chomping down a white pine branch that had fallen in the water in front of my blind and then carrying off the remnants to its lodge. The problem seems to be in cutting down the tree in the first place, a gummy, sappy process no pleasanter for the beaver than one might imagine. But any pine branch downed by breakage appears to be palatable to them.

Beaver dams eventually breach and their pond drains. This breaching often occurs in spring flood where its construction of mud and loose sticks is overwhelmed by pressure from the enlarged volume of water. Once the breach starts, the rush of water enlarges the gap and bends back the arms of the dam in a characteristic way. This appearance can be used to tell a natural breach from a man-made one. Humans looking to protect their property or roads from flooding, or others looking to expose the lodge to get at the animals themselves, may also breach the dam. In these cases, the breach will usually lack the bend of the arms of the dam on either side, and debris from the gap may be piled next to the breach.

The bottom of the pond is composed of rich sediment that was transported over the years down to the pond by rainwater that leached nutrients from surrounding hills. After the dam breaches and if the beavers don't repair it, the pond drains exposing the rich, muddy bottom to the sun. The newly drained pond quickly becomes a beaver meadow full of vegetation, such as sedges, grasses, herbs and eventually woody growth that begins the process of succession back to forest. During this stage the drained pond attracts many animals for the energy in its lush growth. Bears commonly feed in these spots as do deer, and small animals use its dense cover for both concealment and food. As the vegetation on the bed of the old pond grows up to sapling stage, one of the trees that often invades is the aspen, or what New

Englanders call poplar, an early successional tree. This happens to be one of the beaver's favorite food sources, and its growth sets the stage for reclamation by the next generation of dispersing beavers that happen upon the old pond bed.

Tracks and Trails

Due to the danger to the animal of moving beyond its watery sanctuary, surprisingly few beaver tracks (Figure 6.24) will be found around a pond, and those that are found are usually hind prints. The beaver's larger hind foot is the paddle the animal uses to swim around its pond and readily overwrites the much smaller front print. Then the tail, dragging along behind the animal, often erases all but very deep hind prints, at least in non-snow seasons. Your best chance to find a

Figure 6.24. A near perfect set of beaver tracks in a loping gait. This is about as fast as a beaver can move on land. The tracks were headed toward water suggesting that the animal was fleeing danger.

229

legible trail is in the winter when, for one reason or another, a beaver has come out of the safety of its pond for an excursion ashore. In this case the beavers that I have tracked seem to hold their tails up off the cold and abrasive snow. Perhaps this is to avoid wear and tear or perhaps it is an effort to conserve body heat. I once saw a beaver on a hot summer day asleep on a log with its tail in the water, so perhaps the naked tail can regulate the internal temperature of the animal by carrying off heat from the blood of the otherwise fur-insulated animal. In any event, the resulting trail in snow usually shows a line of hind prints in a walking gait with the feet pointed somewhat inward and with only occasionally a front print peeking out from underneath.

One oddity I have seen a number of times is beavers' seeming unwillingness to use culverts under roads to move between impound-ments. Perhaps they are afraid of getting stuck in the tunnel, or they are suspicious of an ambush on the far side as they emerge. The first beaver trail with good tracks that I ever found many years ago was across a dirt road in the winter where the beaver had deliberately avoided using a broad culvert directly under its trail. On a few other occasions I have found legible prints on or about a lodge in the snow of early winter. I suspect this to be evidence of young beavers that have sensed vulnerability in their first lodge and have been forced to emerge in order to reinforce their work.

Scat

Beavers are caprophagous, that is, they reingest their own fecal pellets. The first pellet is extracted from the anus and consumed once again. Apparently it takes two trips through the beaver's gut to get much nutrition out of the woody material it normally eats. The second scat is the one found in the shallow water around the base of the lodge or other spot where the beaver spends a lot of time. The end result looks like a little yellowish dumpling less that an inch across. Brought to the surface on the blade of a canoe paddle, the collected pellet will dry quickly and in a few days disintegrate into a little pile of sawdust. Although most of the time beavers defecate only in water, in early summer these little piles can be found sometimes quite far from the edge of a pond or river. Although I have occasionally discovered places where a beaver has dropped a pellet when it is out of the water, these early summer pellets were usually left high and dry by the receding

Figure 6.25. Beavers normally defecate in water. This dropping, although found on dry land, was deposited during spring flood. Note the leaves blackened by acidic water.

spring flood (Figure 6.25). In the North Country, the shade of the leaves upon which the dropping lies will usually tell if this is the case. The acidity in the water of northern bogs "burns" anything that is left in it for long, turning brown leaves a shade of dark gray.

Scent Mounds

Scent mounds (Figure 6.26) are common around the edges of beaver ponds where the resident beavers have felt the need to mark territory or advertise their presence. Piles of debris from the bottom of the pond are pushed up onto the shore and anointed with castoreum from the animal's castor gland. This oil has a rather pleasant odor that, I have read, was once used by the perfume industry as a fixative to extend the durability of expensive scents. Some people can pick up the odor of castoreum while standing anywhere near a pond while others may have to get their nose close to the mound to detect it.

Figure 6.26. Scent mounds can be small piles of mud and debris such as this one, or as much as two feet high. They are usually anointed with castoreum as markers of territorial residence.

A beaver scent mound can be a few inches tall to occasionally a couple of feet high, reinforced with mud to hold it together. Someone has suggested that these piles are a signal to an itinerant beaver that either the pond is inhabited by a mated pair and so the beaver should keep moving on, or that one or the other sex is absent from the pond and new blood is welcome.

Cutting

When beavers cut brush or saplings near their water, the resulting cut surface has the familiar diagonal orientation created by other rodents. A look at the width of the chiseling on the end of a cut will easily tell it from the cuttings of smaller rodents. These cuts are generally very neat with little splitting of the wood (Figure 6.27) as opposed to the cuttings of porcupines, the ends of which often show stress splitting (page 275). This splitting results from the porcupine's need to control what it is cutting when the animal is high in a tree.

Since beavers cut while standing on the ground, the branches are under no such stress and the result is a finely chiseled diagonal cut.

"Sits-on-Lodges"

I recall tracking a lone coyote out to Church Pond off the Kancamagus one winter. Coyotes often use forest bog-ponds for a variety of reasons. Sometimes they gather on them on moonlit nights for a convocation in which they romp, sniff each other, scratch around, play and howl at the moon. The winter winds clear much snow from the surface of these ponds, piling it up just inside the border of spruce and fir along the shore. This either leaves the ice away from the shore clear or wind packs the snow there into a firm surface, both conditions for efficient transit of one's hunting range. Often the visiting coyote will approach an inhabited lodge on a pond and scratch at it, seemingly aware of the beavers within and no doubt their edibility (Figure 6.28).

Figure 6.27. Compare the perfection of this beaver cut to the splintered branch in Species Notes: Porcupine.

Digging through the covering snow the animal soon encounters the frozen mud with which the lodge was reinforced. At that point the coyote may make a few desultory scratches and then pass on to other pursuits. Only a few times have I seen a lodge at any season that has

Figure 6.28. Once a pond freezes, predators can move over the ice to the lodge. However, the beavers are safe inside the frozen mud of their fortress. Here a pair of coyotes has approached the lodge, pawed at it but without a serious attempt to dig into it.

been dug into. Once this was done in spring by a bear on a bank den and another, also on a bank den, by local dogs in the winter. I suspect the bear work was accomplished by a hungry animal just out of hibernation that sensed the beaver kits within. The latter, much harder winter dig was done by creatures that knew their livelihood lay elsewhere and could afford the waste of energy.

My Church Pond coyote, however, had discovered another use for beaver ponds. Heading along shore, it approached a free-standing lodge and, instead of scratching at it, climbed up to the top and sat down. At the summit of the lodge-cone beavers leave a vent, a vertical column criss-crossed with sticks as with the rest of the lodge but not mudded. This allows the exhaust from their own respiration to filter up through the branches and dissipate into the outside air. This column of waste gas can be observed on foggy mornings as "smoke" rising from the top of the lodge. My coyote had decided not to let all that warm air go to waste. By sitting on top of the lodge the animal achieved an elevated perch from which to survey the pond while having its backside warmed by the borrowed heat rising from below. Interestingly, when the coyote descended on the opposite side of the lodge from my approach, it left a trail of gallop patterns as it hurried off into the bog brush at the edge of the pond. Once into concealment, the animal had stopped and turned to stand with the long axis of its body oriented toward the lodge. The crystals of snow kicked up by the energy of its gait out on the pond could still be blown around, showing that the trail was very fresh. Coyotes don't gallop in winter unless a sure meal is nearby or unless they feel the need to escape. On this remote pond in the dead of winter there is not much to scare a coyote, except perhaps a human tracker intent on following its trail.

Species Notes: Bobcat

Behavior

Bobcats are fascinating animals to track. Their trails are so expressive that one can "see" into them for varied behavior more than is usually the case with wild canids, whose personalities tend to be more phlegmatic. The bobcat's expressiveness is due in part to its anatomy. These wild felids have long, supple legs with which they not only traverse the landscape horizontally, but also climb vertically on trees and cliffs. Typical of tree-climbing mammals, the pads on a cat's foot are deep and rubbery for shock absorption when they jump down from whatever they have climbed. As a result, any movement of the body above translates into a distortion of these pads and their relative arrangement. A shift to the left, for instance, may exaggerate the size of the medial toe pads in the right print and cause the bulbous secondary pad to spread and round on the lateral side. Acceleration forward tends to enlarge the primary pads and deemphasize the secondary, reducing its relative size. The impact of a jump down spreads all the pads, greatly exaggerating the size of the print. Delicate foot placements on a firm surface and in a gait with little vertical force tend to do the opposite, reducing the size of the pad impressions. The former may give the impression of a much larger animal than it really is; the latter may give the opposite impression (see pages 40-43).

In west-central Massachusetts where I live the easiest way to find bobcat trails is to look for remote patches of mountain laurel. These dense thickets are prime habitat for snowshoe hares, providing them both food and concealment from predators. Ironically, because of this they also attract predators like bobcats. These animals are best at stealth hunting, prowling these thickets though which they are small enough to insinuate themselves and yet large enough to overpower any prey found within. Tracking bobcats in the Quabbin region, then, is a process of following a direct trail between thickets. Once the cat enters one of these clumps, it circulates through it hoping to catch a dozing hare. The initial reaction of a snowshoe hare to danger is to hold still, hoping that its white coloration in winter will conceal it from the source of that danger. (In the summer a drab brown pelt allows it to blend in with the lower trunks of the laurel and the dead leafage on the ground.) At some point, if the hare senses it has been detected, it will resort to its other defense, speed, fleeing over trails it has packed down in the snow. The bobcat is no match for a snowshoe hare in a tail-chase. Instead, the cat must act first, charging before the hare makes up its mind to flee, and then the cat must catch it within a few steps.

After a fresh fall of soft, uncrusted snow hares will begin moving about, leaping from spot to spot over and over again until little "hare islands" of foot-packed snow are developed into an archipelago chain along its trails. With repeated passage, these islands extend toward one another, eventually connecting into a long packed trail that the hare can use both for negotiating its feeding area and escaping from predators like bobcats that regularly invade their thickets.

The worst mistake a hare or squirrel can make is to create a curve in its course of flight. An experienced bobcat will anticipate this and cut the angle to head the prey animal off. I have an interesting video clip sent by a friend showing a bobcat stalking a squirrel that was scavenging seeds beneath a bird feeder. The cat simply laid down in the snow until the squirrel turned its back, whereupon it charged, but not at the squirrel. Instead, this veteran of past winters anticipated the squirrel's escape route and ran over the snow with lightning speed toward the nearest tree where it caught the squirrel just before it was able to scamper upward.

Bobcats attack other prey besides snowshoe hares and squirrels. They also prowl the margins of beaver ponds, where the thick edge growth provides concealment for everything from birds to snakes and frogs. Once I was in my video-blind on a favorite remote pond in Quabbin when I noticed grackles raising a fuss in the grasses down the shoreline. The disturbance moved gradually toward me, with the grackles mobbing something moving through the sedges. I waited, following the bird's progress. Suddenly, into a shaft of sun among the pondside hemlocks stepped a beautiful bobcat, twitching its black and white banded tail in irritation at the avian harassment. It paused for a moment in the light, looked in my direction and then disappeared into the shadows of the forest beyond. It seems all of my sightings of bobcats have been this fleeting.

Bobcats are shy prowlers of thick cover, only occasionally using the human infrastructure of roads and paths where they can be observed. Most of their trails cross our own rather than following them the way coyote trails commonly do. There are two exceptions that are always worth investigating for evidence of this secretive creature. One of these is plowed roads in the winter. Bobcats are not as well adapted to snow as their close cousin, the Canada lynx. In deep, soft snow following a storm, a bobcat will generally lay up next to a hare or rabbit run hoping dinner will run into its mouth so that it won't have to struggle clumsily through deep snow to catch its next meal. Apparently this tactic isn't successful often enough to keep the cat fed so, after a couple of days of this, hunger will drive the cat to begin actively hunting. Forced into action, the bobcat will use any packed snow available to get around to likely coverts. These may be the packed trails of everything from deer to porcupines. They may also use remote plowed roads. Quabbin Reservation has a network of unused dirt roads in the forest around the reservoir. The workers at the reservation plow a loop over these so that they can drive their service vehicles in the park in the winter. Many wild inhabitants of the forest use this cleared road to get around in their ranges, as well, including red foxes, coyotes, porcupines and bobcats. Most of the time, as I have suggested, bobcat trails cross rather than follow these roads, but in difficult conditions they will use them, risking observation, as the course of least resistance and therefore greatest winter economy.

In summer a good place to look for bobcat evidence is on trails across isthmuses where the alternative would be to swim. Not only do bobcats dislike mud and frozen water, but they also dislike the melted version of the latter. I can think of any number of times in the spring, summer and fall when I have tracked bobcats to these crossing points between wetlands. As I mentioned in Chapter III on mud as a tracking medium, there is a public hunting preserve in western Maine called the Brownfield Bog. An old farm road, originally used to hay native grasses out on some high ground deep in the bog, is still maintained by Maine Fish and Game for hunter access. It is also used by bobcats and other predators with large hunting ranges to get around the bog. Several of the bobcat casts that I have in my collection come from the mud puddles that develop in the silty soil from which this road was built. It seems that ease of movement even trumps the characteristic feline dislike for having mud on their feet.

A third place to look for legible bobcat trails is on the sandy shore of a lake (Figure 6.29). Quabbin Reservoir, for instance, is bordered by a narrow beach, the width of which depends on water usage and the intensity of evaporation from the water surface. Both tend to be highest at mid- to late summer. Once the water recedes enough from the surrounding forest to create a continuous beach, many animals use it as a highway on which to get around (Figure 6.30). Otters must cross it to get to favored scat stations just within the woodline. Short-legged porcupines prefer its lack of resistance for easy movement. Raccoons patrol the wrack line for anything edible. Coyotes and foxes, both red and gray, as well as bobcats move along it to one

Figure 6.29. Predators like bobcats may find a deserted beach easier for passage between hunting areas than they find the adjacent woodland.

or another of their hunting grounds. Interestingly, coyote trails usually follow the firmest part of the beach, that part that has more recently been inundated and firmed by waves from wind or boats. More often than not, however, bobcats avoid the easy going and stay higher on the beach, closer to the dense laurel brush that will provide quick conceal-ment for this shy predator as well, perhaps, as hunting opportunities on any unwary hares detected within its shadows.

Cat Smarts

One November day many years ago I was hiking up Mount Car-digan in New Hampshire over a shoulder of the mountain called Firescrew. There was a cover of 2-3 inches of soft thaw snow over bare ground for most of the route, and because of the thaw, dense fog had enveloped the mountain. At the bottom edge of the boreal zone of spruce and fir, a bobcat's walking trail emerged onto the hiking trail heading upward — as was I. Every other track in its direct-register pattern showed a tiny spot of blood indicating a cut foot. This reliable little spot made the trail of the cat easy to discern even in the low contrast of fog and damp, gray snow. As I climbed I began to find it

Figure 6.30. The narrow strand on the margin of a lake may be used as a highway by many animals including bobcats and porcupines, seen here.

hard to pick out the hiking trail from the background and started wandering off it every hundred yards or so. The bobcat seemed to be having the same problem, as its trail would suddenly disappear from time to time only to reappear a short while later. About half-way up Firescrew, a bear wandered out onto the trail as well, also heading upward, but its trail, too, would disappear for a while, then reappear. Upward I climbed, zigzagging through the fog over the switchbacks in the trail, congratulating myself that I knew the landscape no less well than two of its native denizens. Then suddenly I realized that both these animals were using shortcuts! Having four legs rather than two and being in far better condition than the best human athlete (not to mention me), they failed to see the wisdom of moving back and forth over a slope when the direct and most efficient route was as close to the fall-line as possible. Stumbling around in the ground spruce under the summit, I finally gave in and started following the trail of the bobcat instead of hunting around for the man-made trail. Sure enough, the cat's trail followed narrow defiles through the growth straight up the slope and emerged on the densely be-fogged summit of Firescrew.

The bobcat's trail did not show a pause on the top of the shoulder but led off down into the col and then up toward the main summit of Cardigan. Normally I would not fore-track a winter-stressed animal if I thought its trail was fresh, but in this case there was no choice. The stunted boreal spruce-fir growth was so dense that the hiking trail we were sharing was the only route of passage over to the main summit of the mountain, at least the only safe one in fog and snow. About half-way up the summit cone I discovered just how fresh the trail was. The bobcat's tracks ended at a slanted rock slab next to the trail beneath which there was just enough room for a medium-sized animal to squeeze. The cat with the bloodied paw had forced its body under it for shelter and rest. But its trail also came back out from under the slab and charged off at a gallop down into the woods at the side of the trail. Clearly, I had just pushed the animal out of its resting spot and forced it run, expending precious energy in the process.

Powerline Incidents

Above (page 68), I referred briefly to an incident of hands-and-knees tracking in the Quabbin region of central Massachusetts while describing some of the attributes of snow. I elaborate on it here for some of the attributes of tracking bobcats over hard surfaces in the winter.

Tracking can be characterized as the art of seeing what almost isn't there. One January day, early in my tracking experience, I was excited to find the trail of a walking bobcat crossing the Gate-35 road at Quabbin. Under the leafless hardwoods near the road there was an inch of soft snow over bomb-proof crust, showing the bobcat's direct-register walking gait. Since the trail looked as if it had been made no more recently than the previous evening, I decided to fore-track it southward away from the tamarack bog from which it had emerged. At first the trail was easy to follow. The cat had maintained its walking pattern, only occasionally shifting into an aligned overstep trot. As it neared the road around the reservoir, however, the inch of soft snow over the frozen granular in which I had been following it suddenly disappeared. The newer snow had hung up in the hemlocks and sublimated into the atmosphere without ever reaching the ground. Under these trees the bobcat had walked on the firm crust alone and left behind very little for this young tracker to go on.

I had taken some measurements earlier that showed a step length varying from 13 to 17½ inches, the kind of variation that a stiff-legged canid would never show and even in a trail of vague prints a good indication of a cat. So I knew within fairly narrow parameters approximately where the next step should be. Starting with the last recognizable track as the cat entered the hemlocks, I fell to my knees and began searching 13-17 inches ahead. The light was filtered-sunshine that provided little shadow relief, so finding depressions in the frozen granular was impossible, especially in the shade under the evergreens. However as I searched with my eyes within a foot of the snow I noticed something. The cold night air of the evening before had caused moisture escaping from the snowpack to crystallize into tiny needle structures on the surface. If I looked very carefully within my search parameters I could find places where these crystals had been crushed. If I then backed off a couple of feet and looked at each spot, I could see a faint yellowing of the surface where each foot had been placed. This discoloration was faint at best, like the color of dilute lemonade against the otherwise pale gray snow, but it was recognizably there, apparently a function of some refraction of light on the altered surface of crushed crystals. But was what I was seeing really there or a product

242

of my expectations? I poked a squaw-wood stick into the snow each time I thought I had found a track and after twenty minutes or so stepped back and looked at what I had done. Sure enough there was the familiar zigzag walking trail of the animal I had been following, headed toward the plowed road.

The object of all this effort was to discover whether the bobcat I was following had moved directly across the road and into the forest on the other side, or whether it had walked down the plowed road a few yards to a powerline cut and used it to continue on its hunting route. Soon the zigzag line of sticks in the snow provided the answer. As the line approached closer to the road, the trail of sticks curved to the right, toward the powerline. Now I did not have to search the woods directly across the road, but instead could concentrate on the powerline, where after a short search I was able to re-locate the trail in the open.

I have since learned that these powerlines are favorite hunting and travel lanes for a number of predators, but particularly bobcats. This is especially true in the winter if the line runs east-west and if it is bordered by evergreens. The powerline cut itself is a brushland providing dense cover and food for many different prey animals like rabbits, mice and voles. Furthermore, the evergreens along the north border absorb sunlight in their somber green foliage. This short-wave solar radiation is re-radiated at night as infra-red, long-wave radiation, that is, as heat. This heat thaws the snow along the evergreen border, melting it back to bare ground faster than in the surrounding woods, and even before it melts to ground, the dampened snow refreezes overnight, forming a firm surface for a predator like a bobcat to walk on. Powerlines in winter, then, are another good place to look for bobcat sign.

A Bow-legged Bobcat?

Another powerline incident comes to mind that sheds a little more light on bobcat behavior. About a mile eastward on this same highline cut one winter I was following another bobcat's trail as it, in turn, followed the faint trace of the service road, really just a jeep track, under the lines. Along the edges of this "road" was extremely dense cover of bramble and laurel. Following the cat's trail I was able to locate places where the odor of ammonia showed the cat scent marking three times within 300 yards by leg lifts on this cover. The step lengths in 4-5 inches of softened frozen granular varied from 14-16 inches, about average for bobcat, and the direct-register walk showed a straddle of about 5 inches. The track width of 2¼ inches suggested a fairly large animal, perhaps a male. At one point the bobcat's trail disappeared into the dense cover and reemerged a few yards farther on.

But as it emerged, its trail, which up until then had followed the highline, now cut across it on a perpendicular, heading for the bordering woods. Here the step lengths declined to 12-14 inches (Figure 6.31) and the straddle widened to as much as 8 inches! Had I picked up a different animal? The trail itself now lacked the neat methodical character of a winter-adapted animal, looking as if it had been made instead by a drunk. I pondered, and then obeying the simplest and most important of tracking maxims, I looked more closely. Next to one of the cat tracks as it had emerged from the brush was a tiny tuft of silky fur (Figure 6.32). And along the sides of the trail as it headed toward the woods I detected faint linear marks parallel to the cat's trail. The brush of a powerline is too dense for a hare, one of whose defenses is to outrun its opponent, but it is just right for a cottontail rabbit whose only defense is to hold still and hope that its mottled pelt blends with the dense, tangled background. Apparently this one's did not. The bobcat had the rabbit in its mouth, a leg or an ear perhaps dangling down to touch the snow, and was walking with this awkward load toward the trees at the edge of the highline where overhead cover would allow it to dine in peace away from observation by other predators and scavengers. In order to avoid tripping over the dangling parts of the rabbit, the bobcat spread its legs to either side, leaving the extraordinarily wide straddle.

Once inside the woodline the cat had laid the rabbit down and fed. After the bobcat left the site, two other animals visited. A red-tailed hawk that may have witnessed the kill from one of the powerline poles left a feather perhaps as it was accosted by a coyote that also visited and probably carried off whatever remained of the hapless cottontail.

At another spot along this same powerline a couple of years ago I found a bobcat scat station. Such places are located along the habitual hunting routes of these cats. I know of at least eight of these in my area that have been used over and over for many years. This one

Figure 6.31. The tracks identify this as a bobcat's walking trail, but with exaggerated straddle.

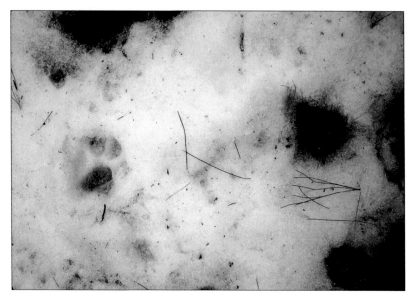

Figure 6.32. A tuft of fur explained the wide straddle. Its silky feel showed that the cat was carrying a cottontail in its mouth.

was under a powerline pylon surrounded by dense brush. It afforded several scats as the cat or cats worked up and down the nearby service track. All were left in the open, uncovered, and none were placed in prepared scrapes. The same is true of the scat at two other sites on the side of Mount Wachusett. One is well off the summit road in mixed hemlock and beech woods in a small clearing caused by an eruption of ledge. In two spots in this clearing, always on small areas of exposed ledge I have found perhaps a half-dozen scats over several years. Being on rock there could, of course, be no evidence of site preparation of the sort that one finds when bobcats void scat singularly on their hunting routes. Another site, at Quabbin Reservation, is deep in the woods at a large glacial erratic (Figure 6.33).There I have found scat on top of the rock as well as at its base just in front of a crevice that appears to have served as a lay site (Figure 6.34).

What characteristics do these three scat stations have in common that cause the cats to select them? The only correlation that I can come up with is privacy. The one under the pylon can only be approached through dense, tick-infested brush. The Mount Wachusett sites I described are in a trailess area so far from any road that I doubt

Figure 6.33. Bobcat scat stations tend to be in inaccessible spots. At this one a cat used the crevice at the base of a glacial erratic for a lay site and left several scats nearby including the ones visible on top of the rock. These may be from the same animal twice or may be counter- marking by two animals.

anyone but me has been to either of them in years. And the one at the erratic is in such a dense young woodland full of similar erratics that I have trouble myself locating it when I return occasionally to have a look.

The bobcat that ranges in the region around this last site revealed itself to me on one memorable late morning. An unused dirt road over a culvert between Bassett Pond and its boggy outlet created an isthmus of dry ground across the wetlands. Often I have found the tracks of coyote, porcupine, otter and bobcat using this crossing in one direction or another. One late morning I descended a hillside and came out at a road intersection next to the isthmus where I sat to have a drink and rest. As my gaze wandered aimlessly down one of the roads leading to the intersection, I spotted the motionless figure of a bobcat standing in the road looking intently at me. It had come out of a side path, intending apparently to travel down to the intersection and across the isthmus. But it saw me a moment before I saw it. It froze and I did the same, regrettably forgetting the camera on my chest.

Figure 6.34. Scats found at the boulder scat station.

Slowly it turned its head to the right, pondering its next move — up into the woods or left down into the brushy wetland? In a second it turned and loped to the left into the brush below the road. Then I remembered to breathe.

An Encounter at Barre Falls

On the second day of March, Joe Choiniere and I had an interesting adventure down at Barre Falls. We had decided to go to an area where we had found ermine sign in the past on the chance of getting some more in damp snow that would provide defined prints to photograph. Five inches of snow had fallen just the day before so it was a little soon to expect a lot of new tracks and trails to appear on the refreshed surface, but it was the only time when both of us could get out together.

Walking down an abandoned road we began to find the trails of chipmunks, tricked out of torpor by a warm, snowless winter that had suddenly turned snowy. Then, a hundred yards or so down the road, the trail of a large animal crossed. Loose snow had obscured detail in the tracks, but alternating direct registrations showed a consistent step length of 23 inches with a fair amount of straddle. I followed the trail out a bit to find a place under overhanging hemlock where the tracks would be clearer. Within a hundred yards off the road, I finally got some clear impressions that showed symmetrical splayed, oval prints without nail marks. The secondary pad was small, with the winged-ball appearance of a gray fox. I stepped back and noted the straddle as far too wide for this species so I looked more closely. The plane of the tracks was slanted down toward the toes, de-emphasizing the secondary pad in the hind print that overlaid the front. The "wings" of the gray fox's winged ball secondary pad turned out to be the lobes of a cat's, only partially registering due to the foot posture. This was yet another protean bobcat.

One might not think that a cat could easily be confused with a dog, but the gray fox is a very cat-like animal and, indeed, I had seen these foxes leave as long as 22 inch step lengths in the same aligned direct-register trot that the bobcat on hand was using. That the cat, whose normal walking step length is only about 15 inches, was also using this gait was suggested by the spray of snow crystals in front of each track as the animal's feet lifted out of the snow into a moment of suspension. This was evidence showing higher energy than in a walk. Such a gait is appropriate to the circumstances on a couple of counts. First, its bouncy nature allowed the animal, which was sinking down nearly to the bottom of the five inch snow cover, to vault over the intervening snow rather than plowing through it. Secondly, as we have seen with many predators, the front foot in each direct registration packed down the snow for the benefit of the hind foot on each side, a small efficiency that would pay off in the long run for an animal that travel-hunts.

With the animal identified, we decided to backtrack rather than foretrack it. Given the freshness of the trail, made perhaps that morning, we didn't want to stress the animal in the winter by following it and perhaps forcing it to waste energy fleeing from a lay site where it was resting. Soon the trail led us into an area of trees older than the usual New England managed forest. Here were large red oak, hemlock and pine with a scattering of white oak. Under the high canopy was a strongly developed understory of hemlock, apparently the result of a thinning cut perhaps twenty years earlier that had left enough hardwoods to provide shade for a regrowth of the shade-loving conifers. In this dense understory we began picking up the trails of snowshoe hares, a species of lagomorph that likes just this sort of habitat. These hares, it seemed, were the reason for the presence of the bobcat.

For a half-hour we backtracked the cat, which we decided was a male by the consistent size of its tracks. As it trotted toward us from the west, it maintained the 23-inch aligned DR trot wherever it traversed small openings in the forest but would slow down to a walk as it approached thickets and groves within which a hare might be dozing. Eventually the cat left a few ammonia-smelling urine marks, first on a branch that had fallen from overhead and lodged butt-down in the snow, then more conventionally on tree stubs a foot to a foot and a half high by 5-6 inches across (Figure 6.35) I say "conventionally" because it is the habit of large, thinly distributed predators to mark a class of objects such as these presumably to make it easier visually to distinguish scent posts from the infinitely varied background. These scent marks are used to communicate all sorts of information to other bobcats: territory, health, breeding condition and so forth. Another spot favored by bobcats for scent posts is just under the edge of any overhang on a rock they encounter in their travels

Figure 6.35. Scent marking on a stub. Bobcats often choose a class of objects on which to mark their passage

(Figure 6.36). And indeed we soon located one of these marks at a stone wall that traversed the woodland.

After about an hour of tracking we suddenly found where a second bobcat trail had come in and joined the one we had been following. Another bobcat? I was skeptical since I had many times found bobcats circling around in hunting areas, eventually cutting their own trails and creating such a maze that it was impossible to follow the cat sequentially. I continued on, confident that we were still following a single animal.

The woodland we had entered got better and better, becoming very close to an older-growth forest, with tall trees, wide crowns and the kind of vertical complexity that promotes diverse wildlife. Joe and I were separated out of sight of one another when I noted some bark chips at the foot of a pine that the cat had approached (Figure 6.37). Side-by-side prints showed that the animal had stood on its hind feet and reached up on the tree to scratch down, creating the debris.

Figure 6.36. Here the bobcat has lifted a leg on another favored object, a boulder with an overhanging surface.

Although a common nail-sharpening habit of all cats including housecats, sign of it is not often distinguishable in the field, at least when there is no snow on the ground. Pausing to photograph it, I also noticed a nearby lay site melted down into the snow (Figure 6.38). The heat of the animal's body had melted the white snow down to gray ice, making the cat's form conspicuous against the background. I called to Joe who shortly appeared to say that he had found more lay sites nearby. OK, one lay site would not be that remarkable, but more than one since the snowfall the previous day could only mean that the multiple trails we had been following had indeed been made by two animals!

We poked around some more and began to notice commotions under low-hanging hemlock boughs the tips of which had been caught and pinned down in the new snowfall, creating a bower beneath. What was going on? Then we noticed that the lay sites that we were finding were in pairs, oriented the same way and a few feet apart. Carefully examining each, we were able to make out the hind foot impressions as long troughs in the ice to which the snow had melted down. In each pair of lay sites the paw arrangement showed that the animals had faced each other. Then we looked more closely at one of the commotions and realized that some of the dark specks, which I at first had thought were bark chips, were actually fur clumps (Figure 6.39). Was this predation? We felt the consistency and appearance of the clumps and soon realized that they had apparently come from one of the cats!

Figure 6.37. The male bobcat stood on his hind legs to scratch the tree, then walked off left.

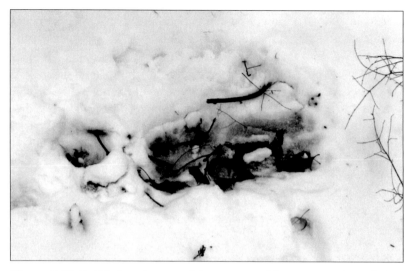

Figure 6.38. In this lay site one of the cats rested between matings, facing right.

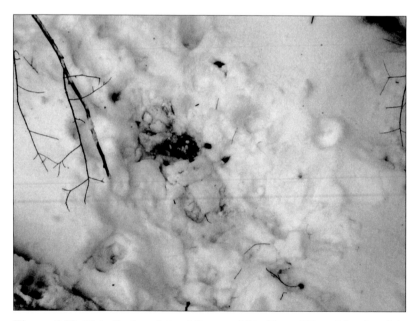

Figure 6.39. Note the imprints of front and hind legs. Bits of the female's neck fur can be seen here at the site of one of the mating commotions.

We paused to ponder a bit and soon came to an exciting realization: what we had stumbled on was a bobcat mating arena! Here a male and female bobcat had encountered one another and engaged in multiple periods of copulation interspersed with periods of rest. Trying not to obscure any evidence with our own tracks, we carefully looked around for more evidence. Soon we discovered another scratch tree similar to the first, as well as a single engorged tick that had been dislodged during the violence of mating and was frozen in the snow of one of the commotions.

Reorienting our minds from backtracking to imaginatively foretracking allowed us to piece together the likely scenario. Each commotion was a mating, where the male had grasped with his teeth the fur on the back of the female's neck while copulating, pulling out a few clumps in the process. In between these sessions the animals had laid down to rest, facing each other a few feet apart before commencing relations again. After several mating and resting periods the two animals set out on circular hunting loops through the female's range. It was one of these loops that had first alerted us to the possibility of more than one animal. After hunting for a while the male departed alone toward the old road where we had first picked up his trail. Unlike wild canids, male bobcats don't form lasting pair bonds or contribute to raising the result of their copulations. Nor, in fact, would the female allow them to. After a brief bit of cooperative hunting, the male moves on to make the rounds, mating successively with every female that has come into condition on his circuit.

Happy with what we had found, we then noticed another trail heading into the mating area. This was a series of four-print patterns in high-energy motion, leading to one of the scratch trees. Snow debris from the energy of the gait had once again obscured detail in the prints, but the outline of each suggested an oval with the long axis in the direction of travel. Cat prints tend to appear round while many canid prints look oval like these. Could this be a coyote? I began to backtrack the animal and was astonished to see intervals of up to 9 feet between the print groupings! At a high stonewall, I found where the animal had perched momentarily on top before leaping out and down. Clearly this was no coyote! Looking over the wall I could see a tight group of prints where the animal had gathered itself for the leap upward. After climbing over the wall I followed back over the trail perhaps twenty yards when I became aware of a familiar 23-inch trotting trail closing on the left. Shortly the two trails converged at a point where the trotting bobcat had suddenly come to a halt, then reversed itself in a series of colossal leaps toward the mating arena. The oval appearance of the prints apparently was the result of snow splash in the direction of motion. Mentally reversing the process I realized that this last part of the episode that we found last was actually the first. The male bobcat had been trotting and hunting from westward through the female's range when it sensed something, perhaps a yowl from the female in heat. Stopping dead in its tracks, it turned around and bounded toward the cry, leaping onto the wall and onward, determined to be the first male to arrive at a receptive female. Finding no competing male at the site, our male had stopped to sharpen its claws on the pine before its first mating session with the female.

An afternoon of eco-tracking had yielded a very satisfying encounter, an incident that could only be reconstructed after the fact by using interpretive skills. Here was mating behavior recorded in the snow by two animals that had no need to alter their behavior because they knew they were being observed by humans. This candid account provided insight available in no other way and did so only because a couple of trackers with prepared minds had stumbled on the scene at just the right time. Indeed, when I went back to the GPS coordinates a couple of days later, all sign of the event was gone. The debris at the base of the scratch trees, the squat void we had found in the midst of the other sign (Figure 6.40), the lay sites, the altered gaits at the scent locations and the copulation turmoils were all concealed beneath a couple of inches of additional snow that had fallen overnight. Our own vague prints were all that I could make out, but even those would disappear in the approaching spring thaw. I was reminded of the art practiced by oriental monks, where colored powders are painstakingly applied to a surface in order to create an exquisitely intricate design, only to be erased at its completion. We may at first regret the loss of these short-lived phenomena, as did I while fruitlessly searching the woods a few days after the event, but their disappearance also makes them more valuable to the mind and perhaps to the spirit behind it.

Bobcat Gaits

Most of the time bobcats hunt in dense cover: laurel thickets, brushy edges of beaver ponds, thick boreal growth or dense chaparral.

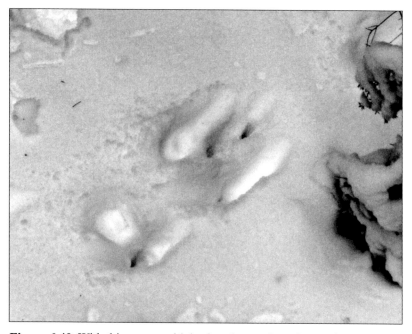

Figure 6.40. With this squat-void the female may have been advertising her breeding condition.

This habitat affects its choice of gaits. While coyotes and foxes put in the miles traveling roads and trails, hoping for coincidences between their route and that of potential prey, bobcats typically are stealth hunters, patiently prowling coverts in hopes of pouncing on an unwary animal or capturing it in a short chase. The canids mentioned above habitually use flat, displaced gaits on their typically more open and even terrain as an efficiency that may help them survive the winter. Bobcats, on the other hand, would find the enlarged frontage of such gaits cumbersome in the dense cover of their chosen terrain, requiring them to force rather than insinuate their way through. The resistance of the bobcat's hunting habitats is further increased by obstructing debris hidden in the dense growth at its feet. A flat gait, however efficient for a dog on open ground, would cause a cat to trip in the thick cover it typically hunts. For the bobcat, then, the reduced frontage and higher arc of stride of an aligned gait is preferred. The two most common bobcat gaits are the direct-register walk (Figure 6.41; Diagram 6.3) and the overstep-walk/trot (diagrams 6.4 and 6.5).

Figure 6.41. A bobcat's perfect direct register walking trail on a snow-covered beaver pond. Note the slightly wider straddle than with either fox or coyote.

Diagram 6.3

The average step length for a bobcat using the direct-register walk gait is about 15 inches. However, the distribution is extremely broad. Reviewing my field notes, I find examples with a step length as short as 10 inches and as long as 23! At what point these direct registrations transition to a trotting gait is hard to tell, since the higher arc of an aligned trot is concealed in the invisible third dimension of the gait, that is, how high the animal bounces through each cycle is seldom readily discernible on the tracking surface. On one occasion at Gate 33 in Quabbin, I was tracking a cat in the winter in a dense regeneration of young pine. At one point the cat actually stretched this DR walk out to 27 inches! I suspect that this was a stalk, the animal having lowered its center of gravity and reached out far in front to place each foot, levering itself forward and placing the following hind foot in exactly the same place. Reducing the number of steps in its stalking approach as well as the area of each footprint reduces the chances of snapping a twig and alerting its prey.

You may have noticed that the average step length of a direct-registering bobcat is the same as that of a red fox using the same gait, that is, 15 inches. However, where a fox's trail shows a nearly straight line of tracks, the trail of a bobcat shows a more pronounced zigzag. Where the straddle of a red fox is perhaps 2-3 inches, or not much wider than its footprint, the straddle of a bobcat distributes around 5 inches.

The footprint of a housecat is similar to that of a bobcat. Both have the same enlarged secondary pad skewed toward the shorter of the two middle toes, the same tri-lobing along the posterior margin, the same oval toe prints and so forth. However, the body shape of the two is somewhat different. Where bobcats have long bodies and legs, housecats typically have relatively short legs for their body length. The step length of about 8-10 inches for the housecat when direct registering approaches that of a small bobcat, but the former, due to its body shape, often understeps, that is, the hind foot may register somewhat behind the front. It is much more likely that a bobcat's hind print will overstep, registering well ahead of the front. Here is an example of that overstep walk/trot pattern.

Diagram 6.4

If the bobcat exaggerates the overstep even more, the trail becomes the familiar sinuous S-pattern common with this species.

Diagram 6.5

This pattern (Figure 6.42) is very similar to the wiggle-walk I described above (pages 201) but may also represent an aligned trot. As I often seem to find the 15-to-16-inch step length in these patterns, which is about the same as a normal walking gait, I suspect the invisible third dimension of such an aligned trot accounts for the combination of large overstep without a corresponding increase in the step length. That is, the bobcat may be introducing a lot of verticality into its gait, bouncing up and down as much as forward. I am not sure of this, but so far it seems to be the only way of explaining this oddity. If there is a high arc of stride, of course, there may on some surfaces be pressure effect around the tracks, that is, mounding or other evidence of impact.

Of course, a bobcat can use almost any other gait in the book if it chooses. It may gallop briefly, either transverse or rotary, to charge prey or to escape (or to arrive first at a receptive female), and it may bound or jump to leap obstacles or gain height. In my experience, however, bobcats rarely lope. Loping is a gait that animals use for efficient long-distance running. A deer may do it, for instance, when run by a coyote, and the following coyote may do it as well. But cats don't have reason to run long distances. They are hunters of the shadows, moving carefully through the brush, ready to pounce at any moment. In between their hunting areas they walk or trot.

Bobcat Scat and Scent Marking

Like other predators with large ranges, bobcats need a way to communicate with others of its species, either to warn off a competing animal from its territory or to find and attract a mate. To do this, bobcats use their scat, urine and scent glands. Many of the glands leave a scent that is undetectable to the human nose, or at least to my nose.

Figure 6.42. This sinuous trail is identifiable as that of a bobcat even without track detail.

However, the first two are more obvious, especially scat, which, if left so as to be conspicuous to another bobcat, is equally obvious us as well.

Bobcats are carnivores, much more so than foxes or coyotes, which will feed commonly on whatever edibles they can find including apples, grapes and other berries in season. Bobcats stick to meat, so their scat almost always contains fur but without the large bone chips that often show on the surface of coyote scat. Lacking the robust molars of canids, bobcats can't crunch large bones. Their carnassials are used just to shear meat and smaller bones which they immediately swallow. While bobcats will scavenge a carcass, especially in winter, they merely lick the meat off the bones with their raspy tongue. The result of these habits and dental limitations is that bobcat scat is remarkably uniform in consistency, at least within a single deposit. In size they usually range larger than red fox and smaller than coyote, that is, slightly narrower than ¾ of an inch, and generally less copious than coyote scat. They are typically tubular and rather blunt, with perhaps a tail at one end, which is typical of carnivorous scat of any predator, the tail being the excreted fur or feathers of the prey. If it is fresh, it will also have the characteristic sweetish odor of such scat.

It is often proposed that bobcat scat is more segmented than that of other predators of similar size. I believe this started with Olaus Murie, whose experience with bobcats seems to have been mostly in the Southwest. Perhaps the dry climate or the preponderance of small rodents in the desert contributed to this impression. In the Northeast I have found this characteristic unreliable, having discovered both segmented and smooth bobcat scat in this region. Furthermore, much coyote scat also shows segmentation of sorts. However, one fairly reliable characteristic of shape that I have found locally and attribute to bobcat is what I call the "ball-and-socket effect" (Figure 6.43). In a segmented scat, one or more of the seams may show what looks as if the rounded end of one segment as it travels through the cat's gut has pushed into the adjacent segment, creating a ridge around the rim of the receiving segment. To me this looks like a mechanical ball-and-socket joint. My friend Kevin Harding believes this forms as the scat is being excreted in pulses. Whatever the cause, I often find it in scat that other evidence leads me to attribute to bobcats.

Figure 6.43. The ball-and-socket segmentation on the lower left and the scat's location in a gap of a stone wall reliably identify it as that of a passing bobcat.

One should be careful about averages. While the maximum width of a bobcat's scat averages somewhere between ½ and ¾ of an inch, the distribution can be quite wide.

Once again, tracking identifications should depend on the sum of the evidence; pinning a scat identification on just the size of the scat alone won't work here or elsewhere. And one can't assume that the lack of large bone chips in a scat identifies it as bobcat, either. While large bone chips on the surface of a scat may eliminate bobcat, a scat without them may be any carnivore feeding on a fresh kill. The bone chips only appear later as the carcass is reduced and its bones are crushed by an animal with teeth to do it in order to get at the marginal nutrition that is left.

What, then, are a reliable set of characteristics for bobcat scat? We have already looked at shape and lack of large bone chips. To those we can add location. Bobcats often leave their scat in predictable places. Scat stations where bobcats habitually leave their scat along their hunting routes have already been mentioned. On the other hand, single scats can often be found in gaps in stone walls and occasionally on tops of the walls themselves. In winter, the snow-covered top of a cut stump is a favorite location. Such spots are also favored by fishers, however, so some care should be taken. In any surface that is loose, such as snow, dirt or dead leaf debris, bobcats usually prepare a scrape by pawing at the surface before they deposit the scat. Since the cat's back is to the scrape as it defecates, it sometimes misses the pawed surface and so the scat is found alongside the scrape rather than in it. This may lead to confusion with canid droppings, since dogs habitually scratch beside their scat. However, they may also leave a urine scent mark, the odor of which is distinctively different from the ammonia smell of cat urine. Prepared scrapes under a scat deposit are one of the best clues to bobcat, assuming that the other evidence agrees. Of course, on a hard surface such as exposed ledge, a rock wall or a boulder, no scrape will be evident and the identification may be less certain. Go by the sum of the evidence and then mentally put a percentage of certainty on your identification.

Bobcats also make scrape-ups where they paw up debris into a pile. The reason for this is not clear since these piles, when I have found them, lacked a urine odor and didn't contain buried scat. Certainly they would contain a scent from inter-digital glands, but I can't detect this odor with my faulty, civilization-corrupted sense of smell. However, I am sure bobcats and other animals can.

As far as the habit of burying scat, I have read that these cats leave their droppings exposed in the interior of their own range but bury them near its borders or in another cat's range. However, I have seen it written the opposite way as well. Personally, I have never come across a completely buried bobcat scat. Perhaps after they bury it, weathering disguises the site from me, or perhaps their tendency to bury deposits is exaggerated. Occasionally I will see a rudimentary effort to cover the scat by pawing up dirt or debris, but this hardly represents a thorough effort to conceal it. I suspect it is the sort of instinctive behavior that one often sees in wild animals where they go through the motions almost ritually rather than effectively.

In addition to scat, bobcats also mark with urine. Bobcat scent marks, to my nose at least, usually have the same ammonia smell as that of a housecat. Rather than placing marks at random around their range, where the chances of another bobcat finding them would be reduced, bobcats select out of the infinitely varied background a set of objects for anointing. The two most common are stubs and overhanging rocks. In my area I often find the ammonia mark on the dead stub of a tree that sticks up above the surface about 1-2 feet (Figure 6.44). These are generally back sprayed, much as a cat does on the tires of a car. The other location is on any boulder the upper edge of which forms a slight overhang. I think bobcats may use these overhangs because the scent is less likely to be washed off by rain. Whatever the reason, any bobcat moving through new territory does not have to check out every detail in the terrain it crosses to find out what other animal is in the area. All it has to do is check out any of these typically selected objects.

Fun Tracking Bobcat

Many of the challenges in tracking bobcats were compressed into one memorable winter day in the Federated Forest. I went to a spot on Fever Brook Road between two large laurel thickets where I had found a bobcat crossing several times before, and sure enough there were tracks across the three inches of snow on the unplowed road. A thaw and refreeze after the trail was made showed in the tracks as a softening of their features. I decided from this that the trail

Figure 6.44. Bobcats like foxes often spray a class of objects. A common one for bobcats is a tree stub about a foot high.

was at least a day old so I decided it would be no stress on the animal to fore-track it. The trail dove straight into the thicket on the east side where I tried to follow it as it wound around among the laurel shrubs. After wrestling with the thicket for a few minutes, I gave up and withdrew, circumnavigating the patch until I found the cat's exit trail. Apparently it had not met with hunting success and now headed off through the woods to the next patch. So it went for the next half hour, from thicket to thicket through a tableland of forest, until the trail headed downhill toward a wetland. Descending this slope, the cat wandered into the trail of deer herd that had been moving up from the lowland. It followed this packed trail downslope as a convenience, no doubt, but for the following tracker it was much less so. Here there was so much commotion from the passage of multiple animals that tracking my animal through it was very artful indeed, depending largely on anticipating where the next track should be and trying to decide whether the track I thought I was seeing was really there or existed only in my wishful imagination. After a frustrating half-hour of this, I thought it might help to widen my perspective and try to think like the bobcat. Having tracked this particular animal several times in the past and gotten into its head a little, I considered where it might have been intending to go. This guesswork seemed to help a little as I was able to locate, but then again lose, the trail several more times. Finally, I seemed to lose it for good, both actually and imaginatively, in an especially large commotion of deer tracks. The only thing I could do was continue to follow downslope over what I suspected would be the cat's route and look for the place where its trail might have diverged from that of the deer herd. After two passes up and down the slope to the last known real track, I finally located the cat's exit point. It had leaped up out of the deer turmoil directly onto a fallen limb, mostly bare of snow, and invisibly walked its length into the bordering trees before jumping off onto trackable snow once again.

Impressed with myself, I followed for about a hundred yards to where the trail, still showing the cat's familiar direct-registering walk, crossed an open seep by jumping from one snowy island to the next. Here I was forced once again to circumnavigate but did manage to pick up the animal's trail on the other side. No sooner had the cat and I got clear of this seep than it found the trail of a porcupine in the snow and decided to follow it as well, as long apparently as it headed more or less in the direction the bobcat wanted to go, which was downslope. Here the tracking was easier. Each of the steps of the much larger bobcat was about twice that of the porcupine and since the cat was heavier as well, the tracks were deeper. That depth had a created a shadow along the edges of the tracks as the sun of the day before had traversed the sky. That shadow had absorbed ambient light and melted the bobcat's tracks down to the bare leaf litter under the snow. As a result, the cat's tracks stood out as dark spots in every other track of the porcupine, which were themselves gray against a contrasting gray shade of the old snow.

Down the slope the bobcat walked, eventually leaving the porky's trail and emerging onto a beaver impoundment. As the ice was firm enough to support me, I followed the cat out onto the surface and soon encountered, in the inch or two of snow over the ice, another bobcat walking trail crossing the one I was following. I suspected the cat was circulating again, hunting through the alder scrub that lined the pond. I refused to be diverted by the crossing trail and stayed on the one I had been following, at least until it suddenly separated into two trails! Both of these were headed in my direction and diverging. Somewhere along the line behind me the bobcat had circled and entered its own trail, using it just as it had that of the deer and porcupine. It had placed its feet so perfectly into its previous prints that I had been unaware that I was following more than one trail. Chagrined at being tricked, I continued fore-tracking one of the diverging trails until it stopped at a commotion at the base of a dead snag standing in the pond. On the snow were many fine splinters, and the surface of the snag showed where the splinters had come from (Figure 6.45). This was a scratch post of just the sort that a housecat uses to sharpen its nails. After taking a photo, I looked around for the exit trail and quickly found . . . several!

Stepping back mentally and widening my perspective once again, I realized that the cat had wound around and around through the alder, crossing and re-crossing its own trails to create an unintelligible jumble. In several places it walked in its own trails, then magically disappeared and reappeared elsewhere. After 20 minutes of trying to reconstruct its route and activities, I gave up and reverted to a circumnavigation strategy, this time making a very wide circle around all of this confusion.

After a few minutes I found the exit trail up in the woods above the bank of the pond and followed it over a low ridge and down to the

Figure 6.45. Fine splinters at the base of this snag show nail sharpening by a passing bobcat. Interdigital glands also leave the cat's scent behind.

edge of another beaver bog. Here the animal walked out over ice too thin for me to follow and gathered itself on a sedge tussock from which it jumped over open water to another and then to another. I stood on the bank and tried to see where the cat's trail emerged onto the ice on the far side, but a surge of water had spread out over this ice as it had settled and this flood had erased the bobcat's trail. I circled around through the woods to an old farm road that crossed the head of this wetland, hoping to pick up the trail again in better snow. But after searching for nearly an hour I decided the bobcat had been assumed into heaven as I was able to find no trace of it on this firmament. As anyone knows who owns a housecat, they can be contrary creatures. It seemed almost as if this bobcat knew it would be followed and made every effort to confound the follower. In this case it succeeded.

Species Notes: Porcupine

An Elegant Theory

Years ago in need of some intellectual stimulation, I evolved an elaborate theory on how a porcupine walks. It was elegant thing, with its own vocabulary no less, to describe what I was sure the porcupine was doing as I closely watched it advancing across the mown field up in the Federated Forest campground. The theory was full of "compensated weight adjustment" and "error anticipation." The fall that inevitably follows pride arrived with the purchase of a good videocamera and a chance to record a porky walking through the woods at an upland pasture near the Brownfield Bog. I took the tape home, eager to establish proof of my elegant theory. Alas, my fantastic intellectual castle was suspended in air with, excuse me, no point of contact with the ground. My theory had only one defect — it was wrong. Absolutely wrong. As it turned out, the video proved that a porcupine walks the same as almost any other mammalian quadruped: the left hind foot moves forward and replaces the vacating left front foot, which then itself moves forward, then the right hind foot does the same with the right front foot. How pedestrian!

Porcupine Tracks, Trails and Other Sign

Having little need for speed, porcupines usually walk (Figure 6.46), only resorting to a few loping cycles as an escape gait. In the resulting

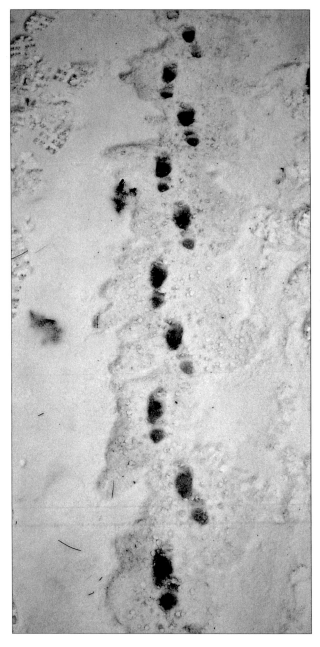

Figure 6.46. An overstep walking trail of a porcupine on thin snow moving away from the camera.

trail the larger hind foot often direct registers, obliterating the smaller hind print. However, just as often the hind will overstep leaving a double registration. Porkies have rather broad chests, which are useful for hugging trees in order to climb them. (When a porcupine climbs a tree, it hugs the curve of the trunk with its front feet and hitches its hind directly upward under its body, levering itself upward.) As a result, the straddle of the front feet is slightly wider than that of the hind so that a walking trail (Diagram 6.6) may look like a diagonally arranged series of commas, each facing away from the next.

Diagram 6.6

A porcupine's footprint is interesting in that it is the only mammal track I know of in which no toe pads register. Although the animal has toes, it has no need for pads since its gait is plantigrade and its weight is supported by both its large rubbery secondary pad, fused to its tertiary (heel) pad in one large "plantar" pad, and by the tips of its long, curved nails. The resulting footprint is unusual in that it shows a large oblong flat area and an arc formed by the nails far forward of that pad. With no toe pads in between the pad and nail arc, the nail marks are easily dissociated in the mind's eye from the main pad as if the two were not even part of the same print. The combination of a large flat oblong pad and an arc of nails well forward of the pad is diagnostic even in vague trails such as those on the loose sand of a beach (see page 70).

It should be noted here that porcupines have five toes and associated nails on the hind foot, but only four on the front (Figure 6.47). Olaus Murie, the pioneer in our art, made a rare mistake in his groundbreaking book, *A Field Guide to Animal Tracks*. Apparently, he mistook a hind print for a front and drew them both for his guide with five registering nail marks. It is interesting to see how many field guides since have made exactly the same mistake.

Porcupines are active year round. In the winter their trails often lead from den to feed tree, usually a pine, hemlock or other conifer. Because these rodents have rather short legs, they tend to plow through

Figure 6.47. Porcupines have only four toes on their front foot and no registering toe pads. Note the pebbled surface for traction while climbing.

soft snow, leaving a sinuous waddling trail that is diagnostic even if recognizable prints are not available (Figure 6.48). In the other seasons, you may find faint porcupine trails even on a hard-packed dirt road by looking for alternately slanting quill marks Under one of the trees at a feeding site you may find sprays of hemlock or other evergreens littering the ground. Since other rodents also cut sprays that litter the winter snow, whether or not the work was done by porcupine is to some extent a matter of weight. No animal does more work than it has to in order to feed itself. Therefore, the cutting mammal intent on reaching end buds will work its way out as far on the narrowing branch as it can before its weight threatens to collapse or break the branch. If a red squirrel is the feeding mammal, then its light weight allows it go far out to nip off only the last few inches of the twig. A porcupine, however, cannot go nearly as far out before the branch begins to sag. Thus, most boughs resulting from porky cutting will be thicker than those cut by smaller feeders.

The mechanics of a porcupine cut are helpful in distinguishing its work from that of another large rodent that clips larger branches as well, the beaver. After the porcupine crawls as far out on the branch as it dares, it extends one paw forward and rests the long, curved claws on the branch. Then it begins a series of diagonal cuts with its front incisors. When the lead paw feels the branch begin to weaken, the animal uses the curved nails to pull the branch back toward its body so that, as it finishes the cut, the branch is under control against its side and does not fall to the ground. This bending backwards stresses the cut end, which then shows splitting when it is recovered by the tracker under a feed tree (Figure 6.49). A beaver, on the other hand, is standing on the ground when it makes its cut, with no need to control the branch at all. Thus, with no stressing, the end of the cut looks clean.

Occasionally you may find a site in the woods covered with the quills of a porcupine, but with no carcass or blood to indicate a kill. Porcupines sometimes fall out of the trees they are feeding in, having misjudged the strength of a branch and crawled out too far. When the animal hits the ground, quills spray out in all directions. If it was lucky, it may have gathered itself and wandered off with only bruises or perhaps a broken bone or two. Once, however, I found the body of the porcupine itself, having died from the fall. Lore has it that fishers may

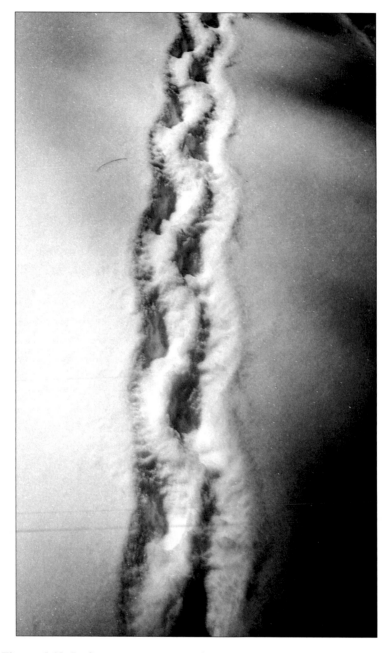

Figure 6.48. In deeper snow a porcupine makes a distinctive wavy trough with long drag marks of nails forward of its footprints.

Figure 6.49. Because of the method a porcupine uses to control the branch it is cutting, bend-back and splitting usually show at the end of a branch.

harass a porcupine and force it far out on a branch to induce such a fall and make predating the stunned or dead porcupine easier than attacking it directly.

Wisdom does not seem to apply to porcupines, and furthermore they have very poor vision so that sometimes they do not foresee what they are getting themselves into. Once, down below Rattlesnake Hill at Quabbin, I heard what sounded like leather rubbing on a rough surface. After a while I realized the sound was from the knubbly foot pads of a porcupine climbing up one of the slabs on the side of this ledgy hill. I watched it for a while as it groped forward, trying to force its way up the incline. Above it the route looked unpromisingly steep, but the animal could not see that far ahead to predict the folly of its route. (A porcupine's eyes seem adjusted to focus not much farther than the end of its nose — a convenience for feeding but not for route selection while climbing cliffs.) As it tried nonetheless to push its way upward, the friction of its pads on the rock finally slipped and the animal tumbled to the bottom of the cliff. I have read that autopsies on porcupines often find poorly mended fractures.

Scat

As with other animals abroad in the winter, the cold-season scat of porcupines tends to be drier and formed more perfectly into pellets than summer scat. These pellets can be found in quantity wherever the animal dens or feeds. The winter pellet is sometimes described as curved like macaroni. I often think of it as a bean, curved and often with an irregular line on the concave side. Since porcupines feed on evergreens including pine in the winter, the resulting scat and urine often smell like turpentine.

Porcupines don't feed on hemlock and pine because they prefer these trees or because they are very nutritious. Just as with deer, they feed on them in the winter because their boughs are the only edible greenery around. As soon as spring arrives, porkies abandon the conifer stands and seek out the flowers of hardwoods and later the leaves of many deciduous trees, especially those that have not developed a lot of tannin or other anti-browsing defenses. During spring and summer the shape of the scat becomes less well defined. It is still in vaguely pellet form, but often the individual pellets are connected by a membrane or fiber into a little train of soft, irregularly-shaped globs.

Dens

Porcupine dens tend to be located in large hollow trees or in rock crevices. Two located near me come to mind. One is an ancient white oak standing isolated in a hemlock wood (Figure 6.50). These trees were maintained in pastures on old farms as shade for livestock. After the farms were abandoned, the pastures grew back, often to pine and in this case eventually succeeded to hemlock. This particular oak dentree is about 5 feet in diameter and hollow, with two openings on either side about 4 feet off the ground. From both of these orifices has spilled, over generations, perhaps millions of individual winter porcupine pellets so that the cone of turpentine-smelling fecal matter is about 4 feet high and spread out on the ground to easily twice the diameter of the tree itself. If I pound on the tree, I can often hear the scraping of the animal's claws inside as it tries to climb higher and out of danger. A kinder trick that Joe Choiniere uses to determine porcupine presence is to insert his digital camera into the hole, facing upward and use the flash to take a picture that is immediately available upon withdrawing the camera. (I never thought of this simple tactic, perhaps because I have yet to fully adjust to the digital revolution where snapping photos no longer carries the expense of film and processing.)

Figure 6.50. Generations of porcupines have lived in this old white oak. Note the large flow of perhaps millions of droppings emanating from the tree hole.

The other den site is in a deep gully under a powerline. The sedimentary nature of the underlying rock in the area has resulted in a jumble of boulders at the bottom of the gully with a complex series of caves and crevices that generations of porcupines have used as dens. In the winter the porcupines in residence create a network of trails in and above the gully, easily picked out from the background snow because of dirt tracked into them when the animals emerge from their subterranean hiding places. Porcupines pee on the run, so to speak, and the yellow-green stains of this turpentine-smelling urine also dot the dirty trails. Look carefully at such winter trails and you may also find the tracks of porcupine predators like fishers, bobcats and coyotes using them to approach the den.

Predation

Although the porcupine is a prey species, its defense is formidable. Loosely held quills embed themselves in any would-be attacker and teach it a memorable lesson. Each quill has layered scales along its surface, the ends of which face backward, so that once it is embedded in flesh a quill cannot easily be pulled out. Instead, unless it is removed, it will work its way deeper and deeper into the animal as a result of that animal's own muscular contractions. DeBruyn relates a post-mortem on a bear cub he had been following that died quite suddenly for no apparent reason. An autopsy revealed that a porcupine quill, embedded during an encounter much earlier and of which there was no longer any sign on the surface of the bear, had worked its way into the cub's body cavity and pierced its heart.

As Nature abhors a perfect defense, the porcupine's formidable one has a defect: all of its quills are oriented toward the rear of the animal, leaving its face and head vulnerable. Porcupines in any given area sort themselves out into the stronger and the weaker. The dominant animals get first choice of crevices in which to hide their heads from predators; the lesser animals have to settle for what they can find. These sub-dominants represent a local surplus for the species, which is available for harvesting by predators. In this way the best genes are preserved and the weaker genes are weeded out of the population. I have heard it said that once a fisher moves in on a porcupine population, that population is doomed; the fishers will wipe them out. This is certainly not the case. No predator but one wipes out its own food supply. A wild predator, tutored by millennia of ruthless selection, unconsciously applies a sort of instinctive wisdom that leads it to farm rather than eradicate its prey. When the vulnerable surplus of a prey species is eliminated and the remaining dominant animals are too hard to kill, the predator moves on to some other prey animal that is enjoying a temporary surplus. This is the way predation works, for the ultimate benefit of both the predator species and the prey species, as well, whatever the cost to the individual animals (see pages 151-162).

A Bloody Lunch

One winter morning, Kevin Harding and I were tracking out towards Bassett Pond in north Quabbin on a couple of feet of fresh snow. We had picked up a coyote's trail and determined to follow it as long as we could or until dusk. Along the way, through a frozen seep-swamp, we came upon an attack by the coyote on a porcupine. The process was difficult to reconstruct, involving a couple of holes in the snow and overlain tracks of both predator and prey. But what was clear was that the porky had been caught on the ground with no place to hide its face. There were blood and quills on the snow, but we were uncertain whether the coyote had been successful. We picked up the coyote's exit trail and followed it to the edge of Bassett Pond where the intense late-February sun obliterated the animal's tracks. As it was nearly noon, we picked out a point across the pond to head for as a sunny, warm place to have lunch. I sat on a log and Kevin stood out a meter or so on the pond surface eating his favorite p-p-and-j sandwich. We lazily discussed the morning's progress, lulled nearly to sleep by the sun and the warmth. At one point I looked down at Kevin's feet and was startled to see that he was standing in a pool of bright red bloody slush. My first thought was "He's hemorrhaging and doesn't even know it!" He followed my eyes down to his feet and nearly jumped out of his skin. We both stood looking at the blood and at each other. Retrieving a stick from the shore I stirred the slush and tried to clear it. I felt something firm under the snow and more digging unearthed (or rather un-snowed) the carcass of a porcupine! It seems our coyote had been more successful than we had thought and had been carrying the dead animal along as we had been tracking it. Heading for the same promontory into the pond that we had chosen for lunch, the coyote had buried the carcass under the snow to hide it from other coyotes and scavengers as well, perhaps, as to refrigerate its next meal.

A Porky in a Hurry

Porcupines normally walk from place to place, as do skunks and bears. All three have formidable defenses that belie the need for great haste, the bear because it is the largest mammal in the woods, the skunk because of its noxious scent glands and the porcupine because of its quills. However, the porcupine's defense has the defect mentioned above that leaves its front end unprotected. Fishers, coyotes and even foxes take advantage of this defect if they can find a porcupine on the ground in the open where it has no place to hide its face. So when they are in the open, moving from tree to tree, and sense that predators are around, porcupines can speed up their normal waddling walk to a lope. As related above (page 116), I came upon just such a trail of a worried porcupine one winter several years ago (Diagram 6.7).

Diagram 6.7

In a 3X rotary loping pattern such as this, the central track in each pattern is the impression of a front followed by a hind direct registering over it. Little detail was evident in the tracks due to a thaw so I at first thought I was looking at a fisher trail as the animal loped through the woods. Since a fisher's front foot is larger than the hind, just the opposite of a porcupine, I reasoned that the animal had been moving from right to left. However, I soon realized that this was not the predator but rather the prey, hot-footing it between trees where it would be relatively safe from the foxes and coyotes whose trails I had been picking up in the same woods. Furthermore, it was moving from left to right, with its smaller front foot initiating each pattern. I believe this loping gait is about as fast as a porcupine can travel, being too bulky and short-legged to manage anything like a gallop or bound.

Species Notes: Fisher

It is appropriate that the notes on porcupine should be followed by those on fisher since fishers are often thought to be the chief predator of the former. In fact, the presence of fishers in New England may have resulted from a forestry scheme to re-introduce the animal into the North Country as a biological control on porkies. It is also possible that a relict population in inaccessible old growth in New Hampshire may have spawned the current population, which has since spread far southward even into the metropolitan area around Boston. As was discussed above (pages 151-162 and 279), predators manage their prey, taking off the surplus of weaker animals and leaving the robust, which they find too hard to catch. When the easier prey of one species runs out, the predator moves on to other prey species that are enjoying a surplus.

In the metropolitan parks of Boston where there are no porcupines to begin with, this "wilderness predator," the fisher, has become a suburban squirrel specialist. Gray squirrels are so abundant in these older-growth parks that fisher numbers have increased to the point where it is much easier to find the trails of these animals there than in the less-populous central and western areas of the state despite the greater forestation in these regions. Figures 6.51 through 6.53 provide a representative sample of the kinds of sign one might look for in suitable habitat.

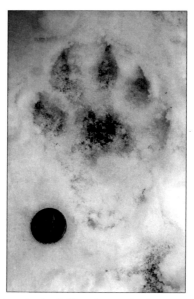

Figure 6.51. Fisher tracks look like those of other weasels: 5 toes arranged asymmetrically and with rather weakly registering secondary pad. Tracks of males are usually an inch wider than those of females.

Figure 6.52. Both bobcats and fishers habitually leave scat on snow-covered stumps. However, bobcats first paw a depression for the deposit.

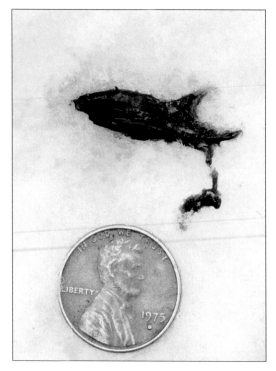

Figure 6.53. Fishers sometimes mark with a "force scat," a small deposit often left at the base of a fallen tree.

As I described earlier (pages 88-89), this "wilderness" of more western parts of the state is actually like a huge tree farm where timber is removed for lumber before it ages enough to develop natural cavities, cavities that animals like fishers depend upon for birthing dens as well as hunting sites. It is significant, I believe, that if I need to find the trail of one of these interesting mammals, I can do it much more quickly in some of the Boston metropolitan parks than in the much larger forests outside my door at Mount Grace.

Suburban Fishers

I first discovered the trail of a suburban fisher in Breakheart Reservation in Saugus, Massachusetts, a suburb north of Boston, in 1992. One of the wilder metropolitan parks, its 600+ acres is composed of hardwood forest, rocky outcrops, several undeveloped ponds and a stretch of the Saugus River that winds between it and the adjacent woodland of an old Boy Scout camp. One early winter afternoon in an inch or two of soft snow near a cove of the Lower Pond I found a trail of the illustrated patterns winding through a pepperbush thicket (Diagram 6.8).

Diagram 6.8

Clearly this was the loping trail of a mustelid, and the print width of 2¼ inches suggested a smallish otter. As there is a lot of water in this park, I was used to finding the trails of that animal commuting among various of them, and as I was close to the "lagoon," as I called a cove on this pond, I jumped to this hasty conclusion. I followed the trail for a while, increasingly disturbed by several anomalies that I was not used to finding in otter trails. First of all, the front prints were about the same size or perhaps even a shade larger than the hind, just the opposite of what I should expect for an otter, which depends on its larger hind feet as paddles. Secondly, the trail wound through the thickets rather than heading straight for the pond. Otters subsist on fish and crayfish, but this animal seemed to be hunting on dry land. Thirdly, this weasel was not sliding. Otters prefer sliding to walking or loping and do it wherever they get a chance. Even on small slopes my animal seemed positively averse to the thrill of letting go of its quadrupedal stance. Could this animal be a fisher?! I had got used to the idea from the literature that fishers were strictly wilderness animals, found only far from civilization under the dense overhead cover of northern coniferous forests. My tracking of them in the past was in the White Mountains far to the north or at Quabbin far to the west. But here I was in a Boston suburb; it couldn't be a fisher. No one had ever reported or imagined such a thing. I must be wrong.

With mounting excitement I continued to follow the animal as it closed on the lagoon. Finally, it approached the cove through a fringe of highbush blueberry. Here, I thought, this otter will drop its disguise, jump into the water and solve the riddle. Just short of a narrow border of open water between the bank and the ice, the animal stopped, then backed up, showing a pause and some indecision in its tracks. Then it moved off to the right, resuming its slow lope around rather than into the pond. No otter would ever be guilty of such a thing. Water is its natural element, summer or winter. This animal had to be a fisher! It was the first I had ever heard of occurring inside the outer beltway of the Boston metropolis, and the reservation in which I found it was just within the *inner* beltway. (This occurrence was so unaccountable that, when I later reported it, I was dismissed by a park ranger as a "confused amateur.")

Little did I realize it at the time that this female fisher was the precursor of an invasion of wilderness species into the park system called the "Emerald Necklace" that circles Boston. In the previous half-decade I had already found red fox, white-tailed deer and river otter, and now this fisher, but in the next few years they would be joined by discoveries of eastern coyote, mink, gray fox and bobcat!

A couple of years later I was commissioned to conduct a mammal tracking inventory in another of the metropolitan parks, the Middlesex Fells Reservation. This park, although larger than Breakheart, was even closer to the city center and so received much more human and dog visitation. Because of this I didn't expect to find much. Wildlife managers for the state derisively refer to such metropolitan reservations as "squirrel parks" and I had accepted that opinion, not believing that a woodland so ravaged by human recreation of all sorts, from off-leash dog-walking to ATVs and snowmobiles, would support much else. It was one thing to find wild animals in my beloved and much wilder Breakheart, several miles farther out, but in the Fells, with the

Boston skyline in full view? On the first day of the census I stepped out of my car and headed on snowshoes into the woods. Within two hundred yards I cut the trail of a fisher.

In the next three months, I encountered several such trails each day as I prowled the woods in this park. It became clear, as my experience increased, that much of what I had read about fishers was wrong. Aside from a few large groves of hemlock, most of the Fells is an oak woodland with a mix of white pine, red maple and a few relict hickories. Nothing about it suggested the dense overhead cover that I had read was a requirement. Furthermore the park was laced with fire roads that the fisher regularly crossed despite the insistence in print that fishers need large tracts of roadless wilderness. And, they were not preying on snowshoe hares. With unpredictable and often minimal snowcover that quickly melts back to bare ground, the Boston metropolitan parks support no snowshoe hares. Nor do these reservations harbor porcupines, another favored prey of fishers (Figure 6.54).

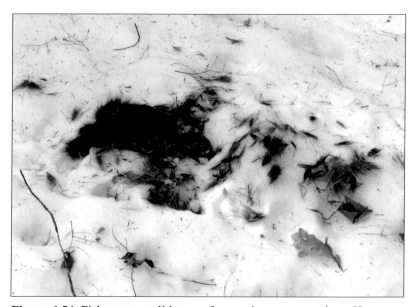

Figure 6.54. Fishers are well known for preying on porcupines. However, they don't eliminate them, as some foresters who reintroduced them to New England had hoped. Instead they "farm" them, taking off only the weaker, less-dominant animals.

What, then, explained the presence of these "wilderness animals" in this seemingly unsuitable habitat? The answer, it soon became clear, was the very prey species used by game managers to deride the metro parks: squirrels.

In my boyhood, growing up in woods only five miles from both the Fells and Breakheart, these woodlands were dominated by rather undersized white oaks and hickories, a shortness of stature I attributed to the poor, rocky soil. With the myopia of youth, I failed to realize that this was regrowth from the disastrous Hurricane of '38, a few years before my birth, which blew down much of the Eastern forest. And with elongation of time, also a characteristic of youth, I failed to realize that this familiar, seemingly static forest around my home was in transition toward something else. One of the few benefits of old age besides a dubious wisdom is that one gets to see almost geologically slow processes like forest succession within one's lifetime and come to appreciate first hand that in Nature everything is changing all the time. Take a walk in that same forest today and you will see that, from the short, warm, sunny white oak/hickory woodland of my boyhood, it has evolved into a tall, shady, moist red oak forest, dominated by very old trees with large boles and broad crowns. With tight state budgets these woodlands have been left generally unkempt, free from the manicuring that renders inner-city parks so sterile. This "benign neglect," to borrow Daniel Patrick Moynihan's felicitous phrase, has resulted in the metro parks taking on many of the characteristics of an old growth forest, a closer resemblance to the pre-colonial forest than at any place or time in the state since colonial deforestation.

These older growth red oak forests provide two things pertinent to fishers. First, they produce large mast crops of acorns. And acorns mean squirrels — literally. The word for squirrel in Norwegian, a modern language descended from the Old Norse that is one of the main contributors to the English we speak today, is *ekorn*, so closely are the seed and the mammal related in the human mind. It seems these squirrels of the metropolitan parks are a naïve lot. As prey animals largely without predators they had got used to the easy life. Only winged predators could take them in numbers, with terrestrial hunters like red foxes stuck on the ground, unable to follow up into the canopy.

And the other attraction for fishers is the increasing age of the forest, not, as was supposed, whether it was hardwood or conifer.

The natural cavities that form in old trees from limb loss provide the birthing dens that the females need as well as places in which to trap the abundance of squirrels.

One day I was tracking another small fisher in Breakheart over a few inches of dry frozen granular. The trail led up a hillside near a hemlock grove where I found an interesting and enlightening story in the snow. The animal's loping trail led to the base of a large oak, where the animal had paused, then loped off at speed to the base of another oak a few yards away, then back again to the first. Back and forth the trail went several more times, then finally left the area toward the hemlocks. I sat on the butt end of a log and pondered. After noticing a lot of gray squirrel bounding patterns in the snow, the answer occurred on me. The fisher had been running between the two trees, watching a squirrel up in the canopy. As the squirrel moved back and forth over the high limbs, the fisher had tracked its progress from the ground, judging whether it was worth climbing one or the other tree in pursuit. As long as the squirrel stayed out on the limbs, the climb might not be worth the energy needed to climb after it. After all, the squirrel is light enough to jump from twig to twig where the heavier fisher would be unable to follow. But if the squirrel went into a cavity in the tree, then the effort might indeed be worth it, as the fisher, with muscular head and killing jaws at the front end of its body, could enter the hole and trap the squirrel. As it was, the squirrel apparently stayed out, through chance or increasing experience with this new danger, and the fisher loped off in search of easier prey.

Fisher Size

In my mammal census of the Middlesex Fells I found something unusual in the fishers that I recorded. Almost all of them showed the print size usually associated with females. Fishers are normally dimorphic with the males as much as half again the size of the females. Traditionally any print that was 2¼ inches across the five toes was regarded as a female, and any 3¼ inches wide was a male. In the Fells, nearly all the trails I found were made by fishers with the smaller measurement. What was going on?

Several theories came to mind. The first was that, since among weasels one male services any number of females he finds within his own much larger range, there simply were fewer males. However, in other, wilder areas of New England I had not found this disparity as pronounced as it was in the Fells — nor, for that matter, in Breakheart. Another idea occurred to me. Male fishers are big and powerful enough to successfully attack raccoons. Indeed, the one trail of 3¼-inch tracks I found in the Fells was in a raccoon's trail. As raccoons are a principal vector for rabies, perhaps they had communicated that disease to male fishers, killing off a disproportionate number of them. However, since raccoon numbers seemed unaffected by the periodic rabies outbreaks, was it reasonable to think that the effect of the disease would be more pronounced on the predator of those raccoons? Perhaps, especially if the large number of raccoons, despite rabies, could be attributed partly to the elimination of competing scavengers by the large number of off-leash dogs allowed by their owners to roam the park. Raccoons are unaffected by this practice because during the day they can climb up to a den high in a tree, only descending at night when dogs aren't around.

A third theory evolved in my mind. In the Fells, the main prey of fishers was found during the survey to be squirrels, naïve squirrels easily caught by, among other ways, trapping them in tree cavities. It is often thought that the disparity of size between male and female fishers is a function of resource partition, the males sticking to larger more terrestrial prey so that the females will have smaller prey all to themselves. But in the metropolitan parks, squirrels, at least at the time, were nearly the only readily available prey. Cottontails were in a periodic crash, and there were no hares upon which to shift their attention. Was it possible that the squirrels were so abundant that competition with the female was mooted? Were the males selecting for smaller size so that they, too, could climb efficiently, enter squirrel-size holes and trap this prey animal in its refuges?

Preferred Gaits and Hunting Tactics

Fishers are not cats and don't fish. For that matter they don't even like water. They are large weasels, their name deriving from *fitch* or *fiche,* an alternate name for the European polecat, also not a cat, which apparently resembled the fisher to a colonist's eye. As described above with regard to mustelid bounding (pages 141-147), most members of the weasel family, including fishers, rely on just a few gaits to get around. As with any other mammal they have the ability to walk, usually from either fatigue or the need for stealth. Mostly, however, they choose either to lope or bound. The 3X or 4X lope is what they employ on firm, even surfaces such as crusted snow (Figure 6.55). With its relatively low arc of stride and delivery of force exclusively by the hind legs, this gait is a good compromise between speed and efficiency for a predator that hunts on its feet, putting in a lot of miles around its hunting range. In snow where this short-legged animal sinks down more than 3-4 inches, it must resort to bounding, which with its higher arc of stride allows the animal to overcome the resistance problem by rising at a steep angle clear of the obstructing snow (Figure 6.56).

Unlike otters, which slide at any opportunity both for business and pleasure, fishers hate to slide. I recall tracking a very large male in the Douglas Brook drainage at Passaconaway one winter. I spotted this animal crossing the Bear Notch Road early in the morning, and with its fur erected for warmth, it looked more like a wolverine than my mental picture of the smaller fisher. Its trail led down into the bed of the brook, where it had loped along in a remarkably uniform gait, and then crossed the stream on a fallen log to avoid getting wet. After a half-hour or so of tracking, during which I followed it to a lay site under a log where it had fed on a cached hare, I found where the animal arrived at the brink of a steep slope down to the bed of a feeder brook. The slope was about 50 feet long with an open runout. This is a circumstance in which an otter would delight, probably sliding down and coming back up like a kid with a sled to enjoy the ride over again. My big male fisher, however, would have none of it. To descend the embankment the animal repeatedly plunged its front feet forcefully into the snow in front of it, braking its progress all the way to the bottom. Fishers, it seems, are very serious adult animals, not happy-go-lucky thrill seekers like otters.

The route of this fisher was instructive. I tracked it for what was left of the day and saw that its trail curved slightly on average, describing the beginning of a large circle that would take the animal perhaps several days to complete. Its progress was dogged and its gait homogeneous. I had read somewhere that fishers change their gaits frequently. But, here and elsewhere I have seen that this is not often true. My big male on this afternoon employed the 4X lope almost exclusively. It performed a sort of bound down the steep slope,

Figure 6.55. Because it is not sinking deeply into the snow, this fisher has opted for an efficient 3X rotary lope. From the bottom of the frame the order of tracks is right-front, then left front under left hind, and finally right hind.

Figure 6.56. When a fisher sinks in more than 3 inches or so, it will resort to a 2X bounding gait with repeating slants. Each two-track pattern is a direct registration of two hind into two front prints. With its high arc of stride this gait allows the fisher to vault over intervening snow rather than plowing through it.

as I just described, and at one point walked a few yards when its trail crossed that of another animal that had aroused its curiosity. But otherwise it was as if the fisher had awakened from sleep, tested the snow surface to decide on the appropriate gait for the day and then remained in that gait as long as it could.

Circuits

The arc of this fisher's route got me to thinking about the instinctive wisdom of having a hunting circuit. One might think frequent visits to the same terrain would soon strip the route of potential prey. But I have seen the same circling so often with fishers and other predators that apparently this is not the case. Some advantage must dictate a circuit rather than a linear course. Certainly, conflict with adjacent males might very well follow if blocks of territory were not parceled out to various individuals. Furthermore, during mating season a circular route would bring the male into contact with more than one female, the territories of which are smaller than those of the male. Male fishers are promiscuous (excuse the anthropomorphism) with males impregnating as many females as they can find, departing each in turn to take no part in the raising of the resulting kits.

However, another reason occurred to me, perhaps equally pressing. A hunting circuit over which the predator passes every few days allows the animal to learn all the hiding places of its target prey. And the fact that the animal comes by so rarely allows those prey animals, busy with avoiding starvation themselves, to forget after a few days about the potential for a violent end in the jaws of the fisher. In the event of lack of hunting success, a circuit also allows the fisher to revisit old standing-deadwood snags (Figure 6.57) as well as hidden caches such as the one I found under a log at Douglas Brook. This may be the partially eaten carcass of a hare buried in the snow or concealed in a squirrel or raptor nest high in a tree, above the scent plane of scavenging foxes or raccoons.

This sort of caching is a common practice of fishers. I remember once tracking a fisher in Breakheart that was "root-humping" through the woods in a 4X lope. This is a tactic I have seen foxes, coyotes, bobcats and fishers use, where the animal heads directly toward the base of a large tree and then passes over its root hump before heading directly for another. Done often enough, the predator will eventually discover a squirrel foraging on the other side. And so

Figure 6.57. Standing deadwood benefits not only bears but also fishers. Both species tear at rotting trees in search of overwintering beetles and larvae. Here a dirty fisher trail leads away from a hulk in a 3X transverse lope.

this fisher did; at the far side of a large red maple were the foraging prints of a squirrel, looking more or less as shown in Diagram 6.9.

Diagram 6.9

A blush of blood stained the snow, and the fisher's trail headed straight off toward an old burn that had grown back to poplar. Over logs and rocks it headed on a beeline, its loping pattern showing occasional drag marks where some part of the hapless squirrel touched the snow. A couple of hundred yards on, the animal's trail emerged from the brush, headed to the base of a white pine and then ended. Looking up, I could see an old raptor or crow nest next to the trunk on one of the branch whorls. There the fisher could eat in peace out of sight or scent of any hijacking scavenger. Circling the tree I found the fisher's trail departing from the base, now unencumbered by its load. If the next circuit of its hunting range were unsuccessful, it could return to the cache as it passed by in a day or two and finish the meal.

Hijacking of hard-won prey is common among predators. I have followed fishers to fox caches and foxes to fisher caches. Once, also in Breakheart, I was tracking a fisher for a video I was producing when I came to a place where the animal had departed from its straightforward route and began circling, investigating everything its senses led it to. After a couple of circuits, the trail came to a dig in the snow. Feathers of a waterfowl were scattered about; it seemed that the fisher had found and exploited the cache of a red fox. Foxes have small stomachs and so must store any uneaten remains of larger prey they manage to kill. As the nearest water was a quarter of a mile away, the fox that left this duck buried in the snow apparently felt obliged to carry it that far to conceal it. Unlike gray foxes and fishers, red foxes can't climb to hide their prey in a handy nest. Instead, they have to cache it near the ground where it is susceptible to detection by other terrestrial scavengers.

Menage a Trois?

One mid-winter day I was prowling around the Conway River bottomland through some young spruce/fir regeneration when I found an interesting trail in the inch or so of soft snow over heavy crust. Much of it looked like this (Diagram 6.10).

Diagram 6.10

Nearby, the 4X loping trail of a fisher, a female by her track width, meandered back and forth through the trail illustrated above. A look at the individual tracks in this bizarre trail showed them to be those of fisher, as well, but, at about 3¼ inches across, appropriate to the male of the same species. I puzzled as to what to make of the pattern?

I followed the two trails to a point just past the root hump of a large tree where the riddle was solved. I was following the trails of three fishers, a female and two males (Figure 6.58)! The first of the two males for some reason had taken a sudden right turn around the trunk while the second animal, fooled for a moment, had continued on for a pace or two before recovering and turning to the right as well, back onto the trail of the other. Apparently, the first male was trespassing on the territory of the second, who was determined to pursue the interloper either to the border of his property or, catching up with him, to deal with him decisively.

The trail of the female seemed fresher than that of the other two, showing that she had come upon the scene after they had passed. Furthermore, while the trails of the two males were oblivious to hers, her trail showed an indirect awareness of theirs — more evidence that she had come afterwards. Apparently, it was important for her to give the impression that she was there solely to hunt the snowshoe hares whose tracks were abundant in the vicinity and couldn't care less about the males. However, her trail disclosed the subterfuge. Its course revealed a keen awareness of their passage: her general route, regardless of the seemingly unconcerned meanders, loosely but unmistakably followed theirs. As late-winter mating season was approaching, she did not wish to seem too interested, but apparently also wished to be on hand and within the awareness of the victor.

Figure 6.58. A male fisher following another male with a female's trail nearby. These trails are mixed with red fox on the left and snowshoe hare on the right.

Species Notes: Eastern Coyote

Coyotes are jacks-of-all-trades, predators, scavengers or vegetarians as the situation requires. This adaptability has helped to make them a very successful species that can attack the center of the food web or nibble around its edges as needed. Furthermore, two or three hundred years of ingenious persecution by humans has only resulted in the evolution of a super-coyote: smart, athletic, resourceful and deceptive. The trickster of Indian legend has grown to top-animal in several branches of the web.

The eastern coyote is the western coyote, only more so. In attempting to fill the vacant wolf niche as a predator of big game, the eastern version of this animal is bigger, with a thicker muzzle and correspondingly stronger jaws. It possesses, as it turns out, an influx of wolf genes apparently acquired as it filtered through Canada on its way to the eastern half of the United States. These genes are prominent in the eastern coyote's anatomy as a result of convergent evolution; because it is trying to fill a wolf niche, it has come to look more and more like the animal it replaces.

But a great wolf it is not. Although it runs in packs when it is hunting or scavenging large concentrated masses of fat and protein, it has not learned to function at the direction of an alpha animal. It may use its numbers to cut off retreat down a shoreline when trying to drive deer out onto an ice-covered lake, and it may hunt together in small groups. Many years ago I tracked such a group of three animals, presumably two adults and an adolescent helper, as they hunted in the woods. Two animals walked conspicuously on a ridge top while the third moved along a defile below waiting to snatch prey fleeing from the other two. But beyond this limited cooperation, pack activity seems to be primitive. I suspect that coyotes run in packs less as a hunting tactic than as a convenience in breaking out a trail in deep snow and as a way of avoiding conflict among themselves. If two coyotes that have never met arrive at a dead deer, trouble will probably follow resulting in injury to one or both. Any injury in the winter, because it requires energy to repair and by the same stroke inhibits the acquisition of that energy, is tantamount to a death sentence. But if the two coyotes are related members of a "family-pack," they will already have sorted out their dominance relationship. The senior will eat and the junior will wait. This and not wolf-like hunting tactics, I suspect, is one of the reasons coyotes consort so often in the winter.

Preferred Gaits

Coyotes are capable of nearly every gait in the book. They often direct-register walk in soft snow, gallop in order to attack or escape, lope along when driving a deer and trot when traveling on a smooth, even surface. Rather than the bouncy aligned trot of the bobcat, however, coyotes normally use the flatter displaced trot (figures 6.59 and 6.60) described above (pages 113-114). Bobcats hunt in dense cover where the narrower frontage of an aligned trot allows them to insinuate their way through thickets and vault over obstructions, whereas the coyote typically hunts more open terrain, often using the human infrastructure to get around. With less need to restrict its frontage or bounce over hidden obstacles, the coyote employs the greater efficiency of displacement. Diagram 6.11 shows the usual pattern that results.

Right front / Right hind   Left front / Left hind

Diagram 6.11

Here, the speed increases from left to right with all the front prints retarded and all the hind prints advanced. Furthermore all the front prints are lined up in a nearly straight line on one side of the trail and all the hind on the other. To achieve this, the animal orients its spine diagonally to the direction of travel so that each advancing hind foot passes by rather than over the print of the front foot on the same side of the animal. In an aligned gait the hind foot must pass directly over the vacated front print, hence the bouncy appearance of this gait with its waste of energy in the vertical direction. In the displaced trot, or "side trot" as Elbroch and others call it, however, the animal's body barely rises at all through each cycle, minimizing the vertical vector and so achieving greater efficiency. In the winter when every animal including the coyote is starving, this small efficiency repeated thousands of times in an evening and morning's travel, may mean the difference between life and death.

Figure 6.59. Prints of an eastern coyote in a slow, displaced trot pattern, front print on lower right. Note the light nail marks and tight pad grouping typical of this species.

Figure 6.60. A coyote's displaced trotting trail on thin snow over the ice of a beaver pond.

This pattern of prints arranged with a repeating slant from group to group may superficially resemble the slant-2 bounding gait of a mustelid. However, the weasel's pattern is composed of direct registrations of hind-over-front while the coyote or red fox pattern is composed of single registrations of one front and one hind. Even in trails where print features are vague, one can easily see the size difference between the larger front print of the canid and the smaller hind. In a mustelid trail, the two direct-registering tracks will be the same size.

A variation of this displaced trotting gait which may be equally efficient, I call the "straddle trot" (Diagram 6.12; Figure 6.61). Here is what its trail looks like.

Diagram 6.12

In this trot the same efficiency is achieved without angling the spine. Once again the hind feet pass by, rather than over, the front prints, but rather than lining up all on one side, they alternate sides. To use this gait the coyote places each of its front feet in a nearly straight line and straddles this line with the hind feet so that the latter appear in a broad zigzag. As you can see in the above illustration, the front straddle is very narrow, scarcely wider than the individual prints, while the hind is considerably wider. In a vague trail you might think that the alternating slants suggest the overstep walk of a raccoon, but a closer look at how the prints line up in the canid trail will show this contrast between front and hind straddles. In a raccoon's trail (Diagram 6.13), the front and hind straddles are the same width.

Diagram 6.13

Generally, the many coyotes that I have tracked over the years almost always prefer the displaced trot to the straddle trot. I have a sense that the aligned straddle version is more often used by young or otherwise small coyotes. I cannot say why this seems to be the case; it is yet another imponderable to ponder.

Figure 6.61. A coyote trail using a straddle trot. Note that the retarded front prints are all in a nearly straight line while the advanced hind prints have a wider straddle. In similar raccoon patterns both front and hind have equal straddle.

One cannot assume in following the perfect direct registrations in snow of a walking coyote that this is the trail of a single animal. Those single holes in the snow may have been made by two feet but may just as likely have been made by four, or six or even eight or more feet. Coyotes habitually follow each other around in the snow. The ease of this arrangement when traversing deep, soft snow is one reason for a social organization into packs. On the other hand two animals following each other seems to occur more during and after midwinter mating. I have wanted to assume that such a trail represents the male chivalrously breaking trail for its mate, but I have no evidence to support this. If a single trail of multiple animals is followed long enough, one or another of the followers will give away the trick by departing for a few steps to lift a leg or investigate something that has attracted its attention.

Especially in late winter, coyote tracks will be found among those of groups of deer. Since deer in deep snow often travel in numbers from bedding to feeding sites, coyotes can use these beaten down trails to get around themselves to rabbit coverts and other places with potential for capturing small prey. The coyotes get a triple reward for such behavior. First they save energy, secondly they gain access to small food, and thirdly, they get to keep an eye on the deer themselves, looking for signs of weakening as the condition of the herd declines toward spring.

Scat Placement

Like other predators that roam over large areas, coyotes use both scat and urine to communicate territoriality or breeding condition — as well, possibly, as many other things. Wild mammals generally have a very well developed sense of smell, so acute that humans have difficulty imagining what such a mammal can discern from a scent that often we cannot even detect, sight-reliant creatures that we are. Although wild predators have scent glands all over their body, on their face, tail and even between their toes, humans usually can only detect by odor either scat or urine marks.

Coyote carnivore droppings can be copious (Figure 6.62). They usually average about ¾ of an inch to 1 inch maximum diameter and are often in two discreet items with one blunt end and a tail of excreted fur at the other. The largest one I've ever found followed a saturation hunt to reduce the Quabbin deer herd. Coyotes feasted on

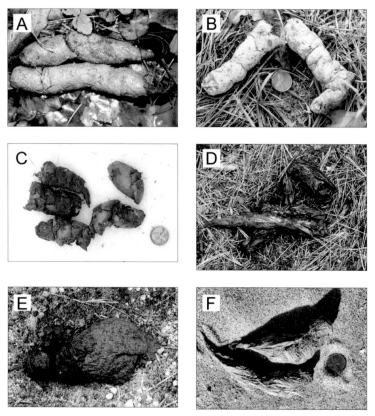

Figure 6.62. Samples of eastern coyote scat. **(A)** Fresh scat composed of fur and ground bone. (B) Coyote carnivorous scat often turns white with age. (C) Coyotes often visit orchards after fallen winter apples. (D) Scat composed of green grass to stimulate bowel action. (E) Diarrheic scat composed at least partly of berries. (F) Little was left of nutritive value in a deer carcass, resulting in this dry winter fur scat.

piles of guts left in the woods, the end result of which in one instance was a scat on a cut stump that totaled 11 inches long! Coyotes have powerful carnassial teeth with which they can chew bone. Therefore, large bone chips in an appropriately sized scat in the Northeast can safely be attributed to this animal. However, when feeding on the organ meat of a fresh kill, the resulting scat may have no chips and be the consistency, if not the color, of a dollop of oatmeal. Coyote scat may contain fur, grass, apple skins, grape seeds and skins and other berries. If they are living close to humans, their scat may contain garbage. I was puzzled one winter when I found a red scat in a sanctuary in central Massachusetts, until I found out that an abutter had been feeding coyotes with pizza left-overs. In another case in Concord, Massachusetts, I discovered a purple scat. When I fractionated it out of curiosity, I found a piece of a cordovan wing-tip golf shoe, the polish from which had dyed the scat. I tracked another coyote to a lay site in the same upscale town and found an empty container of Hubbard Farms cheese spread. Presumably Velveeta would be too *de classe* for Concord's coyotes!

Many years ago I noticed I was often encountering some odd coyote scat in very early spring composed entirely of yellowish fur. At first I though the animals had been feeding on old, dry winter caches from which nearly all the meat had been removed. But the color perplexed me: what prey animal had yellowish fur? Then a friend related a story to me about hiking in the fog on Mount Wachusett in this month when he spotted a coyote walking ahead of him. It looked like a ragbag, he said, with clumps of molting fur hanging from its pelt. I realized then that the scat I was finding was the coyotes' own molting fur that they were biting off and swallowing.

Coyotes are well known for marking the intersections of human hiking trails with scat. They also may leave it on a raised portion of such a trail or some elevated object along it, such as a cut stump. Like any dog, a coyote will usually scratch beside a scat or urine mark, apparently to spread the scent. This scratching can usually be told from the prepared scrape of a bobcat by being created after the deposit rather than before it. A coyote scat will never be found in the scrape as a bobcat's will. And one does not often get to say "never" in the tracking game.

Coyotes also use urine to communicate. The odor of most coyote urine is mild musk mixed with burnt hair, at least to my nose. In quality, it is not unlike the musky urine of red fox, but milder. My quick and dirty rule for distinguishing the two is: if you can smell it standing up, it's a red fox; if you have to crouch down to smell it, it's a coyote.

Oddly I have yet to find coyotes marking their frequent "convocations" with either scat or urine. These are locations to which coyotes call their related pack members every now and then for a family reunion of sorts. It always involves an excited turmoil of tracks as the pack renews its bonds (Figure 6.63).

Figure 6.63. A coyote convocation. Beaver ponds are favorite gathering sites in winter, especially on moonlit nights.

Incident on a Beach

Down on the Moore Island channel one day in early fall I discovered on the narrow beach the remains of a Canada goose. For the most part, all that was left was the ribs sticking up like claws from the supine backbone. There was a lot of commotion in the sand, not only around the carcass but also up and down the beach on either side. I approached carefully, hoping not to erase anything in the track record that might help me reconstruct the scene. On the beach were the trails of both bobcat and at least one coyote. There were also drag marks in the sand on either side. Gradually the incident became clear. A coyote had lain in the shrubs just back of the beach until one or more geese had floated near. Galloping patterns in the sand showed where the predator had suddenly charged across the narrow beach and plunged into the water, seizing the goose by the neck and dragging it back up on the shore. After killing and feeding on it, the coyote had either left it on the beach and retired or it was joined by at least one more coyote. Here ownership seemed to have become an issue as the carcass was dragged back and forth on the beach to a distance of about 50 feet either way. Quickly the carcass was reduced and, then, apparently abandoned, at which point a line of tracks in the sand showed that a bobcat had walked down the beach and approached the remains. The bobcat also seemed to have fed on what was left of the goose as the ribs sticking up from the backbone had been licked clean rather than cracked off as a canid would do. As was pointed out elsewhere, cats lack the teeth to crunch large bones. Instead, they use their raspy tongue to lick the meat off, leaving them cleaned just like the rib bones of this goose.

After investigating the scene, I worked my way down the beach a little farther and found where a coyote's trotting trail had come in my direction, stopped and turned back down the strip with the coyote running in a gallop with as long a pattern length as I had ever seen (page 134). While the gallop patterns on the beach that have been left by the attack were perpendicular to the shoreline, these paralleled it. Since the damp grains of sand in the ridge around each of the tracks had yet to dry and crumble, I took this to mean that one of the coyotes had been returning to the scene when it had spotted me and fled.

Since that incident I have found several more cases where coyotes have lain in wait for a Canada goose to drift near before plunging into the water for a kill. A fisherman with whom I chatted one day along Great Brook in Carlisle, Massachusetts, said that, while fishing one morning, he had witnessed just such an attack. And a few years ago I found where a coyote had used the same tactic on a common merganser on the margin of Spot Pond in Stoneham, Massachusetts. In my boyhood, Canada geese were strictly migrants locally, the calls of a flock passing high overhead a thrilling prelude to the coming winter. Since then, however, some of the population apparently has decided that feeding on grass on suburban golf courses is an easier life than flying all the way to Hudson Bay. Since their droppings on the greens of country clubs have drawn the ire of golfers everywhere, the duffers may take some comfort in this newly developed skill of coyotes.

This incident at Moore Island provided an immediate and intimate look at the lives of the two species of mammals involved. It shows the adaptability of the coyote to a new food source and the competition for food among members of the same species, and it demonstrated a reversal of roles between the normal predator and the normal scavenger, the bobcat and the coyote. Finally it shows the dependence of scavenging on effective predation. None of this incident was witnessed directly, but could be reconstructed by inspection of the evidence, the track record, by a prepared mind.

The Trickster

I have a dozen coyote stories, but this one shows the keenness of the animal. It is said that you can fool a coyote once, but never twice the same way. I learned this first hand one spring many years ago at Bassett Pond in north Quabbin. I had just bought my first videocamera and a new blind, as well, with which I hoped to take wildlife footage for use in interpretive programs. I had placed the blind the day before on the shore of Bassett Pond on a cove across from the dirt patrol-road that circles the reservoir. Right away, I started to record common mergansers that used the pond for early spring courting. Then a great blue heron glided in, and a pair of wood ducks cruised the quiet water of the cove. A little later two ladies walked down the road on their morning constitutional with a dog. When they were well out of sight, I caught movement to the left far down the patrol-road. After a few seconds a coyote appeared in the yellow morning light, trotting up the road. I started taping as it came over a small rise and then slowed to a walk while lowering its nose for ground scent. It's checking to make sure the two ladies and dog have already passed, I thought. Then it picked up its pace again until it was directly across the cove from my position. As it neared the right limit of my vision, it turned its head in my direction and spotted the blind. The sudden recognition of this new feature on its familiar landscape seemed to knock it a little off its stride as it disappeared behind some roadside brush.

That was it for the morning. Eastern coyotes were still fairly new in Massachusetts at that time and this was the first clear look I had ever gotten of one. In one stroke I now had not only seen one but videotaped it as well. Quite a morning! This wildlife photography stuff isn't as hard as they say, I thought.

After reviewing the videotape at home on a monitor, I was excited to see that I had actually recorded the animal, but I was unhappy with the panning, which was jerky and tilted. No problem, I thought. I'll just go back again and get some more footage. And so I did just that: morning after morning I arose in the dark, drove to the gate and walked the mile into the pond to sit in that blind in the cold. No coyote. Two ladies and a dog, an occasional waterfowl for diversion, but no coyote.

Finally, one chilly April morning after I had sat in the blind for two hours, I was shivering so badly that I could no longer hold my hand steady on the handle of the tripod or reliably finger the buttons on the camera. In fear of hypothermia I decided to leave the blind and go for a walk to warm up. Backing away from the blind into the woods behind, I circled out to an intersection on the patrol-road through which any animal using the road as the coyote had previously done would pass. I noticed that the sand and remaining snow of the crossroad were clean of tracks as I walked through. From there I bushwacked into the woods and up the side of Fairview Hill. After warming myself with the exertion for fifteen minutes or so, I turned and descended to the road intersection once again. Straight through the crossroads were the unmistakable tracks of an eastern coyote! I cursed my luck, myself and anyone else I could think of. If I had only stuck it out for a few more minutes, I thought, I would have had him. I decided to see if I could backtrack the animal in the spotty snow. The trail came from the direction from which I had approached the intersection after leaving my blind. Under the hemlocks I could pick out enough of its trail in some old snow to determine its line of approach and so managed to follow the animal to a position behind a small rise perhaps a hundred fifty feet to the rear of my blind. There I was astonished to find a sit mark with the outlines of both entire back feet in the snow and a number of holes for front feet as well. Perhaps the animal had seen me enter the blind two and a half hours earlier, and had sat behind this rise shifting its feet while watching my location with a patience bred of countless generations of selection, waiting for me to give up and leave. I recall feeling a little inadequate after this event. How could I reasonably expect to compete with this or any other wild animal, handicapped as I was by characteristics bred of the many protections of civilization?

At best, all I could take away from this was an awareness of those limitations in myself. The coyote had had its revenge. It is not wise to play tricks on the trickster of Indian legend.

Species Notes: Red Fox

A Myth with Amazing Legs

A story I have been told several times about the origins of North American red foxes has had amazing durability. As the tale goes,

Virginia gentlemen in colonial times found the native gray fox unsatisfying to hunt on horseback with hounds because it could easily throw off its pursuers by climbing trees, leaving no scent on the ground for the hounds to follow. To solve their sporting problem they supposedly sent back to England for some red foxes, which lack the physical equipment for climbing effectively. From these imported foxes came the population of this species that subsequently spread across the continent.

There are two problems with this theory, one anatomical and one historical.

First of all, our North American red foxes are superbly adapted to cold. In the winter most of what we see as a robust animal is actually fur: a dense coat of white underfur insulates the animal down to very cold temperatures. Think of its tail as a sleeping bag that the animal drags along after it. When the fox needs to rest, it finds a sunny, windless spot, lies down, draws its paws up close to its body and then covers both nose and feet with this enormous brush. In such a position it is said that a red fox is good down to the upper teens Fahrenheit without so much as a shiver. Indeed, even the pads of a red fox's feet are covered at all seasons with dense fur to insulate them from cold surfaces (Figure 6.64). When I have had the tale about red foxes originating in England repeated to me, it has always seemed unlikely that an animal so overbuilt for cold could have originated in that rainy, temperate land.

However, as suggestible as the next person, I at least allowed for the possibility that this story might have some merit until one day when I was poking around the stacks of the Concord Free Library. There I found a reprint of a book published in England in 1650 by Roger Williams, the founder of Providence, Rhode Island. One of the few white men to show an interest in Indian culture, he had set about recording and translating much of the language of the Narragansett Indians around his settlement. It occurred to me that this book might give me some insight into the native wildlife that existed at that time before the white colonists had had much impact on the land. In it I found words for foxes among the vocabulary. The Indians had names for three, in fact: the gray fox, which they called *pisquashish*; the red fox, *mishquashim*; and a black fox. (The black fox was actually a black morph of the red fox, a detail probably of no concern to the Indians with little need for taxonomic precision.) The Indians would not, of course, have had a name for an animal that did not live among them. Furthermore, the name did not seem to be derivative from anything they may have picked up from the white Europeans, who would have had little impact on Rhode Island anyway before Williams arrived. Therefore, I regard the testimony in his book as convincing evidence that red foxes were not originally imported from temperate England. Given their extreme adaptation to cold, it would seem that instead the species arrived from the North, perhaps over the Bering Land Bridge from Siberia.

Preferred Gaits

Red foxes have a track morphology similar to other canids such as coyotes. However, the surface and the gait can alter their appearance from the expected norm. Figure 6.64 shows the classic front track on a firm surface. Figure 6.65 presents a widened direct registration with spread toes and extended nails, while Figure 6.66 shows prominent nail marks where the animal used its claws to support sudden acceleration.

Figure 6.64. Red fox front print. Note the fur striations, longitudinal symmetry, light nail marks and chevron-shaped bar across the secondary pad.

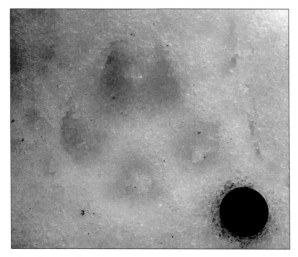

Figure 6.65. The surface can greatly affect the appearance of track, contradicting the usual traits used for identification. Soft spring snow has caused this red fox to spread its toes for support, making the track look more like that of a small dog.

Figure 6.66. Gait can also greatly affect a track's appearance. Here a galloping fox has dug in its front nails for acceleration. Nevertheless, the transverse bar on the secondary pad is still prominent.

Red foxes use all the gaits that coyotes do, and the patterns they leave behind are similar, although proportionally smaller. In snow, where they are sinking in more than an inch or so, they walk, leaving behind an elegantly efficient trail of perfect direct registrations (Figure 6.67). Close examination of a coyote's trail under similar circumstances will show an occasional lateral indirect registration, but that of a red fox almost never. The resulting trail is a line of dots in the snow one almost directly ahead of the next in a nearly straight line with no foot drag to mar its elegance (Figure 6.68). The step length tends to average around 15 inches with a distribution of only an inch one way or the other. It should be noted that in snow where the animal is sinking more deeply, this fox may also launch itself into an aligned trot with the prints still direct registering. The pattern looks the same since the increased arc of stride is the invisible third vertical dimension of the gait and so is difficult to detect in the trail. If a fox is direct registering in the usual walk pattern but each foot is sinking deeper than 4 inches, it may very well be using this bouncy gait as a way of clearing the resistance of the snow between its footprints.

On a firm, even surface, red foxes usually prefer to travel in a displaced trot (Figure 6.69) just as do coyotes. Where a coyote in this gait will present a 24-inch+ step length, however, this trotting trail of a red fox will begin at about 17 inches and increase with speed. I have also found red foxes using a transverse lope (Figure 6.70) traveling down a long slope. As with many other predators, the use of galloping gaits is restricted to attack or escape, while bounding seems to be used mostly out of impatience with deep, unconsolidated snow.

Marking with Scat and Urine

As with other predators that are distributed thinly over large land-scapes, red foxes use scat for communication, placing them in prominent locations on human trails, lightly used dirt roads, trailside stumps, downed logs, exposed ledge in a dirt road and so forth (Figure 6.71). Foxes have very small stomachs compared with either coyote or bobcat and so the scat tends to be deposited in smaller amounts than by either of these two animals with which it might be confused. When the scat is composed of the aftermath of predation, it often has characteristic constrictions along its length (Figure 6.71A). As for other food, foxes also eat grapes, as Grimm reminds us, as well as other berries (Figure 6.71C), fruit and grass. The appearance of the resulting

Figure 6.67. Red foxes often travel the margins of frozen ponds where the snow is wind-packed and where the brushy edges may conceal potential prey. Note the perfect foot placements and narrow straddle in this walking gait.

Figure 6.68. Which one is the dog and which the fox? The perfect stepping of a fox shows an animal on a tight energy budget, unlike rover out for a careless romp.

Figure 6.69. Displaced trotting is used by a traveling fox on any smooth, firm surface. This flat gait is faster than a walk and more energy-efficient than a lope or gallop.

Figure 6.70. The spacing of prints often makes it hard to discern the separations between loping patterns. Here it is nearly at the center of the frame. From upper right the order of prints in this transverse lope is left front, right front/left hind, right hind. Note the slightly larger front prints.

scat will vary but, once again, will be deposited in smaller amounts than even a small coyote. The maximum diameter of most red fox scat is about ½ inch compared to ¾ of an inch to 1 inch for the larger canid.

Urine is also used for communication. If you have had the experience of walking down a trail and smelling skunk for only a step or two after which the odor is gone, you have certainly just passed a red fox scent mark. Returning to the spot and working upwind, you may be able to find the source. Red foxes often anoint a class of objects: in winter, for instance, the red foxes in New England, at least, will go out of their way to pee on any conifer tip they encounter that is isolated and sticking out about a foot above the snow (Figure 6.72). At all seasons they may mark any oddity they find isolated along their trail, such as a prominent tuft of grass along a hiking trail, or perhaps some odd object they find along their way.

While coyote urine also has a musky scent, that of red foxes is much stronger. Once in a metropolitan park, I came across an old cotton work glove that had fallen out of one of the staff trucks and lay beside a lightly-used road often followed by foxes. The glove had been urine-marked as simply a noticeable anomaly in the landscape.

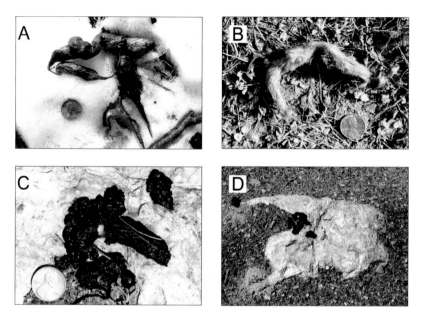

Figure 6.71. Red fox scat samples. (A) Classic fur scat with constrictions. (B) This scat is composed entirely of cottontail fur, probably the result of attempted feeding on a depleted winter cache. (C) A mid-summer scat composed mostly of berries. The tough cuticle of an unripe berry has caused it to survive the digestive process intact. (D) Typical placement is on a conspicuous feature, here an exposed ledge on a lightly used dirt road.

Since I was to conduct ranger training for the park on the following day, I collected the glove and sealed it inside two zip-locked bags, after which I left it in the park headquarters building overnight. The next morning, I arrived early at the headquarters to set up for the ranger session. I could hear the park manager puttering inside, making morning coffee for the coming group. When I opened the door, a gust of skunk odor that had managed to penetrate the plastic baggies and accumulated in the closed building overnight rolled out the door and nearly overwhelmed me. I looked inside to see the manager whistling unconcernedly as he worked at the coffee maker. He looked up at my contorted face as if to say, "What's the matter?" As it turned out, he had a nasal condition that largely eliminated his sense of smell. When the rangers arrived, however, they declined the indoor session and insisted on meeting outside instead.

Figure 6.72. Conifer tips projecting from snow are a favorite class of objects that red foxes anoint with urine scent marks. The track pattern here indicates a leg-lift on the pine seedling.

Normally, unlike skunk spray, the odor of fox urine has a very limited range, at least outdoors. However, during the mid-winter mating period this urine seems to become more concentrated. At this season on a damp, still day where red foxes are advertising heavily, the woods may be filled with this odor on the scale of a skunk disturbed by a dog.

Habitats

Red foxes have a reputation, more or less deserved, for being partial to pastoral landscapes. And indeed, they are usually the fox you will find hunting near farms, skirting pastures and prowling hayfields in search of voles. However, this reputation can be taken too far. Red foxes are very adaptable animals, as David MacDonald's *Running with the Fox* celebrates with the English version of this animal. Although gray foxes are better suited to deep woods, they only recently have managed to extend their range into the more boreal regions of the Northeast. There, red foxes still can be found deep in the woods. I have often come across their trails on old lumber roads in the forests of the White Mountains where their routes seem to connect abandoned log yards and lumber camps. These sites host a regrowth of dense underbrush, cover for prey such as voles and grouse that these foxes favor.

One memorable winter day I decided to hike up to Boott Spur, an exposed shoulder of Mount Washington, the highest and in terms of conditions the most forbidding peak in the Northeast. Down fairly low on the mountain I picked up the descending trail of a red fox in nearly a foot of unconsolidated snow. As I climbed higher into the boreal zone I realized that this fox was using the same hiking trail to descend the mountain that I was using to climb it. But from where? What possible interest could this animal have had in the alpine zone of a mountain famous for the worst weather in the country? Upward I climbed, kicking my snowshoe tips into the steepening snow of the narrow hiking trail, necessarily erasing the fox's neat walking patterns alternating with occasional lopes and bounds. After a couple of hours I emerged from the ground spruce into the alpine zone of low-growing plants clinging onto the mountain for life. As I hiked higher up the barren terraces toward the top of the spur, I could still find occasional prints in the soft snow that had collected behind the boulders that marked the trail. Another hour and I emerged onto the windswept summit of the spur where the fox's trail curved off toward the main summit of the mountain and disappeared onto glare ice. I looked across the col at the main summit of Mount Washington, gleaming in the winter sunlight, a magnificent view, or would be were it not for the collection of summit buildings: summer snack bar, small museum and all-season weather station that collectively stood out on its top like a carbuncle. I followed the disappearing curve of the fox's trail, coming at me straight from the summit. Suddenly, the fox's business so high up on a mountain made sense to me. Where there are humans there is human food, and where there is human food there are left-overs for mice, and mice are food for foxes. That was why this fox was headed down from the top of the mountain. It was trekking the three miles down from an exposed summit that it had climbed four-thousand-feet up to in the first place — after mice! Are there not enough mice, voles, rabbits and grouse in the valley that this animal had to climb all the way up here to kill a few? In the end, I decided that this representative of its species was so fit and so well adapted to cold that such an arduous and exposed trek was of no special concern. I might find a winter climb up to Boott Spur to be an athletic adventure, but to the fox it seemed to be a ho-hum matter of course. And indeed I have since seen a photo, taken though a window by the staff of the weather station, themselves hunkered down against the incredible winter conditions of that summit, of a red fox unconcernedly prowling around outside the buildings. Apparently my fox's trail did not mark a one-time visit.

A mind-bending question arises from this incident. How did the fox know that there was a man-made installation on this summit that might harbor food? After all, the top of Mount Washington is 3 miles from, and 4000 feet of elevation above, the nearest human settlement, with the last mile and a half across a barren, windswept shoulder of the mountain. Would this really be the sort of place a fox or any other wild animal might accidentally encounter in its travels? Apparently so, and that by itself reveals how poorly we can appreciate, limited by our own perceptions, a wild animal's different experience of the world around us.

Species Notes: Gray Fox

If one were to dye a red fox's russet fur and white-tipped tail a grizzled gray with a black stripe down the back of its brush ending in a black tip, it would be hard to tell from outward appearance, size and anatomy, from a slightly large gray fox. However, the two species are not even in the same genus, their anatomical similarities being a good example of convergent evolution. Despite their outward similarities, the different origins of the two species show up plainly in their sign. The tracks of a gray fox are more easily confused with those of a small bobcat or very large housecat than they are with those of a red fox (figures 6.73, 6.74, 6.75, 6.76). Gray foxes have a special talent that red foxes lack: they climb trees, and this ability is reflected in their feet. In order to facilitate climbing they have sharp, semi-retractable claws like a cat, and in order to ease jumping down from the same trees, they have deep, rubbery, oval pads, also like a cat.

The similar outward appearance of the two foxes is reflected in the narrowness of both species' walking straddle, for both have a very narrow chest and pelvis. However, the trail of a gray fox is not quite so straight nor are the direct registrations quite so precise. This latter detail shows a less than perfect adaptation to snow for the gray fox, whose core range and presumed origin is farther south than the red's.

320

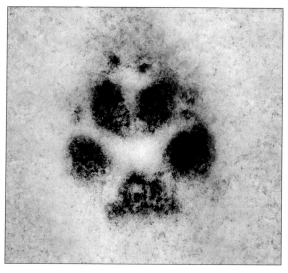

Figure 6.73. A very clear front print of a gray fox shows the cat-like oval toepads and roundness of the whole track. The 'winged ball' of the secondary pad is also visible.

Figure 6.74. Even though this is a direct registration, the very characteristic appearance of this track is dominated by the hind print.

Figure 6.75. A double registration shows the combination of round front with narrow hind that is a common identifying factor for gray fox.

The rapid deforestation for agriculture, begun in colonial times, changed the dense forest into a pastoral landscape. Almost the only stands of trees left were in woodlots preserved for building materials and firewood. Because in such a changed landscape, the gray fox's special talent, the ability to climb, was of little use, it was out-competed by red foxes for which a pastoral setting was just fine. Add to this the species' persecution as vermin — according to Thoreau the last gray fox in Concord, Massachusetts, was shot for bounty in 1810 — and the gray fox was soon extirpated from central New England northward. However, since the decline of agriculture and the regrowth of the northern forest, begun in the mid-1800s, this animal has slowly been moving northward, reclaiming some of its ancestral range. I first encountered its tracks in Breakheart Reservation north of Boston in

November 1995, a walking trail parallel to that of a housecat, and on Mink Pond in Estabrook Woods in Concord, west of Boston, that same winter, the animal having left a walking trail in spindrift on pond ice.

A hundred miles north in the Conway River bottomland of northern New Hampshire I had been used to tracking red foxes on both sides of the stream for quite a few years. Then, in the same winter of 1995, I began finding the trails of a gray fox on one side of the river along with those of the resident red foxes. Within two winters the red foxes disappeared from the east side of the stream, totally replaced by more and more frequent trails of gray fox. That same winter I discovered another trail of gray fox in a walking pattern across the parking lot for the Rocky Branch Trail south of Mount Washington. However

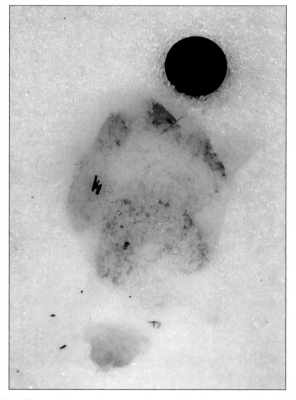

Figure 6.76. This understep by a gray fox makes the toes of the hind look like the secondary pad of a bobcat. The radical toe-down posture of the hind foot hides its secondary pad and completes the deception.

long it had taken this arboreal fox to return, it seemed they were now repopulating with a vengeance.

It appeared odd to me that I should find the trails of this species in south-central New England for the first time coincident with finding them 100 miles north in the White Mountains. Then I realized that on the southern exposure of a mountain tipped up to the low winter sun, snow tends to compact into firmer footing and then melt away faster than in the flatlands to the south. As gray foxes are not nearly as well adapted to winter as red foxes, they seem to have learned to use such terrain features to their advantage, facilitating their spread into colder climes. Around 2005, I discovered another gray fox, a road kill, in the boreal mountainous region of Grafton Notch in Maine, another 75 miles farther northward.

The behavior of a gray fox contrasts with that of the other wild canids of the Northeast in ways besides subtleties of gait. One winter I was tracking out in the Passaconaway intervale when I found side-by-side trails of an eastern coyote and a gray fox in a campground that was closed for the season. Both approached and then crossed the roadbed of the Kancamagus Highway, a grandiose name for a two-lane road through the mountains, and then crossed it. On the far side, each animal had a choice: either follow a dirt road access to a trailhead that led back into the forest or parallel the access road through a dense conifer forest. Not surprisingly the coyote, notorious for using human infrastructure, chose to trot down the road in the open. The gray fox, on the other hand, avoiding the easy going of the plowed road, instead forged its own trail though the unbroken snow of the adjacent woods.

Preferred Gaits

As I have suggested, gray foxes use the same walking gait as a red fox (Figure 6.77), but with a step length usually from 9-13 inches, compared to the 15 inches of the slightly larger red fox. However, like a bobcat, a gray fox seems to be able to stretch its step considerably. At some point in this stretch the gait probably becomes an aligned trot with direct and indirect registrations. The step length in the resulting trail can extend all the way up to around 20 inches, a variation I have never seen in red fox. Also indirect registrations are much more common in this fox's walking trail than in the precise trails of red foxes. I have found these as under-stepping double registrations as well as slightly displaced indirect with the following hind foot shifted slightly to one side and ahead of the front print.

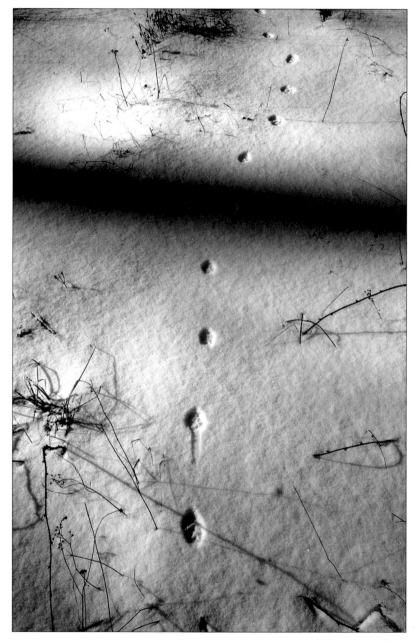

Figure 6.77. The walking trail of a gray fox is not quite as precise as that of a red fox. The straddle is a little wider and the step-length is usually shorter.

Trotting gaits may look the same but with a correspondingly longer step length. Once in Breakheart I found a trail of direct registrations with step lengths of 17", 18", 16", 14", 18" and 19" in ½ inch of soft snow over hardpack. Even though these were all direct registrations, the wide variation in step lengths suggests the presence of a third dimension to the gait. An animal's walking gait is "flat," with three feet in contact with the ground at all times and therefore necessarily very little arc of stride. As a result, that step length is governed mostly by the animal's anatomy, which is constant from step to step, providing for consistency from one stride cycle to the next. The arc of stride of an aligned trot, on the other hand, with only two feet in contact with the ground at any moment, can be much greater and can vary from stride to stride depending on how much vertical energy the animal introduces into each cycle. Consequently these strides, ungoverned by anatomy, can be much less consistent. This is what you see in the variation in the above series of step lengths. So even though the third vertical dimension in a stride is difficult to see in the two-dimensional track pattern, its evidence is sometimes detectable. More clues that the above sequence was an aligned trot rather than a DR walk appeared a little farther along on this animal's trail where it ventured out over rather delicate pond ice covered by the same half inch of snow. Here, wishing to eliminate the vertical vector of force with an equal and opposite reaction that might plunge the animal through the ice, the fox's step length in the same pattern declined to a very consistent 13 inches, the normal walking, anatomy-dictated step length of a male gray fox.

In looking through my field notebook, I note that I have rarely found much displacement in the trotting gaits of gray foxes. In coyotes and red foxes, the side trot is the preferred means of getting around on firm, even surfaces. The most that I recall was the trotting trail of a gray fox straight across a clearcut above Jackson, New Hampshire, on firm crust with a sun-softened half-inch surface in which to record tracks. Diagram 6.14 shows the pattern, with the size and shape differences between front and hind somewhat exaggerated.

Diagram 6.14

326

Once or twice I have found a straddle trot similar to that of coyotes but with proportionally smaller dimensions (Diagram 6.15). One of these that I find in my field notebook shows the usual straight line of front prints with a front straddle of only 3½ inches (slightly wider than usual for red fox) and a zigzag line of hind prints on either side of the line of front prints and advanced. These hind prints showed a straddle of 5½ inches. The step length varied from 16-18 inches.

Diagram 6.15

The shorter step length and the grouping of round front with narrow hind prints, which is so often diagnostic in identifying this species, distinguish it readily from the much larger tracks and longer steps of a coyote performing a similar straddle trot.

Marking

While the scat of both red and gray foxes can easily be confused, given that they are animals of similar size and diet, their urine scent marks are easily distinguished. While red fox has the musky smell of "sweet skunk," that of gray fox usually smells to my nose much like fisher urine. The English language is poorly provided with words for scent, but to my nose the normal urine mark of both gray foxes and fishers smells like chemically-treated paperboard of the sort that clipboards are made of. Once during mating season I found a conspicuous yellow mark that smelled like a sneaker badly in need of washing (Figure 6.78). The gray fox that I was tracking at the time was fairly large, leaving DR walking prints at the 13-inch end of the distribution. I took both the size, its conspicuousness and the odor to indicate that this was a male on the prowl for a vixen and producing more pungent urine in the effort. For the most part, however, gray fox marks are very lightly scented, pale yellow marks at the base of a tree or other vertical object. They are so inconspicuous that only a change of track pattern is likely to give their presence away.

Like other widely dispersed carnivores, gray foxes also mark with scat, a representative collection of which is presented in Figure 6.79. Like red foxes, the scat is deposited in small amounts, but I have never seen it with the constrictions so often characteristic of red fox.

Figure 6.78. During mid-winter, mating male gray foxes mark more conspicuously than at other seasons. The urine also has a stronger odor.

Behavior

Tracking gray fox is very interesting. The trails of this animal are quite expressive, allowing the tracker to interpret the fox's actions, crawl inside its skin, and see the world through its eyes more easily than is possible in the trails of other animals, like coyotes, whose metronomic trails reveal relatively less. In this respect and in several others, this fox is more like a cat than a dog. It shares the bobcat's habit of walking along fallen logs, stone walls or any other level raised linear object (Figure 6.80). Unlike either of the two other New England canids,

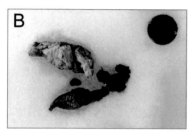

Figure 6.79. Gray fox scat samples from New Hampshire. (A) Fur scat. (B) Winter scat can have a dry, blond, clay-like appearance. (C) Berry scat.

it can climb trees quite easily, and in order to keep its nails sharp for climbing, they have been withdrawn from contact with the ground in many of the tracks the animal leaves behind. The pads on its feet are deep and easily deformable, adapted for jumping down from things. Normal dog feet show wedge-shaped medial and lateral pads in a print, but those of gray foxes are oval, again like those of a cat. Even the impression of a secondary pad is easily mistaken for a felid. Cats typically show tri-lobing on the posterior of this pad as a distinguishing characteristic. The shape of a gray fox's secondary pad, both front and hind, appears as the celebrated winged ball, a central round disc with a smaller lobe appended to either side. As a result the posterior of this fox's secondary pad shows as a very cat-like tri-lobing. For identification, then, the combination of round front print with a narrow hind compared to the long ovals of both feet of coyotes and red foxes and the round-round of cats distinguishes the gray fox from the others.

A Day Tracking Gray Foxes

The following is an entry in my journal about a day spent tracking a pair of gray foxes and a female bobcat.

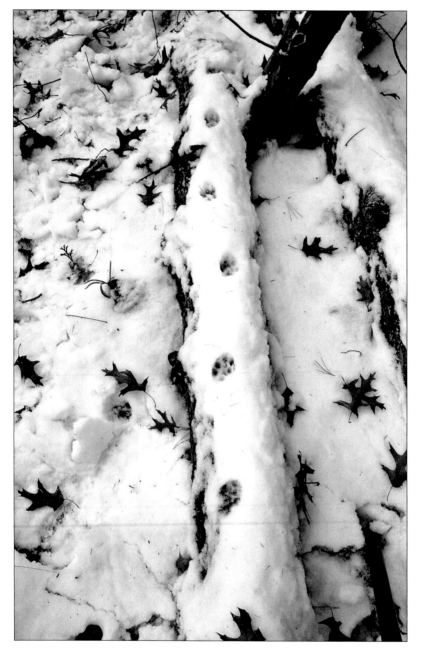

Figure 6.80. One of several cat-like traits of a gray fox is its habit of walking the length of raised linear features like this fallen log.

"December 1

"Weather at the Federated Forest : temp in upper 20s and windy from the NW. Snow depth: 0-3 inches. A warm day yesterday and fairly warm last night after snow three days ago: a good surface for track registration."

Kevin Harding and I were at Quabbin looking for the trail of the female bobcat that regularly crosses the Fever Brook Road just south of the first S-turn. Failing that, we scouted around east of the road until we came upon the trails of a pair of gray foxes near the jeep road over to the Monson Turnpike. Being only the first of December, with mating presumably a month away, I thought it unusual to find a pair although it is possible that they were siblings of that year still hunting together. And, indeed, their behavior was surprising. Right at the spur off the jeep road where we first found them was a great commotion of tracks — circling, bounding and leaping. Although there were some snowshoe hare tracks in the area, the activity did not seem to involve chases. Only two things provoke this kind of foolish behavior, so energy wasteful as the winter sets in: either this represented the high spirits of young animals not yet sobered by the privations of their first winter, or it was the antic behavior associated in most species with the sudden coursing of sexual hormones.

The commotion of tracks was so intense that we could make no sense of it and had to circle the area widely to pick up the exit trails. Once they were located, we tracked them southwestward for the rest of the short winter day. Unlike the female bobcat I have tracked in the same area, these animals stayed on the wooded slope above the wetlands that are favored by the cat and traveled in a fairly straight route toward the southwest. This contrasted as well with the bobcat's usual prowl from one laurel thicket to the next. We came upon no scat but did find several scent marks, which had the usual paperboard odor, but so faint that, with his chronically stuffed nose, Kevin had trouble detecting it at all. One of the scent marks was on the windthrown branch of an oak with leaves still attached, another on the moss of a cut stump.

Throughout the route we both remarked on the cat-like behavior of these foxes. Every downed, snow-covered tree they came to, one or the other would step up onto (Figure 6.81) or walk the length, and on a couple of occasions one would briefly leap up onto the vertical trunk of a living tree before jumping off. (I have watched fishers do

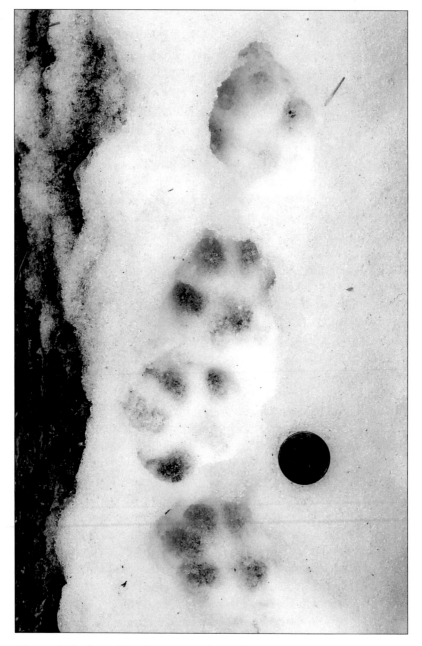

Figure 6.81. One of the foxes paused on a fallen log. Hind prints are at the top and bottom of the frame.

this apparently as a way of elevating themselves for a better view of the surroundings.) Outside of the commotion sites (there were two) the gait for both throughout was a direct-register walk. Occasionally they would perfect step in each other's tracks, but this was exceptional over the length of the route; generally one followed the other rather haphazardly with frequent diversions to one side or the other before rejoining. Every now and then they would root-hump, passing close to the trunk of large trees in the hope of surprising something foraging on the other side. I have seen many other predators do this, as well, including fishers, coyotes, bobcats and red foxes.

Eventually the wetland drainage to the east closes on the hillside and cuts across to the west, interrupting the ridge. From there it flows into a beaver pond that I have always called the "eastside pond" to distinguish it from another pond on the other side of Fever Brook Road. In order to continue their southwestward route, the foxes were forced to cross this drainage somewhere along the cut through the ridge. To do so they located a narrow log, only 4 inches wide, across the brook and gingerly balanced their way across to the other side. As we lacked their agility and weighed a lot more, and as I discovered I had left my pack tripod somewhere to our rear, we abandoned the tracking and retraced our steps. While picking our way back from the wetland, we were happy to find the trail of the small, female bobcat that we had begun the day looking for. Apparently she, like the foxes, had been searching for a crossing point on the brook.

Once we retrieved my delinquent tripod, we continued back to the car, drove down the road paralleling the ridge over which we had tracked the foxes and parked near the eastside pond. We got out of the car, walked around to the hatch, suited up in our tracking gear and went back into the woods to relocate the bobcat trail. Once we did, we followed her to a beaver dam on the inlet brook where she, and we, crossed. On the other side, however, we soon lost the trail, found it again and then lost it again as this little female searched out every snowless patch along the east side of the pond that had been melted off by the late afternoon sun. Hanging onto her trail by our teeth, we just managed to follow her along the east shore to some diffuse blood in a snow patch where she had scraped at the earth beneath. Poking around, we discovered the remains of a small mammal. Since all that was left were some fur and an entrail, we were unable to identify it.

At that point we lost her again. After spending a long time attempting to read her mind to determine her direction and still not being able to locate her trail, we decided that the cold of the previous night must have frozen the margin of open water at the pond edge enough to support her weight and she had used this snow-free zone to continue southward. Jumping ahead, we searched the ice at another small inlet brook at the south end of the pond where a skim of snow remained. There we located not only the female bobcat's trail but that of a raccoon, as well, and the trail of another, larger bobcat coming down the brook parallel to her ascending trail. She had gone up the brook over the ice to where the flow was sufficient to open it, whereupon she hesitated, backed up a few paces and then turned right into the tussocks to the side where her trail disappeared once again. The male's trail had come boldly right down the drainage onto the pond.

As the sun was now down over the pines west of the impoundment, we decided to call it a day and left the small drainage to circle around the south and west shore to the car. Once up in the woods we came upon the trail of our two missing foxes! And mixed in with their trail was that of the missing bobcat. Apparently she had crossed the tussocks bordering the small inlet brook and climbed up onto the bank herself, just as we had. Choosing to follow the foxes with what was left of the waning light for no better reason than that they were headed more conveniently toward the car, we tracked them around to the pond outlet where they, and we, crossed at an old mill site, the abutments of which had been reinforced by the resident beavers. Once across the dam we followed them up onto the road and back toward the car where their trail passed right under the front bumper! Had we walked around the front of the car when we first arrived, rather than directly back to the rear hatch, we would certainly have picked up their trail at once.

With some late-day searching northward back along the sides of the mostly bare road, Kevin found where the trails left it and reentered the woods, moving back in the general direction from which we had followed them in the morning. Apparently, we had found the south end of their hunting range. Where the north end was we could not tell. When we first picked up their trail in the morning it had come out of a dense laurel thicket in a wetland on the northwest side of the road, reason enough, we thought at the time, to foretrack the animals rather than thrash around in the swamp thickets.

334

Gray Foxes in Arizona

In the red rocks region of north central Arizona where the Sonoran Desert butts up against the Coconino Plateau there is only one fox, the gray. The desert kit fox does not range this far north nor does the red fox, winter adapted creature that it is, range quite this far south. In the juniper and chaparral of this region gray foxes, having little competition from anything but bobcats, are very common and their sign is easy to find. It was here in the Southwest that Olaus Murie seems to have encountered them, as he describes them as the "fox of the rim-rock country," and the pen and ink drawings of their tracks seem to correspond more closely to the Southwest variety than to the tracks found in the Northeast. Although some of the tracks I have found in Arizona show the roundness and winged ball of the front foot, others show a more elongated print with a larger, more amorphous secondary pad (Figure 6.82). This may be more a function of the surface upon which the tracks are made than of the foot. Much of the area is covered either with talcum-dry sand or hard sandstone, neither being substrates found in New England. Tracks in mud around cattle "tanks," as the locals call man-made water holes, look more like the ones I am used to seeing in the Northeast in similar substrate. Scat of gray foxes are also common and easy to find (Figure 6.83). Some are quite colorful as these foxes feed not only on animal prey but also on whatever berries are available. Most of these "berries" are actually dry capsules rather than the juicy fruit that the term implies, but apparently contain enough nutrition that many animals, including foxes, consume them.

Two encounters, while tracking with my friend and host, Kevin, shed some light on the behavior of this secretive predator, the first of which occurred in a dry wash southwest of the town. We had been searching fruitlessly for the tracks of mountain lion for several days, but the drought at that time had dried up all the water holes that attract these cats both for the water and the ungulates such as peccaries that come to drink. While working our way up the wash we suddenly heard the complaining cries of a scrub jay coming from a juniper thicket at the edge of the streambed. The cries persisted as we cautiously approached. Usually this means a predator is around. If it were a hawk, then the jay would squawk and either succumb into silence or, if luckier, the jay would fly away. Instead, the persistence of the cries suggested a mammal predator that the alerted and elevated bird could afford to

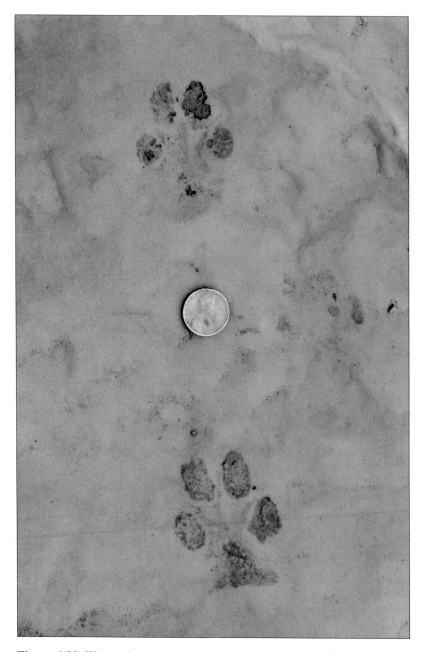

Figure 6.82. Wet tracks on hard sandstone in Arizona. A different substrate yielded longer and narrower prints than on mud or snow.

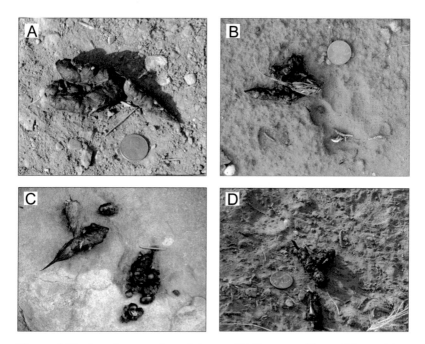

Figure 6.83. Gray fox scats from Arizona. (A) Fur scat with small bone chips. (B) Mixed berry and grass scat with an indirect registration of prints. (C) Mixed fur and berry scat. (D) The scat of desert animals often takes on the color of whatever berry is ripe at the moment. The two hind prints are just visible at the top of the frame.

stick around and scold. At once, I suspected either bobcat or gray fox. There was a slight breeze up the wash that caused a single bluish feather to float out of the thicket. This confirmed a close-call but ultimately unsuccessful attack on the jay. I motioned for Kevin to approach the thicket directly while I circled it, camera ready, in hopes of a sighting. At our approach, the jay shut up and disappeared. All was quiet as I rounded the thicket. But there in the sandy bank was the fresh trail of a gray fox coming out of the clump and slipping off up the drainage.

A second encounter took place at Wet Beaver Creek, south of Sedona. This is a spot where moisture was trapped up against a bluff as a series of stagnant pools that resisted drought. Between two of these pools perhaps two hundred feet or so from the nearly vertical basalt rimrock, I knelt to examine a peccary scat station. Between me and the vertical face was a chaparral-covered slope that concealed the base of the cliff. Suddenly, there was a flurry of wings and, as I looked up, a covey of Gambel's quail, the handsome little quail that graces the logo of the Arizona Fish & Game Department, blew up from a narrow, shaded terrace located half-way up the face and disappeared into the late afternoon sunlight over the rim. At first I thought I had spooked them, but then I spotted a ghostly gray form nimbly ascending the shaded face from the brush below, floating upward as if on wings. In a moment it disappeared into the growth on the terrace. It and not I had spooked the quail, and now the gray fox that had blended so well with the gray-brown basalt was climbing up to the site of its missed opportunity.

Two things struck me about these encounters. From the first I learned the quick shyness of this predator. There was no pause with a backward look, a habit I had seen in departing coyotes. It sensed us (the wind that carried the feather away was to our backs and so carried our scent to the fox at the same time as its attack) and was gone, presumably disgusted by our interruption. From the second incident I got a sense of the usefulness of being gray among the basalt rimrocks as well as the catlike nimbleness and stealth of the fox, so unlike the behavior or ability of other wild canids. I couldn't imagine either a coyote or a red fox climbing that cliff with such ease, nor using the coloration of their pelts to blend into the background in a stealthy approach.

Species Notes: White-tailed Deer

Structure and Mechanics of a Deer's Foot

There are only two wild "ungulates," or hoofed wild animals, in the Northeast: white-tailed deer and moose. In the West, on the other hand, there are many more: mule deer, elk, bison, pronghorns, peccaries, mountain sheep and mountain goats. Added to the list in the far North are caribou. Unlike all the soft-padded animals we have already considered in this chapter, the foot structure of ungulates is quite different and so deserves some discussion. As it is with all ungulates, the hoof of a white-tailed deer (Figure 6.84) is composed of two clove-shaped toes and two "dewclaws." Figure 6.85 shows the similar tracks of a mule deer for comparison.

The cloves, actually the two middle toes of the animal's foot, are composed of a fleshy center surrounded by a thin, hard cartilage. This

Figure 6.84. Track of a white-tailed deer. The hind print is the smaller one.

is analogous to our own fingers with their pad and nail. Such a design provides both traction on smooth surfaces and edging ability on steep terrain. The pointed end of the cloven hoof is also an effective weapon of defense that any potential predator must take into account. As long as a deer has the energy and health to rear up on its hind legs and strike with its forelegs, wise predators, in my experience in the Northeast at least, will leave it alone.

Figure 6.85. This indirect registration shows a mule deer hind print that is narrower than that of a white-tail, a common clue to identification. However, such distinctions are too subtle to be totally reliable. The contextual habitat in which the sign is found should also be considered.

In addition to the two cloves on its hoof, deer also have two dewclaws. These are vestigial toes, comparable to our index and little fingers, and are positioned on the lateral and medial side of the foot behind the cloves. (A fifth toe, analogous to our thumb, appears on the inside of the foot but so high up that it does not register in tracks.) The function of these dewclaws and their position on the hoof is involved in the interpretation of a deer's trails.

This animal's foot is actually the entire structure from the cloves up to what on the hind leg looks like its elbow. This "elbow" is actually the deer's heel, which on the hind leg juts out at an obvious rearward angle. On the front leg the heel is located quite high up near the torso. The springiness of a deer's gait results partly from tendons that run down the back of the foot and attach at the base of the cloves. The elasticity of these tendons stores and releases energy just like a rubber band. As the foot lands, the bony structures immediately to the rear of the toes collapse, articulating the front of the foot and stretching the tendons. During this collapse the delicate tendons are brought close to the ground and actually down to the ground in high-arc gaits like bounding. One of the functions of the dewclaws is to protect these tendons from contact with the ground. As the hoof leaves the surface at the end of the stride, the stored energy in the tendons is released, providing the stride with considerable spring, minimal in a walk but significant in the higher-energy gaits.

The various running gaits, viewed from the side, describe a series of arcs. As the deer is descending on any one of these arcs, its front hooves land well out in front of the animal's center. On slippery ground there is a tendency for the animal's forward momentum at this point to cause the front hooves to slip forward. Here is the second use of the dew claws. Like any other toes, they are under at least limited muscular control and can be rotated in the plane of the foot to bring their sharp edge perpendicular to the direction of travel. In this position, they are used as brakes on any slip that might occur on muddy or loose ground. The dewclaws on the front hooves of a deer are slightly closer to the cloves than on the hind. This proximity brings them into play more quickly in the event of a slip.

Unlike the position of the front feet when they strike the ground in a running gait, the position of the hind when they touch down is directly under the animal's center and the force they apply is down as much as rearward as the animal accelerates upward and forward. In a bounding gait, through tall grass for instance, the force can be close to straight down. The rear hooves' position under the center of the animal's body, rather than forward of it as with the front feet, reduces the likelihood of a slip. Therefore the dewclaws can be mounted farther back on the foot where they function as two points of a stable four-point contact off which the animal can propel itself forcefully into its next arc. Their rearward position may also cause the tendons on the rear of the foot to stretch to a greater degree than on the front, providing more spring to the gait as the stored energy is released.

I have read in hunter lore that the sex of a deer can be determined in its tracks by the hind straddle. One way I have heard it is that a doe's trail will show a wider straddle due to the need to provide space at the pelvis for the birth canal. However, another account contradicts this, saying that the buck will have a wider straddle to allow room between its hind legs for its external sexual organs. Obviously both could not be true and apparently neither is. As explained above, a deer's foot extends up to its "elbow," and a rear end view of either a buck or a doe shows that the heels of the two hind legs nearly touch. Below that point, feet splay outward slightly, much as do the feet of most humans. The straddle in a deer's trail, then, is dictated by the degree of this splay in its feet, not by the structures above the heel, whether birth canal or sexual organs, and this splay apparently varies somewhat from individual to individual regardless of sex. The amount of splay in the deer's hooves may have more to do with the relative weight of the animal. As is often the case with humans, the heavier the body, the wider the splay.

Preferred Gaits

Walking

A walking deer leaves behind the same zigzag arrangement of registrations (Figure 6.86) as most other mammals. The step length tends to vary around 23-25 inches, but the distribution is wide since deer have long legs and since walking is the usual way deer move when browsing slowly from shrub to shrub as well as when traveling in line from bedding to feeding area and back. Since the median step length for an eastern coyote is around 22-23 inches, their trails can be confused with those of deer when track detail is missing. Even in melted out tracks the appearance of the two can be remarkably alike. First, the arrowhead outline of indistinct coyote tracks in snow is about the same size and shape as a deer's print. Secondly, a coyote's indistinct track is often identified by the lack of lateral and medial nail marks. In their absence, the two central registering claws on a coyote's foot can resemble the points of a deer's cloves. In snow, however, coyotes direct register nearly perfectly and in most snow conditions

Figure 6.86. Although a walking deer's normal step-length is similar to a coyote's, the trail shows a wider straddle. Note the "hourglass effect," common with deer tracks in deep snow.

their tracks line up in a nearly straight line. Deer don't travel the long distances that a predator must, so they do not have the instinct for gait efficiency in winter that coyotes do. As a result, their tracks commonly indirect register. A walking deer in a winter of deep snow is usually in a trail of its kin packed down between bedding and feeding areas. If the deer is startled or chased out of this trail, you may find its trail alone but probably in a higher energy escape-gait such as a bound.

343

Trotting

For most mammals, trotting is a traveling gait used most often by lone rather than herd animals. While deer can trot, they seem to have little use for it; walking, loping and bounding seem to be their mainstays. I have, however, seen film footage of agitated deer doing a sort of "hesitation trot." In one clip, this was part of a competition between bucks during the rut, and in the other it was used by a doe out of patience with her uncooperative fawn. In both cases this behavior seemed to represent aggression I have never seen any patterns that I could definitely attribute to this gait, but it seems to be exceptional behavior and not likely to be met with in the everyday trails of deer

Loping

Information presented above about loping, galloping and bounding pertains to the loping gait in deer. As the mechanics of these running gaits are complex, you may wish to review that section before you proceed.

One summer day up in the boundary cut between the Federated Forest and Quabbin, I heard persistent barking by a coyote that seemed to be coming closer. I stepped back into concealment and watched. In a few moments, a deer came loping comfortably along a hiking trail through the brush of the cut. A few yards behind it I could just make out above the bushes the black and yellow fur on the back of a coyote that obviously was running the deer. Conventional lore would say that the coyote was testing the deer for fitness. If this fitness were found wanting, the deer would be dispatched somewhere down the trail where the prey animal finally tired too much to defend itself. However, the evidence before my eyes suggested something else as well. The gait of the deer was an easy lope, not the bounding or galloping of a panicked prey animal escaping imminent death. And the coyote's gait, from the undulations that I could make out above the brush, matched that of the deer with no effort to hurry it along by closing on it. As the pair disappeared over a rise, it occurred to me that this brushy cut was good cover for rabbits and other small prey animals. It seemed to me that the coyote's running of the deer was primarily to prey on any rabbit that the deer might startle out of its form. Sensing the approach of a large running animal, a rabbit would scamper out of the way and then stop to look back at what had startled it, unaware of the following coyote until it was too late. This is yet another example where in

Nature the obvious or proximate cause of wild animal behavior may not be the actual one or may miss an additional, more hidden motive. Too often I have seen these perceptions devolve into an either-or argument. My sense of the elegant efficiency of Nature suggests that in this case the coyote may have gotten double benefit from its action both testing the deer and nabbing the rabbit.

As I indicated earlier (page 144), there is a lack of semantic consistency in gait terminology among the various track identification guides on the bookshelf. In several of them the distinction between loping and galloping is blurred, with one or the other terms used for both. Rather than assign an esoteric definition for the word "lope," I prefer to use the one in common use, that is, an easy running gait. As a reminder of information in Chapter IV: "Patterns and Gaits," in a lope nearly all the power is supplied by the hind legs. The front legs are used as props for the front end of the animal to keep its forequarters from falling to the ground as the animal descends from each arc in the gait cycle. In this respect the front legs are like the spokes on the front wheel of a bicycle, with first one and then the other contacting the ground along the trail to support the animal. As the deer moves forward over its two front feet, the hind feet pass on either side of the planted front so that, for a moment, the two sets of legs, viewed from the side, are criss-crossed under the animal's center. The hind feet then strike the ground and propel the animal forward into a low arc at the end of which the front feet contact the ground, first one and then the other, propping the front half of the animal, leaving either a transverse or rotary pattern behind. One of the distinctions between a lope and the higher-energy gaits is that, in a lope, there is less suspension of the animal above the ground. By the time the hind legs vault the animal forward and up, the front feet are already on or nearly on the ground. By the time the front feet leave the ground under and behind the animal's center, the hind feet are already in contact with the surface or nearly so. This minimum of either extended or gathered suspension contrasts with a bounding gait, for instance, where, especially with deer, there is a great deal of extended suspension as the animal arcs high through the air.

The pattern of a lope can be either transverse or rotary. In the case of deer it often looks like one or the other of the patterns in Diagram 6.16.

Front   Front   Hind Hind

Transverse   Rotary

Diagram 6.16

These patterns can be switched back and forth at any time in the animal's progress. Which pattern a deer chooses probably depends on what it sees in the lower half of its vision as it moves forward, selecting the order of footfall to avoid debris on the ground.

Galloping

Deer lack the chest musculature to sustain a gallop. Compare a deer's chest with that of a race horse and the difference is obvious: horses have bulging "shoulders" with which they can impart energy to the gait with the front legs. In a lope or bound, the hind legs provide the propulsion whereas in a true gallop both front and hind legs do. As far as I know, the only wild North American mammal that can sustain a gallop is the pronghorn. This ungulate is a creature of the open plains whose method of escape is not to run a few meters to cover as with deer but, since there is little cover on a plain, to simply run. Its pronounced chest musculature announces its ability to do so at high speed and for long distances.

Bounding

Most often used as an escape-gait for a white-tailed deer, bounding has a much higher arc of stride than does either loping or galloping. With much of a bound's energy devoted to upward motion, it is inefficient in terms of ground speed. However, in the high-resistance terrain of eastern woodlands where there are likely to be fallen branches and rocks on the ground, a gait with a lot of verticality allows the animal to vault over such debris. There may be one other advantage to a bound as well. Deer have co-evolved with a number of predators including man, the only one of them traditionally equipped to kill at a distance, with stones, spears, arrows or bullets. The verticality in a bounding gait causes the target to move rapidly up and down in the sites of a rifle, confusing the shooter's aim. I have also on occasion noticed a tendency for some deer, caught in the open, to bound away in a zigzag rather then in a direct line. I suspect that this, too, is an adaptation to spoil the aim of the human hunter.

Like loping and galloping, the bounding pattern of a deer may be either transverse or rotary. The pattern (Diagram 6.17) resembles the other two but with a more compressed pattern length compared to inter-pattern distance. (The inter-pattern distance shown here has been shortened so that the patterns will fit on the page.)

Diagram 6.17

Note that the hind prints tend to be quite close laterally, sometimes almost side-by-side. This arrangement, when translated to the temporal rather than spatial aspect, means that the feet land almost simultaneously. Both hind feet on the ground at the same time allows the animal the option of dodging either right or left, depending on what it sees in front or fears behind. It also allows the animal to perform these dodges unpredictably rather than in a repeating pattern, a further aid in confusing aim.

Stotting

Stotting, or "pronking," is a gait used in the West by mule deer. It is a coincident rather than a rocking-horse gait, that is, both ends of the animal rise and fall together. It looks like a quadruped on a pogo stick. As the animal descends from each arc of its stride, it lands on all four feet together. These feet collapse downward storing energy in their posterior tendons, which is then released simultaneously by all four feet, vaulting the animal into its next arc. A look at the grassy "parks" of central and northern Arizona shows one advantage of this gait. Those lovely level grasslands that look as if you could easily gallop over them to the horizon are filled with concealed chunks of basalt disgorged by ancient volcanoes. These rocks are only visible when you are almost right over them. In the midst of each high arcing vault a mule deer can see where these chunks are and subtly redistribute its feet in mid-air to avoid them on landing. Since this sort of terrain is not the usual habitat for white-tails, they have not developed stotting in their repertoire of gaits.

Blended Gaits

In reconstructing an event in the imagination it may be difficult to distinguish loping, galloping and bounding. We separate these gaits into discreet bundles, but the deer does not. Each of the gaits can be blended seamlessly into the others (Figure 6.87). In sorting them out some clues may help. Any gait with a high arc of stride, such as a bound, involves considerable impact as the front feet contact the ground. In some surfaces this can be read as impact debris distributed evenly around the print. While assisting in a tracking survey at Thousand Acre Swamp many years ago, I recall coming upon the trail of a bounding deer in about three inches of slushy snow over ice. Each print was surrounded by an aureole of rays caused by slush globs squirting out from under the hooves as they landed.

Loping, on the other hand, has a lower arc of stride and is a relatively low-energy gait among the several running gaits. This is what makes it useable over long distances. As a result, one is unlikely to find a lot of impact debris around the front prints. Furthermore, since little backward force is supplied by the front legs, there is unlikely to be significant mounding behind their prints. Since most of the forward push in a lope is supplied by the hind legs, one can expect to find what mounding does occur just behind the hind prints. In a gallop all four feet supply forward push but the low body position of this gait results in surprisingly little impact. Therefore one is likely to find mounding behind all four prints but less impact debris surrounding them.

Interpreting gait energy from the appearance of the prints is one of those skills that improve with practice. The above is a thumbnail description meant to provide a start, but as surfaces vary infinitely and interpreting what is going on well above the feet by the appearance of footprints is an almost mystical art, one should allow for anomalies that contradict expectations. That said, facility at fine interpretations, another way of saying "seeing the animal in motion in the mind's eye" is the most satisfying result of any tracker's training and the end toward which this book aims.

Behaviors

Yarding

Much is made in the literature about deer yarding behavior. When snow approaches belly-deep, deer herds may be obliged to live in a constrained network of trails that are gradually packed down in areas that provide both browse and protection from wind as well as providing back radiation of solar energy. These are often among spruce/fir

Figure 6.87. In the real world neat categories break down. This rotary pattern is somewhere between a lope, bound and gallop. The two hind prints are at the top of the frame, nearly side-by-side to propel the deer into the next arc of stride.

regeneration, where the needles, however lacking in nutrition, are at appropriate height for feeding. The dense boughs also slow the passage of wind, reducing convective heat loss, and snow hanging up in the tops of the evergreens lessens the amount that piles up on the ground beneath.

In shallower snow or in places where the surface snow is underlain with a crust thick enough to support weight, deer herds are more mobile. Here the herd will pack down long trails between bedding and feeding areas. These trails become highways for other animals as well, including predators and scavengers of deer. One winter day I was driving near Lake Umbagog in northern New Hampshire when I spotted a coyote motionless in the woods. I did a U-turn and drove back to the site. There I got into a staring contest with the animal, which was crouched on a packed deer trail perhaps 50 yards away. It was a very cold morning and the animal's fur was erected so that its white bib looked like the beard of a canine Santa Claus. The animal eyed me until it could take no more and fled back up into the woods over the deer trail. It paused at the top of a small ridge, back-lit in the morning sun, which picked up the red highlights in its fur. One more look and it fled, loping on the deer-packed trail over the ridge and out of sight. Coyotes spend much of the winter in these trails. In another instance of double-benefit, they use them to get around to sites where they can find small prey animals while at the same time keeping an eye on the deer that created the trail, ungratefully watching for signs of weakness as winter progresses toward spring.

Bedding

Deer beds (Figure 6.88) are locations where the animals lie down, regurgitate and chew their cud. These sites may be occupied for as little as a half-hour during the day or for most of the night. In the winter they are often located under dark evergreens such as hemlocks that absorb the sun's heat during the day and re-radiate it at night in all directions including downward to the bedded deer. The overhanging boughs also reduce radiational heat loss to the sky on clear nights. Beds may be located down in the bottomlands or up on ridges. Bottomland sites may be used in cold, windy conditions to reduce convective heat loss. But many more sites will be located on the "military crest" of a hill or ridge. This is the contour line where the deer is out of sight from below but only needs to raise its head and

Figure 6.88. A deer bed on a ridge. The animal's shoulder impression is on the lower right with its legs folded close to its trunk at the top.

orient its ears downhill to detect the approach of any potential predator, including the human hunter. If danger is detected approaching uphill, the deer can slink off over the convex summit without giving away its presence. In summer, ridge or hilltop beds may also provide a breeze to carry away some of the biting insects that make all of us, both deer and tracker, miserable.

Other Behaviors

There are many other signs of deer presence and behavior. The stumps of cut hardwood trees grow numerous leaders, the buds and leaves of which are at browsable height for both deer and moose. The cuts these animals make when feeding on the end buds tend to be perpendicular to the stem and ragged, as opposed to the neat diagonal cuts of rodents and lagomorphs. Another sign is the "buck rub," a place where a male rubs its antlers up and down on a thin sapling, stripping the bark, which then often hangs off at the top and bottom of the wound (Figure 6.89). "Buck scrapes," on the other hand, are places where the buck paws up the surface of the ground and urinates (Figure 6.90). In my neck of the woods, most of these scrapes are located

351

Figure 6.89. A buck rub is made by a deer scraping the base of its antlers up and down on a smooth sapling. Unlike bark scraping for food, the tails of bark will be both at the top and bottom of the rub. Other ungulates like moose and elk also make such rubs.

under an overhanging hemlock bough, the end of which is often broken and hangs down "pointing" to the scrape. Both rubs and scrapes are mating behavior, intended by bucks to advertise their presence to passing does. Quite different in appearance is a shrub that has been

Figure 6.90. In the Northeast a buck scrape is often on a deer trail with an overhanging hemlock bough.

used to rub off the velvet covering of maturing antlers. In this case the deer plunges its tines into a bush and then thrashes around to remove the felt. The results look like a grenade hit the shrub. The bloody velvet is rich in nutrients and is quickly consumed, perhaps by the deer itself, certainly by any small mammal that follows its nose to it.

Winter is the starvation period when the appropriateness of a wild animal's genes is put to the test, resulting in a weeding out of bad genes and the survival of the best. Food during this period is hard to come by for every animal, predator and prey alike. In this season ungulates such as deer rely on woody browse (Figure 6.91). For a terrestrial browser to take advantage of this source, sometimes risks must be taken. On several occasions I have tracked deer out onto the frozen surface of a beaver pond. Such places are danger areas for this animal since, if it is discovered by a coyote that chases it, the deer will slip and splay its legs, dislocating its hips. This condition makes even a healthy deer vulnerable to the pursuing predator. The deer must balance this risk with potential reward in the form of the ends of

Figure 6.91. Deer browse shows a ragged perpendicular cut, often with a tag. The diameter of the twig rarely exceeds an eighth of an inch and browse height is usually below four feet. Larger ungulates like moose and elk browse higher and often nip twigs up to a quarter of an inch thick.

the branches of a beaver's cache frozen into the ice next to its lodge. With their lower parts in water, these branches tend to hold nutritious leaves on their tips well into the winter.

Red oaks have become the dominant tree in much of the mature northeastern forest. The acorns of this tree are a plentiful supply of fat that can be used by animals to survive the winter, or they would be were it not for this tree's tactic of introducing bitter tannin into them as a defense against consumption. These little globules of fat are intended to sustain the tree's embryo as it grows until it is large enough to produce a leaf with which to process the energy of the sun directly. Few animals, then, will feed on red oak acorns as they fall. But tannin is water-soluble and so, after the nuts lie on the ground for awhile, melt or rain water leaches the tannin out and they become palatable to mice, squirrels, porcupines, bears and deer (Figure 6.92). For their part, deer will sweep away shallow snow in order to extricate these acorns hidden among the leaf fall of the previous season. Other animals abroad in the winter do this as well: porcupines will dig away an amazingly large area of snow to get at them, as will bears and turkeys. The work of these animals can be told from deer by the clearing method. When a deer is digging out acorns, it supports its front weight

Figure 6.92. Feeding on acorns and other nuts by deer is inefficient. However, it provides at least a little energy to keep the animal going during its annual winter testing period.

on one front foot and sweeps the snow away with the other using a diagonal stroke. Often this results in a triangular shaped clearing, sometimes with a front footprint in the center of the triangle. Since deer lack the nimble paws of a squirrel, their feeding on the extracted acorns is crude. They simply crunch the nut with their molars, swallowing what meat is expressed and letting the shell and much of the remaining meat fall. Bears also are sloppy acorn eaters, leaving similar sign, while turkeys leave nothing behind but their droppings. Look for tracks and snow-clearing technique.

Deer and porcupine also have a taste for false truffles, fungi that commonly grow under a few inches of earth at the base of hemlocks. Once again the deer's dig, unlike that of the porcupine, often shows a triangle shape at its head, a comparatively large debris field and occasionally a readable print in the spoil.

The Deer and the Forest

It has long been supposed by ecological historians that deer were scarce in the pre-colonial forest. Lacking written accounts, they seem to have made the assumption that since deer, unable to climb trees, browse only at ground level, then they would have done poorly in a landscape dominated by the sort of forest they imagined existed before Europeans arrived with axes.

As we have already seen in the section on the Eastern Forest (pages 80-83), this partly follows from a misunderstanding of what an ancient forest looks like. The pre-colonial forest was anything but a relentless cover of tall trees; rather it was a forest of mixed age with many sunny openings promoting the regrowth of vegetation at the browsing height of deer — about knee-high to mid-chest on a human.

A second factor that promotes deer, even in a forest of even-aged trees, is the weather. Although in such a managed forest there is little trunk decay to bring live branches down to ground level, there are periodic ice-storms. In the woodlands of eastern Massachusetts, following a severe storm several years ago, the forest floor was littered with limbs torn off of tree-tops by the weight of accumulated ice. Wherever these limbs landed there was a natural redistribution of energy through the forest column. Around these limbs with the next snow-fall could be seen many tracks of deer that had been drawn to them in order to browse the winter buds. So helpful was this storm to this species that there has been a marked resurgence of deer numbers in the hardest hit areas. Such would have been true in the pre-colonial forest as well. And the same effect would have followed the periodic tropical storms that have rolled up the East Coast both in modern and, presumably, in ancient times.

The second natural factor for larger deer populations than has been traditionally assumed is the often neglected beaver effect. Pre-colonial beavers made the same sort of forest openings that modern ones do. The edges of those clearings would quickly have grown back shrubbery at conveniently browsable height for the benefit of deer and moose. Indians trapped beavers, of course, but not until Europeans arrived to pay them for skins did the natives overtrap these animals, causing beaver clearings, so beneficial for wildlife, to disappear from the landscape.

Species Notes: Moose

Moose are a circumpolar species of hoofed animal that inhabits the taiga, a ring of boreal forest and bog that circles the globe in the northern latitudes. In recent years, the range of this huge animal has begun to extend southward so that its sign is increasingly common in northern hardwood forests in the upper temperate zone and even down into the so-called "transition forest" dominated by oaks and red maples. The reason for this expansion, in central New England at least, is a combination of factors. The first is the resurgence of beavers (Figure 6.93). Trapped nearly to extinction for its valuable fur, the beaver has made a dramatic recovery in its traditional range in recent years due to the opprobrium in the fashion world directed at wearing fur. Today in New England, beaver ponds are a common sight dotted around our woodlands anywhere there is water to be impounded. This resurgence has benefited moose because these large animals, in order to regulate muscular control, need a lot of sodium, which they find in the aquatic plants of beaver ponds. In fact, their long legs and long snout are adaptations for wading out into beaver bogs, submerging their heads and cropping these sodium-rich plants. Another source of sodium is road salt. In the Great North Woods moose wallows are spaced along every highway that uses salt in the winter. When the salt drains off the road with rain and snow melt, the resulting brackish water collects on the edges of the roadway, inviting attention from all the local moose.

Figure 6.93. In summer, beaver ponds provide the sodium-rich aquatic plants that moose both need and are anatomically designed to feed upon.

Unfortunately, moose and highways are a poor mix and a great many collisions have occurred near these wallows.

The other reason for the expansion of moose southward from their boreal habitats is the re-maturing of much of the hardwood forest in central New England. The economically valuable trees in this forest have grown to cutting height and as a result many clearcuts now dot the landscape, providing a bumper crop of hardwood sprouts from the cut stumps (Figure 6.94). In the fall, when the energy of pond growth recedes to the tubers in the mud and the ponds themselves freeze over, the moose turn to the buds and tender tips of these sprouts for nutrition to sustain them over the winter (Figure 6.95). Obviously, then, the places to look for moose sign are beaver ponds in the summer and regenerating hardwood cuts in the winter.

Deer also browse stump sprouts, but normally only up to about waist height while moose browse higher. Deer also tend to browse narrower twig ends of a half-centimeter or less in diameter while moose browse up to the thickness of a pencil. Otherwise, the appearance of the cuts is the same: squared-off and ragged. Neither of these ungulates has upper incisors. To cut off a twig, they pinch it between their lower incisors and their hard upper palate and then tear it off. In the southern part of their range, maple sprouts are the moose's favorite while in the North, streamside willow thickets are a magnet for these large herbivores.

One conspicuous sign of moose presence is bark gnawing, where the animal stands in front of a tree and repeatedly scrapes its lower teeth upward on the bark, usually at a shallow angle to the long axis of the trunk. Moose do this on "sweet trees" up to saw timber size, that is, as long as the bark is smooth (Figure 6.96). Maple saplings and mountain ash are favorites for this treatment; presumably the moose are getting not only sugar from the bark but also nutrition from the cambium layer. Sometimes this gnawing is mistaken for bear work, as if the bear stood at the base of the tree and scraped the bark downward with its claws. A close look at the marks of a moose gnaw will

Figure 6.94. In winter, moose invade clearcuts and burn sites to browse hardwood regeneration. Here moose have breasted pin cherry regrowth to reach the higher tips.

Figure 6.95. Moose twig browse looks like deer, but the browse may be ¼-inch thick and at head-height or higher.

Figure 6.96. Sweet-barked trees such as these mountain ash are gnawed heavily for sugar. Maples are also favored.

Figure 6.97. Winter scat when moose are feeding on dry, woody browse.

show, however, that most of the tails of bark at the end of each gnaw are at the top of the scar rather than at the bottom as would be the case if a bear were scratching downward. Bears do shred the bark of certain trees in the spring when they are hard up for food, but these are usually spruce, the pulpy cambium of which is nutritious. They also claw down on mark-trees to advertise their presence. A close look at these scars on a tree trunk should suggest the difference, however. Bear claw marks are narrow at their deepest point while moose and deer leave broad, flat marks.

Scat

Moose scat ranges from collections of large pellets in the winter when their diet is dry (Figure 6.97) to clumps of cohered pellets in the summer when they are feeding on succulents (Figure 6.98). In the transitional seasons while their gut is adjusting to a change in diet, their scat gets looser, sometimes resembling a plop of coffee grounds from an inverted fry pan.

Behaviors

Warning: The moose rut begins in September in the Northeast. This is a period when the bulls are irascible and should be treated with respect by the tracker. The animal may seem unconcerned by your presence until some threshold is passed and the animal suddenly

Figure 6.98. Summer scat when the animals are feeding on succulent plants.

charges. Drooling is a sign that the animal is feeling stressed even though its comportment doesn't seem so. Dogs are perceived by moose as wolves, their natural enemy, and may provoke a violent response.

One spring a few years ago I was lying down behind a fallen log next to a beaver pond videotaping wildlife when I heard splashing up the shore. Through the mountain laurel I could see the felted antlers of a bull moose. Great, I thought. This animal will walk down the shoreline in front of my camera and I'll get a great panning shot. Not so. As soon as I detected the moose, he detected me. Instead of walking down the margin in front of me, he trotted up a beaver trail and came around behind, trapping me against the log. An angry bull moose standing over you with bloodshot eyes is much larger than one seen from the safety of a vehicle or a house. However, once he examined my cowering figure behind the log, he apparently decided I was harmless and calmly passed on, leaving me with some footage of his departing rear end — somewhat shaky footage.

Like deer, moose bed often both at night and, in order to ruminate, during the day. Their beds are like those of deer but larger. In Figure 6.99 the animal was facing left with its large snout disturbing adjacent snow.

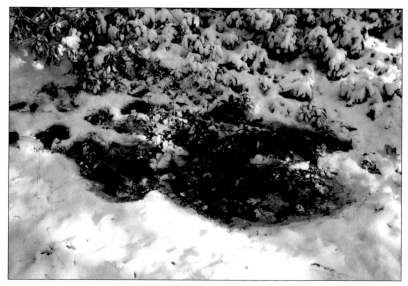

Figure 6.99. Moose beds look like those of deer but are larger. Here the animal's head was on the left with its back to the camera.

A moose's antlers have no known function other than as part of the mating process. Presumably cow moose are impressed by a large rack and accept males accordingly. Bull moose also use them in shoving matches with other males for breeding rights. It has occurred to me that they may also serve to keep bulls and cows in the same general area for breeding, that is, any area that supplies them with calcium such as beaver ponds. The cows need this mineral with which to grow the bones of their calves while the bulls need it to build antlers. It's an interesting speculation but one for which I have no proof. Once mating is over, moose drop their antlers in early to mid-winter and begin to grow new ones. The rarity of finding these in the woods is a testimony to the industriousness of mice and other rodents that chew them up for their calcium.

Species Notes: Otter, Mink and Other Weasels

Sign Far from Water

Both minks and otters are normally associated with lakes, ponds and streams. However, sometimes a perplexing trail that resembles either species may be found far from water. Neither of these mammals, contrary to common opinion, spends its time exclusively on a single water body until its prey source is exhausted. Instead, they hunt a pond or stream for awhile, sometimes a day, sometimes longer, catching whatever is most easily caught, which only rarely are wily, fast-swimming native trout. ("Stupid" hatchery trout, lacking the instincts of a wild genetic heritage, are another story.) Both these species of semi-aquatic weasels, like other predators, farm their prey rather than eliminating it, taking off the vulnerable surplus and leaving the more elusive fish to procreate, producing at a later date more prey with genes selected for survival in the wild. Once the readily caught have been skimmed off, the predators move on, sometimes traveling fairly long distances overland. When one finds an otter doing this, its trail is easily confused with that of a fisher, and when you find a mink trail overland, it can be hard to distinguish from the trail of a long-tailed weasel. This can be especially so in early spring during the mating season of mink. At that time, the males spend a lot more time roaming the countryside in habitats more appropriate for weasels and often well away from water, searching out females.

An interesting feature that I often find in mink behavior is their penchant for feeding in private, perhaps more so than the other weasels. Figure 6.100 shows where a waterfowl that was killed in a nearby river was dragged overland and into the hollow of a tree before being plucked and fed upon. A curled scat can be seen on the leaf in the foreground.

All members of the weasel family are carnivores that will take any live prey they can catch. Weasels, fishers and martens are all terrestrial predators that are restricted to warm-blooded prey, although all will feed on berries from time to time. This latter habit is sometimes attributed to young animals that are having a hard time capturing live prey. Otters are more restricted in their diet, most of which is fish and crayfish, while mink is the swing species, adept at hunting both warm-blooded and cold-blooded prey. This distinction was brought home to me one Thanksgiving on a camping and climbing trip to the high peaks of the northern Adirondacks. Around Lake Colden and Avalanche Lake, I could find no otter sign at all. These high ponds had been so contaminated by runoff from acid rain that they support no fish. However, a quarter of a mile up Opalescent Brook which feeds the ponds,

Figure 6.100. Minks like to consume prey in private. Here a waterfowl was pulled into a hollow tree near the shore of a river. Note the curled scat on the leaf in the foreground.

the familiar slides of otters were conspicuous. The odd thing about this experience is that, around the dead ponds, mink sign was common. Although these tarns were no longer suitable hunting ground for fish-dependent otters, the minks, being more democratic in what they will eat, apparently were released from competition with the larger otters and so increased in number. This irony of the smaller but more flexible species profiting by the misfortune of a larger relative that is more restricted in its diet was readily apparent in the track record.

Among the larger weasels, otter trails are easy to distinguish because as soon as the trail comes to a descent, the otter will slide. Fishers, on the other hand, clearly don't enjoy the experience and do so only for short distances. Once I found an otter trail leaving the Saugus River in Breakheart, crossing a frozen quarry pond and heading uphill. I followed its 4X transverse loping patterns upward, across a park road and then up to the crest of a ridge nearly a half mile from the river. Had I picked up this trail in the snow anywhere above the river, I might have confused it with the similar trail of a fisher. Granted the hind print of an otter is larger than its front, contrasting with the slightly larger front print of a fisher compared to its hind. But this difference is often subtle in soft snow and easily missed.

At the top of the ridge, however, any confusion immediately ended as the appearance of the otter's trail changed greatly. On the way down the slope toward the Lower Pond, the animal slid in the snow at every opportunity, dodging trees and rocky outcrops, moving right and left to find a clear slope downward. As it approached the pond it adjusted its route to the right, apparently aiming for a particular spot. Near an old boat pier a hemlock bough hung over the edge of the pond. This bough absorbed afternoon sunlight which it reradiated in all directions, including downward, keeping the ice weak beneath it. The otter clearly knew this spot and had followed this route before to reach it. At the bank was a fringe of thin black ice with a ragged hole about a foot across where the otter had entered the pond.

As this anecdote suggests, otters slide at every opportunity, not only downslope, but on the flat surface of a frozen pond and even on a slight uphill grade if the surface allows it. For them, sliding is an energy saver, but it also is clearly fun. I read a revisionist theory in a tracking book suggesting that otters don't slide nearly as much as they are reputed to. My own experience contradicts that as I have found they will do so at every chance. Life is easy for otters; they are superb swimmers, semi-torpid winter fish are easy to catch under the ice, and they have excellent waterproof fur to keep them warm. As a result, like other species with time on their hands, they play. And play with an otter often involves sliding, which they will repeat on any suitable snow-covered slope, climbing up and hurtling down like a kid with a sled (Figure 6.101).

Minks also slide, but for these nervous little animals the act is strictly utilitarian (Figure 6.102). Once I followed a mink trail coming

Figure 6.101. Otter slides. Fun!

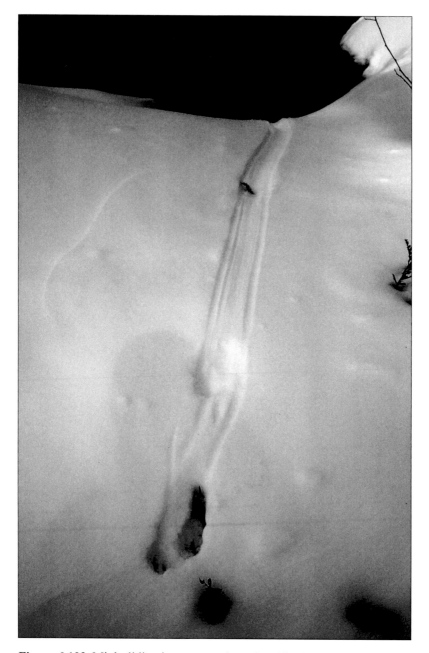

Figure 6.102. Mink sliding is more consistently utilitarian than that of otters and is used strictly as a convenience and energy-saver rather than for fun.

up out of a brush swamp in western Maine and leading to the top of a hill. On the reverse slope was crust, slightly softened by the sun. Finding this surface suitable, the mink started a slide that zigzagged downhill for over a half-mile to a wetland where the trail ended in a hole in the ice. Because the width of an otter slide in snow is around 7-9 inches, whereas that of a mink is about 4 inches, they are easy to tell apart. Distinguishing a mink slide from that of a long-tailed weasel is harder, since a male weasel's size approaches that of a female or otherwise small mink. Both long-tailed and short-tailed weasels slide in favorable snow conditions, but when I have found these trails, they have been very short. Furthermore I have never found their trails to end at a hole in ice. I can find nothing in the literature that suggests that either of these small weasels even swim. Certainly the nature of their fur, compared to the waterproof fur of minks and otters, suggests that they are unlikely to do so in the winter.

A mink, on the other hand, will dive into a freezing stream on even the coldest days. Once while I was skiing along a forest road below Prospect Farm in the White Mountains, I spotted a mink in the brookbed below a bridge. It was nosing around under a rock, searching out some concealed supper with its nose when it apparently sensed my presence. Without much alarm, it simply loped down the snowbank and out onto the frozen ice at the side of the brook where it dove right into the rushing water. The air temperature at the time was about 17 degrees! When you think about it, the temperature under the water can be no colder than 32, so that diving into the water might actually reduce heat-loss for the creature. That is true only so long as the mink's coat protects it from penetration by the water to its skin where, even more readily than cold air, it would conduct heat away from its surface and turn the little animal hypothermic in short order. It is for this protection that minks have that luxuriant oiled coat so desirable to furriers.

Absenting slides, another quick way to tell otter from fisher trails is to look for more than one animal. Otters are social mammals, often hunting together and traveling side by side as they traverse dry land from water to water. While more than one fisher's trail may be found in the same area during mating season in late winter, at other seasons a discovered trail will be that of one solitary and anti-social animal. Minks are solitary animals as well, so their trails and slides will be those of a single creature. Notable exceptions are a male following a female's scent trail or following the trail of another interloping male.

Patterns and Gaits of Smaller Weasels

Mink running trails in snow are distinguishable from those of only slightly smaller long-tailed weasels by their regularity. Although they are nervous little predators, the mink's greater bulk helps it to conserve body heat better than a long-tail, whose small size and long body shape require a red-hot inner furnace to sustain life. If mink trails express nervousness, a weasel's trail communicates a frantic search for its next meal. A mink usually leaves a fairly direct trail of loping or bounding patters from point to point. A hunting weasel (Figure 6.103), on the other hand, will dodge right and left with irregular step lengths and frequent disappearance under the surface where it follows its nose down into the depths of the snow in search of mice and voles. Just as a mink may seek out entry to water in the winter to keep itself warm, weasels get a secondary benefit from hunting in soft, insulating snow to the same end. The long, narrow body of any member of the weasel family is a poor shape for heat retention. While the larger weasels like wolverines, otters and fishers have a sufficient ratio of body mass to surface area to be somewhat immune to this, the smaller weasels do not. Minks must escape now and then into relatively warm water while they are hunting, and weasels can dive down vole tunnels into soft snow for insulation, even reportedly lining their dens with the fur of their victims and then curling up in a ball to minimize surface area.

As described earlier (pages 140-141), in surface conditions where a mink or other weasel sinks deeply, and "deeply" for a short-legged animal may be a matter of a couple of inches for the smaller members, it will modify its stretched-out and relatively flat loping gait to a more vertical bounding one. The pattern will change from the familiar 4X pattern to a 2X DR pattern, that is, from groupings of 4 individual prints to patterns of 2 tracks, each track consisting of one front print and one hind print registering over it. These 2X groups will be arranged on a repeating slant unlike the alternating slants of many raccoon trails but similar in pattern to the displaced trotting trail of a canid. Of course, even the smallest fox or raccoon will leave tracks larger than the largest mink. Because this "repeating slant 2X' bounding trail

Figure 6.103. Weasel trails tend to be more erratic and nervous than mink trails. Here a weasel has used a 2X bound to approach a stonewall, the crevices in which may hide a potential meal. Note the "dog-boning," also a weasel trail characteristic.

of a mink tends to have remarkably regular inter-pattern distances, some writers have supposed a regularity of gait among minks in other seasons as well. However, video that I took in the spring showed me that a mink trail's regularity in the winter is a function of the smoothing effect of snow on the landscape. At other seasons, when downed logs, rocks and forest debris litter the ground, the short-legged mink varies its gait widely, walking, bounding, loping and jumping as the degree of terrain resistance demands. Only on a smooth, level plane, such as a strip of wet sand on a beach, will the mink revert to the regularity of its winter gait habits.

A small mammal's reliance on bounding on any surface that is not smooth and firm results in much more impact on its front feet than on its hind. Descending from a high arc, the front feet of a mink or weasel, bearing the weight of its head, contact the ground first. The hind, following in to the same vicinity, are relatively lightly loaded. The push-off that follows as the animal launches into another arc involves little impact and the spread of the foot is contained by the walls of the snow around it. The result is a characteristic appearance in the tracks of the smaller weasels with a broadly spread front and a narrow hind. I often look at the relative splay of the two central toes to distinguish front from hind.

The three smaller weasels, the long-tailed, short-tailed (ermine) and least, all may turn white in the winter by molting their brown-colored fur and replacing it with white. At the southern end of their range where snow is sporadic and unreliable, they may not change at all. There seems to be less disadvantage to brown pelage against snow than there is white pelage against a dead-leaf background. The fact that neither marten nor fisher molt to white in the cold months supports this. These larger weasels have enough body mass that heat-retention is less a problem for them, and so they opt for darker winter fur that is more absorbent of solar radiation while blending with tree trunks and other dark objects in the winter landscape. Hollow white fur is less conductive and so is better insulation than brown fur at higher latitudes where snow is certain. And insulation for a small, long-bodied mammal is critical. But in the temperate latitudes with a higher winter sun, dark fur is a better absorber of its radiance, and if there is no snow, the little weasels would be less conspicuous against a brown background than they would be if they had turned white.

Whether they turn white or not can affect the appearance of their trails. When a weasel turns white, additional fur grows on the soles of its feet. The result is a very diffuse snow track with vague pad impressions. If the weasel stays brown, its pads remain relatively naked through the winter and so leave distinct outlines. Time and again the identification of white weasels depends on gait patterns — the 2X DR repeating slant bounding pattern on soft surfaces and the longer 4X SR loping pattern on firm surfaces. Both these patterns are constant through the winter regardless of the amount of fur on a small weasel's feet.

Other Sign

Otter haul-outs or "scat stations" (Figure 6.104) are sites along the edge of water where otters come ashore repeatedly to defecate

Figure 6.104. An otter haul-out at the input brook of a beaver pond. Also look for these on peninsulas and isthmuses.

and urinate, have a good scratch by pushing themselves around on their bellies and then slip back in the water to continue the hunt. I have always called these "rolling sites," a term that I inherited as a beginning tracker and which I have continued to use despite the fact that I had never actually seen an otter roll over at one of these locations. Recent literature at my disposal distinguishes between haul-outs and rolling sites, reserving the latter for places where otters roll around on snow or the bare ground in order to distribute oil in their fur. And indeed, I have encountered sites that I attribute to otter where the vegetation has been flattened but without scat, although urine marks are sometimes present.

Before they defecate at a scat station, otters often scrape together some vegetation on the ground such as dead leaves, grasses and pine needles, and then deposit scat on top. Once you know where to look and what to look for, these sites are easy to find. The site itself often appears as a cleared space where rubbing by the otters has ground the leaves and needles to a fine duff. Near this cleared area may be found several scats of variable ages and levels of disintegration. When otters are feeding on fish, initially the deposit is in the form of a black tube. As it dries and bacteria consume its mucus content, the scat will whiten and begin to disintegrate into a mass of scales and small bones. In the summer when crayfish are available, the initial tube will be orange-brown, later disintegrating to a pile of reddish and white chitin. Although live crayfish are dark green with a few reddish highlights, once the chitin is run through an otter's stomach acid, it turns reddish, much as does a boiled lobster. Otters chew their food well and so the individual bits of exoskeleton will be quite small. In addition to scat, one may also find urine burns, sometimes called "brown-outs," where the otters have peed on vegetation that then turns yellow or brown. When the target of the urine stream is moss, these patches are quite conspicuous.

As to location, otters habitually cross peninsulas. Look closely for a run coming up one side, going down the other, often with a patch of pulverized duff and some scat at the high point. Another favorite spot is at the entrance point of a brook to a beaver pond as well as at its exit. These sites are used until they fill up with scat mounds, at which point they are abandoned for another suitable site until the first is cleaned by rain or snow melt. Otters are picky about these sites. I recall one of the first I ever found, along the bank of the Saugus River that was used intermittently for several years until a large maple fell just behind the site. Apparently the otters feared that the fallen log might hide a predator that could leap out from behind it and do them mischief. The otters promptly abandoned this favorite site for another only a few yards downstream that provided a better view of the surroundings.

Finally, one may occasionally find at an otter haul-out a deposit of white mucus, transparent when first deposited but emulsifying with exposure to rain and time to a white rubbery consistency (Figure 6.105). One of my videotapes seems to show this material being expelled at the end of a defecation, suggesting that it may be a mucus lining for the intestines perhaps to protect them from sharp fish bones as they pass through. Other explanations for this material have also been proposed, but the fact that it is excreted at the end of the deposit has always suggested to me that it is a protective lining.

Mink scat tends to be of two types — some resulting from feeding on fish or crayfish and others from predation on mammals or birds. The latter tends to be black, braided and looped; the former looks like a deposit of a tiny otter that has been feeding on the same prey. I haven't often found mink leaving multiple scats at repeatedly used scat stations. Most of the time I find individual droppings anywhere

Figure 6.105. Otter mucous excretion that has turned white with age.

Figure 6.106. Weasel and snowshoe hare droppings. Note the characteristic curve of the weasel scat.

along water, but especially atop root mounds of downed trees and on muskrat domes near the bank. The latter location is appropriate since minks are a principal predator of this rodent.

Weasel scat is almost exclusively derived from mammal/bird predation (Figure 6.106). It is even smaller than mink scat, generally measuring less than ¼ inch maximum diameter. As with mink, weasel scat is often deposited in a loop or crescent that may be blunt at one end and with a long tail. This shape seems to be a function of the short legs of a weasel, which brings its anus close to the ground so that the scat curves as it is being dropped. Weasel scat is easily overlooked. I recall seeing on an asphalt road a small item that I took at first for pocket lint discarded by someone walking the road. Only close inspection showed it be composed of long strands of white fur, perhaps belly fur from a mouse.

Species Notes: Hawk and Owl Sign

Wing Marks in the Snow

As will be shown in "Tracking Tip #5: Timing is Important" (pages 404-406), wing marks in the snow can easily be misinterpreted. There are several ways of identifying an avian predator, including first identifying the prey. Hawks are sight-hunters, that is, they must see their prey in order to attack it. Unlike owls they cannot detect a vole at night or under the snow by sound. Instead, they concentrate on prey that is about during the day and visible on the surface. If a kill-site or plucking site contains the remains or other evidence of squirrels, rabbits or songbirds, all of which are diurnal animals, then a hawk was the likely predator. Owls, on the other hand, mostly hunt by night, relying on their incredibly acute hearing to find their prey. A vole just under the surface of the snow makes noise as it squeaks and scurries through its tunnels. These help a perched owl zero in on the invisible creature and swoop down upon it in a long, silent glide. Thus, wing marks in the snow without the tracks or other evidence of surface-type prey suggests owl work rather than hawk (Figure 6.107).

The manner of the attack may also help to distinguish hawks from owls. As a general rule, hawks land on their victim. Once the hawk has the prey in its talons, it "foots" it, that is, suffocates the

Figure 6.107. An owl strike. Lack of small animal sign shows that the prey was detected under the snow by sound alone, a skill of owls but not hawks. Wing breadth suggests that this was a barred owl.

377

animal in its grip while piercing it with its talons to sever blood vessels. As a habit, many hawks "mantle" during this period, spreading their wings to cover the victim. This is probably residual juvenile behavior where a young bird in the nest tries to exclude its siblings from prey brought in by the parent birds. This mantling, as well as the act of landing on its victim and flopping around as it tries to control and kill it, results in a commotion of wing marks and disturbed snow at the site. There may also be blood if a severed artery leaked to the outside. Although some hawks may begin feeding on the ground, more often than not the predator will fly to a "plucking perch" to feed. This may be a dead branch of a standing tree, a cut stump or just clear ground where it can see around it while consuming its prey. Sometimes the same area will be used repeatedly. I recall finding several such plucking perches with globular felted pellets of more or less the same age among stumps in a small lumber cut atop Ladies Delight Hill in western Maine a few years back. A pair of goshawks that may have been planning to nest in the area had been hunting the locality and using stumps as tabletops.

Unlike most hawk kill-sites with their disturbed snow and commotion of wing marks, an owl kill is usually quite clean. There may simply be a punched hole and one set of wing marks. Having stronger feet than hawks, they tend to snatch their prey and fly off in one smooth motion. This is not always the case as shown in one of the photos (Figure 6.108). If the snatch was not made cleanly or if the prey is large or strong enough to resist capture, the owl may have to land on the animal, riding it while it seeks to kill with its beak by severing neck vertebrae. The resulting turmoil of wing marks will be similar to a hawk kill-site. Here you must look for other evidence. If the wing marks are clear, you may be able to count the number of primary feathers. Most buteos (soaring hawks) such as red-tails, have five primary feathers. These are the long flight feathers at the end of the wing that look like fingers when they are spread (the broad-winged hawk, a common little woodland buteo, has only four). Members of the accipiter family such as Cooper's hawks and goshawks, which also often take prey on the surface of the snow, have six. Owls, on the other hand, typically have eight to ten of these feathers. In counting feather marks be aware that the wing outlines are created on the downstroke as the bird lifts off. On this stroke raptors often seem to

Figure 6.108. Normally owls strike and fly off with their prey in a single motion. However, if they miss, they may land on the snow and create a commotion that looks similar to a hawk attack. Here the large number of primary feathers imprinted in the snow and, once again, the lack of small animal tracks in the vicinity indicate that this was the work of an owl.

fold together the first two or three of the primaries, perhaps to give a stronger striking surface on the snow as they vault back into the air. This can make an owl with 8 primary feathers look like a hawk with 5. However, close inspection on good snow may reveal the edges of the missing feather marks.

Here is another word of caution. Ruffed grouse often roost under the snow, using the insulation of fresh powder to protect them through the winter night. In the morning they may explode from these roosts directly into the air, leaving behind perfect wing marks that can look remarkably like the punch-and-fly pattern of an owl. Note, however, the relatively large opening in the snow, much bigger than the fist marks of an owl. A search of the hole will show melted snow around the form of the grouse and may also show droppings the bird excreted during the night. Most of these droppings will be granular tubes, narrower than a pencil, with whitewash on one end.

Hawk and Owl Pellets

Many birds, including all raptors, eject pellets. A pellet is the indigestible remains of prey that is regurgitated from the gizzard, the muscular stomach that serves to "chew" the prey and mix it with digestive juices.

Hawk and owl pellets are usually easy to tell apart. Hawk pellets tend to look like smooth, felted globs or teardrops with few bones on the surface (Figure 6.109), while owl pellets show all sorts of skeletal parts including jaw bones and skulls (figures 6.110-6.111). I have read a theory that suggests that the stomach acid of owls is weaker than that of hawks, so that small bones in an owl's gizzard are not dissolved during digestion. However, I suspect that the different appearance has to do more with the manner of consumption. When an owl feeds on its kill, it typically either swallows small prey whole or sections bigger victims into the largest chunks it can fit down its gullet — bones, fur, feathers and all. The bones survive the stomach acid

Figure 6.109. Hawk pellets often have a smooth, felted look. Blue jays were imitating a goshawk in the vicinity of this one, and fresh blood at the scene suggests a recent kill. Before feeding on its new prey, it regurgitated a pellet from its previous meal.

Figure 6.110. This great horned owl pellet was found under an habitual roost in a pine grove where the owl had perched out of the sight of harassing crows.

Figure 6.111. A smaller owl pellet, perhaps from a screech owl. Since owls swallow prey whole, many bones usually show on the surface.

and appear in the expelled casting. A hawk, on the other hand, usually strips flesh from the bones and then discards the skeleton. Enough fur or feathers are eaten to make a pellet, but few bones will be included. Illustrated are several examples: the bony owl pellet, based on size, may be from a screech owl while the smooth, felted pellet was probably from a goshawk. The great horned owl pellet was found under a roost; its maximum diameter was over an inch.

The height of an habitual owl perch above the ground can often be estimated by the distribution of the pellets as well as white-wash on the ground: the broader the radius, the higher the bird. In the case of owls, look for a stout branch above that offers both support and concealment in foliage.

Before a raptor at a plucking perch begins to consume its prey, it often ejects a pellet of the remains of its last meal. Therefore, the pellet that is found at a plucking site has nothing to do with the current kill.

Keep in mind that other birds besides raptors eject pellets from their gizzards. Any bird that swallows vertebrate or invertebrate prey is likely to do so. This includes fish-eating birds like mergansers and herons that swallow their prey whole, crush the victim in their gizzard, pass the digestible parts to their stomach and then eject the shells and bones. The resulting pellet can look a lot like otter or mink scat. However, each pellet deposit will be a single discrete ejection while scat often has two or more discreet segments.

Plucking Perches versus Kill Sites

Sites where a predator consumed its prey are often confused with kill sites. Most predators, including both raptors and mammals, eat what they kill at a different location from where the kill was made. A bobcat that kills a rabbit in the brush around a clearing may carry its victim into nearby woods for greater security while it feeds. A mountain lion may drag a deer carcass a considerable distance over amazing terrain resistance in order to feed and, more importantly, to hide the remains for later consumption. A fisher will often carry a squirrel or rabbit to the base of a tree and then up to an old raptor nest where the predator will feed and rest. Then it may cache what is left of the prey there high in the tree above where a raccoon or red fox might sense it and steal the fruits of its labor. In one instance reported to me recently a fisher chose an active Cooper's hawk nest for this purpose, forcing the parent birds to abandon the nest.

Commonly, raptor "plucking perches" in dense forest are simply a clear space of ground. Since a hiking trail is such ground, it is chosen surprisingly often. The most common remains that I find at such sites are blue jay feathers. Blue jays play a most dangerous game with accipiters like Cooper's and sharp-shinned hawks. If they discover one in their range, they will harass it relentlessly. I took some video-tape years ago on a beaver pond at Quabbin where a sharp-shin was subjected to this treatment. The hawk was perched at the top of a dead snag while several blue jays bounced up the lower branches toward it. Suddenly the hawk would swoop downward to make a pass at one of the jays, which would dodge around the far side of the snag. After the hawk returned to its treetop pinnacle, the jays would once again work their way upward to within striking distance of the sharp-shin, whereupon another attack would ensue. This went on for 20 minutes. Judging by the frequency with which I find blue jay feathers at plucking perches, the blue jays do not always escape harm in their dangerous game.

When one finds a pile of feathers, careful examination will tell whether the predator was a raptor or a mammal. In removing the larger quilled feathers, it is well known that a fox, for instance, will often but not always bite through the feathers with its carnassial teeth while a raptor, lacking teeth, is obliged to pull the feathers out of the carcass. Look for these cuts versus pulls among the feathers. One can examine the large feathers for grooves across the shaft. Indentations that are close together indicate a raptor's beak. Also, the canine teeth of a mammal may have made holes in the broad part of the feather. Finding a single feather along a hiking trail might not mean much. A bird flying through branches often loses one or two in the process. Also, birds molt periodically, during which feathers fall out at random locations. Once again look for evidence of toothing or beaking to distinguish these from the results of predation.

If you find more than a couple of feathers in the same area, then you are probably near either a kill site or a plucking site. Considering recent, local weather may give you a direction to it. For instance, if a Canadian front has come by in the past few days with accompanying strong winds from the northwest, then a search in that direction from

Figure 6.112. A river bank at low water is a good place to look for bird sign such as these bald eagle tracks with a faint raccoon print at the top. Look for raven and crow tracks, as well. Being black they absorb a lot of solar radiation and so need to fly down to drink fairly often.

scattered feathers may turn up a site. The dispersion of the feathers, furthermore, will give you an idea of the distance to the site: the more dispersed the feathers the farther away it is likely to be. Heavy rain or snow tends to mat down feathers at a feeding site or plucking perch so that they no longer can be blown around. Several feathers floating loosely along a hiking trail, on the other hand, suggest a recent occurrence. By thinking about evidence, then, you may be able to reconstruct the direction, distance and age of the event.

If you are searching for bird tracks, a river bank at low water is a good place to look. There, if you are lucky, you may find bird sign such as the bald eagle tracks shown in Figure 6.112. Notice the faint raccoon print at the top of the frame. The tracks of these animals that habitually skirt the margins of water bodies are often found in the same locations. One should also look for raven and crow tracks here since, being black, these birds absorb a lot of solar radiation and so need to drink a lot.

Species Notes: The Great Foolers

Rezendes used to say that, given the right conditions, any animal can make its tracks and sign look like those of any other animal. This may be hyperbole, but the point is well taken. Consider the confusion described earlier in this book between a gray squirrel's track pattern and the track of a fisher or the perception of a deer's print as that of a duck! There are two species, however, that create so many mistaken identifications that I regularly describe them to classes as the Great Foolers. Even after 27 years as a tracker, I still have to keep my guard up against embarrassment by these two.

The first, abundant in most parts of the country, is the raccoon. As anyone knows who has had a supposedly animal-proof trash container opened by these enterprising animals, raccoons have marvelously dexterous feet. (I almost wrote "hands," so nimble are its front feet at manipulating objects.) While the comparatively wooden feet of a coyote or fox are good for little else than traveling and pinning prey to the ground, and even a bobcat's more mobile feet are capable of little more than snagging prey and climbing trees, the raccoon is actually able to handle things. For instance, it is famous for feeling through the muddy margins of beaver ponds for edible organisms hidden there, then "washing" the tidbit with its "hands" before eating it. There is a revisionist theory insisting that the washing action has some other obscure explanation. Having examined much raccoon scat over the years that contained of a lot of grit, I suspect that the original explanation is the right one. Perhaps eating dirt is as disagreeable to a raccoon as it is to us.

For the tracker trying to make an identification of a partial track in thawed snow, for instance, all this manual dexterity proves a challenge. Raccoons can extend and spread their toes for support on soft mud (Figure 6.113), or contract and close them to reduce thermal loss (Figure 6.114) when crossing a frozen pond. This latter leads easily to mistaking them for fishers, small otters and even canids or cats (when one of the closed toes fails to register). With any uncertain identification, you should keep a shadow of doubt lurking in your subconscious that you may be on verge of being tricked once again by this great fooler.

Raccoon scat can also be deceptive. Raccoons, like turtles, seem to be able to grow as large as the food supply permits. Because of

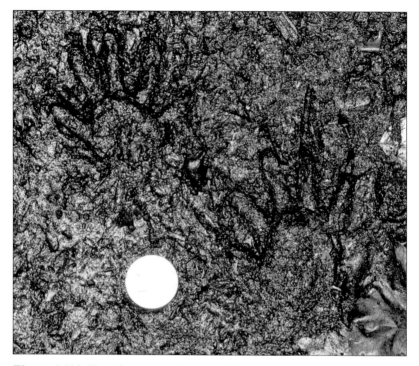

Figure 6.113. Here the raccoon has spread its toes for maximum support on the soft mud of a streambank. Regardless of the toe arrangement, the rounded leading edge of the secondary pad is a clue to this species.

this, their scat can also vary in diameter. Since they are omnivores, the resulting scat can also vary a lot in appearance, and the resulting tubes, ranging from ½-inch to ¾-inch in maximum diameter, can easily be mistaken for those of a scavenging canid or felid. Several clues may help distinguish them, however. First, raccoons often follow watercourses, leaving scat on the root hump of a bankside tree, frequently, in my experience, one that leans over the water. Another favorite location is at the base of a large tree, the upper branches of which may have been used for a day-den. Look for a broken top on the tree or a flattened whorl of branches or even an abandoned crow or raptor nest, all favorite day-bed sites. Another clue is the omnivore nature of the species itself. Since raccoons often excrete at scat stations, such a site with deposits of many different colors and consistencies is itself a sign of this species. Carnivores, such as bobcats,

Figure 6.114. Raccoon prints with toes closed, smaller front on the right.

also leave collections of scat at specific points. However, these will be composed of the fur or feathers of whatever prey animal is currently in surplus, and thus the resulting scat is likely to be more uniform.

When the mid-summer berry crop ripens, raccoons and just about everything else feed on this resource. The resulting raccoon scat is plops of seeds and soft material large enough that I have seen it mistaken for bear droppings derived from the same food source. Look for some tubular form in the otherwise shapeless mass, the differences in diameter suggesting the depositor. Look as well for the kind of damage a bear often leaves at a feeding site: matted down vegetation as well as broken stalks and branches.

Warning: Avoid fractionating raccoon scat. It can contain the spores of a heart roundworm that can make you very ill, even causing death in some cases. This is particularly so with dry scat, where disturbing it can cause invisible particles to be released into the air where they can be inhaled.

The Other Great Fooler

The second of these especially deceptive species may come as a surprise: the snowshoe hare. With large fur-covered feet and a distinctive bounding pattern, hares are usually an easy identification. However, the front and hind toes of this hare can be spread widely (Figure 6.115) or contracted narrowly (Figure 6.116), depending on surface conditions. As a result, the feet themselves can leave very different impressions as shown in the illustrations. Furthermore, in heavily crusted snow where only a part of the familiar lagomorph bounding pattern is discernible, or in cases where the hare stayed in

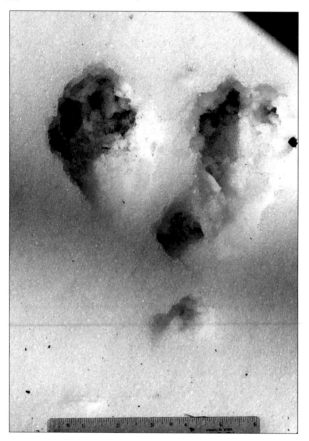

Figure 6.115. On deep, soft snow the same hare can spread its toes to almost unbelievable size. When the front prints don't register, these tracks are often mistaken for lynx, cougar or a large canid like a wolf or domestic dog.

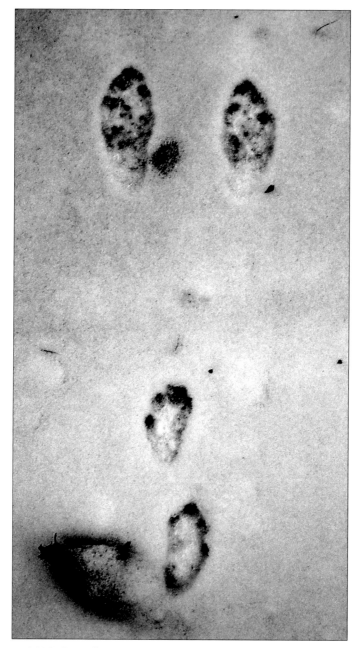

Figure 6.116. On a firm surface a snowshoe hare closes its toes, leaving compact impressions. Such a toe arrangement conserves body heat.

one place feeding for a while, shifting its feet around and obscuring the pattern, identification can be more difficult.

The most common error with this species is to mistake its prints for those of a canid. This can happen in two ways. If the surface is firm, the front feet may penetrate and leave tracks while the larger and more lightly loaded hind prints may not register at all. Spread front toes of a hare can look amazingly canid-like. However, the medial and lateral pads will not show a pie-shaped outline, instead appearing oval and often coming to a fur-point at their forward end. The other confusion with tracks of the dog family can occur where the fur-insulated hind feet remained in one place while the animal fed or rested. In this case, the body heat of the animal can melt toe indentations into the surface. If looked at carelessly, these toe impressions may show a central mound with star-like radiations, just like a canid print. As with any identification and interpretation, it helps to back off one's perspective a little and look at what the animal was doing while it sat. If twigs have been nipped on a neat diagonal just in front of the prints, it wasn't a coyote! Taking in an even broader context, is there more of this kind of sign in the area? Evidence of a lot of heavy browsing on plants of appropriate height (given variable snow depth) may alert you to a more probable identification. "Caveat tracker."

Species Notes: Cougar

Cougars go by many names: mountain lion, puma, catamount, panther and painter. They are a singular new-world species, although arguably there may be a number of sub-species depending on who is doing the taxononomic analysis. For the most part, this cat is a denizen of the West although persistent reports suggest that it is expanding its range slowly eastward. Since it is the only large cat north of Mexico and one of a handful of animals that may be dangerous to humans, much myth surrounds it and enlivens the imaginations of people who believe they may have seen it outside its traditional range.

My own experience with the animal is restricted to the Southwest, specifically a region of north central Arizona consisting of habitat ranging from chaparral desert to alpine forests. The cougar is quite democratic about its choice of habitats. The theme that runs through all of them is that the habitat must support fairly large prey animals such as deer, elk or peccary. In addition, they must include areas of refuge from contact with humans and, more importantly, their dogs. Cougars seem to equate domestic dogs with wolves, their traditional enemies. It may be an exaggeration to say that a Chihuahua can chase a mountain lion up a tree, but it is certainly true that the traditional method of hunting these cats is with horses and dogs. Chased by a pack of hounds, a cougar will tree for protection, making it an easy mark for the following horseman.

In warm environments, ungulate prey need water with which to cool their large bodies. Consequently, cougar sign in the Southwest can most easily be found by concentrating one's search on natural waterholes, on man-made "tanks" for watering cattle and on the routes to and from these features. Often these routes are dry washes that flood suddenly following cloudbursts. The rushing water clears them of brush whereupon they become relatively smooth highways through desert scrub. Furthermore, the sandstone ledges, exposed by erosion, trap water in puddles from which cats and other animals can drink.

Another place that has yielded cougar tracks is the passes between canyon bottoms. The dense shrub-growth of the bottoms provides cover for large prey animals as well as concealment for large stalking predators, of which there is only one. A cat cannot stay for long in any one canyon indefinitely because the prey animals there learn of its presence and become very wary. Once that happens the cat must move on, using the lowest pass it knows to travel to another canyon bottom. Many of these canyon passes accommodate human hiking trails and at these points it is possible for humans to encounter not only cougar tracks but, perhaps regrettably, the cougar itself.

Up on the plateau I have found cougar tracks at places where the table-top mesas are constricted into a narrow ridge. To traverse from one of these sky-islands to the next, the cat must pass over the constriction, leaving prints in the talcum-powder dust of the hiking trails that pass over the same bottle-necks.

Cougars are stealth hunters, lacking the heart and lungs for long tail-chases. Instead, the cat must close carefully on its prey until it is within 30-40 feet before charging. If the attack is successful the cougar, typical of all cats with which I am familiar, will drag its prey to a concealed and secure site for feeding. This may simply be a matter of carrying it farther from a hiking trail, or it may mean dragging it under overhead cover to shield it from the eyes of scavengers like magpies and ravens. Overhead foliage may also serve to dissipate scent, helping to conceal the carcass from patrolling vultures. After feeding the cat may scrape debris over the kill to further conceal it and then return later for another meal.

Cougar tracks are large, about 3-4 inches across the four toes, although for support on soft surfaces the cat may spread its toes to 5-6 inches. In distinguishing them from the spread-toe prints of a male bobcat, note the much larger secondary pad of the cougar. Although a male bobcat can spread its toes out to nearly 4 inches on soft spring snow, this does not affect the size of its secondary pad, which will be much smaller: 1½ inches wide compared to over 2 inches for the bigger cat. All felids have deep, rubbery foot pads designed for shock-absorption as they jump down from tree limbs or ledges. These pads are easily distorted under load and so the tracks of the same cat can differ from place to place depending on the amount of impact in the gait. Bounding, for instance, with its high arc of stride will result in considerably more pad-spread than walking. Track appearance also can differ on hard versus soft surfaces. A wet track left on hard sandstone will look different from one left in dust. It should also be noted that the front and hind tracks look different from one another. Bearing the weight of the animal's head the front toe pads spread more than the hind and the easily deformable front secondary pad tends to show more convexity, with rounded corners and edges (Figure 6.117). The lightly loaded hind show more flat or concave edges and sharper corners (Figure 6.118). Finally, I have noticed that the feature I most often spot first in obscure cougar tracks is the feline tri-lobing at the posterior of the secondary pad, indentations and raised seams that are often very pronounced in this cat's tracks. Keying on this feature allows me to reconstruct the rest of the track, the contours of which gradually reveal themselves as I look at where the rest of the print should be.

Trailing cougars in the redrock country can be challenging. It is much easier to find tracks than to follow them for any distance. Soon the red dust in which the tracks are followed switches to hard sandstone or cobbled stream boulders. At these places one must inhabit the cat's mind and divine its intention in order to pick up the trail again farther along.

A cougar that is travel-hunting along the dry washes to waterholes will use an efficient walking gait with a low arc of stride. Most often this takes the form of an aligned overstep walk, a gait common to many cats (Diagram 6.18). Loping may be used for descents, galloping

Figure 6.117. Bearing the weight of its head, a cougar's front print tends to spread more widely than the hind, with more convex margins.

for attack or escape and bounding to clear impediments. Otherwise, these cats seem to walk a lot. Several examples that I have in my records show a typical step length of 17 inches for a direct-register walk and up to 20 inches for overstep single registrations. The total distribution of step lengths for all walking trails in my records is 14-20 inches. Walking trail widths have been 5-7 inches. Given the relatively small track widths of these examples, most if not all of the measurements may be more typical of females. Diagram 6.18 is a trail speeding up from left to right.

Diagram 6.18

Figure 6.118. A cougar's hind print is less heavily loaded, with more concave margins. In both tracks the tri-lobing on the posterior of the secondary pad is pronounced.

Cougar scat is large; one I found in the Ponderosa pine forests of the Coconino Plateau had a 1½ inches maximum diameter (Figure 6.119). It contained fur, most likely from elk, the only large ungulate in these open conifer woodlands. Like other cat scat, cougar droppings often have a blunt end, some segmentation and perhaps a tail of excreted fur. Whether or not the scat is found in a prepared scrape, as with other cats, seems to depend on the surface on which it is deposited. Stony ground may not be scraped while loose leaf duff or dirt probably will. Like the droppings of other cats, cougar scat will not show large bone chips on the surface. Lacking the teeth for crunching

Figure 6.119. Cougar scat. This one, presumably from feeding on elk, was found about 200 yards from a busy highway.

large bones, cats use their raspy tongues to lick the meat off, leaving their characteristic feeding sign of cleaned bones, the larger of which are intact.

A word or two of warning. Cougars are, once again, stealth hunters. You may be watched by one without knowing it. Most often these cats recognize humans as dangerous and try to avoid them. However, young animals that are inexperienced in hunting may be driven to attack a person out of hunger or youthful lack of caution. It is best to travel with a companion in cougar country since this cat is less likely to attack two people walking together. Secondly, be wary of crouching down, for example to examine tracks, to tie a shoe or to excrete. Such a posture may be seen as an opportunity for attack. Be especially careful at points where human and cougar travel may coincide, such as the canyon passes and mesa constrictions mentioned above but also under bridges and through large road culverts. Prey animals recognize these as dangerous ambush zones and so should we.

Chapter VII

Applications

This chapter begins with a number of tips on techniques that I have found useful over many years of tracking. Many refer back to items previously introduced in the book in various places. Some are concerned with simply finding and identifying animal sign in different lights and backgrounds, that is, with "looking." Others deal with perception and habits of mind in the act of interpreting, that is, with "seeing." The second part of the chapter presents a series of problems designed to test your skills at both identifying and interpreting evidence presented in photographs of real tracking situations. The chapter ends with several accounts of tracking in Arizona where I have often gone in search of mountain lion sign in recent years. Being thinly distributed and almost mystically elusive, this animal calls out the tracker's best skills at finding and interpreting its sign. As an Easterner I would hardly qualify as an expert on cougars, but the principles of the tracking game translate to any natural landscape, even one as foreign to my background as the chaparral and red rock canyons of the desert Southwest. Besides, perhaps there are things to be learned from the experiences of an ardent student about finding and interpreting animal sign as he goes up against one of the most elusive of real-world mammals.

Tracking Tips

Tracking Tip #1: Face the Light

What the eye admits to the mind is, of course, light reflected from surfaces. Detecting faint impressions in the substrate is often aided by taking advantage of the contrast between darker shadow and the lighter areas of an impression in a surface. The rods in our retinas detect the contrast between these in the contours of a track and throw it into relief in the mind. The dramatic effects that the angle of light can have on an object can be seen by holding a quarter at

Figure 7.1.

Figure 7.2.

varying angles to the light from your desklamp. Washington's image changes constantly as you tilt the coin.

If the sun is overhead as at noon in mid-summer, we are likely to overlook much subtle information because there are no shadows. If the sun is lower in the sky, as it is for most of the year in temperate latitudes, and at early morning and late afternoon in the summer, then shadows form on the side of an impression toward the light. Whether or not we detect the impression depends largely on where we stand relative both to the track and the source of light. If the sun is behind our head, then the shadow along the near edge of the impression is hidden from view by the print's lip. However, if we stand facing the light, then that same shadow is on the far side of the print and stands out against the background more clearly.

The photographs in figures 7.1 and 7.2 are of the same site, the upper one with the sun behind and the lower one facing the light. A thorough search of the upper photo may reveal a faint print at the left center, but that is all. The frame of the photograph suggests the presence of something else to its right, but it is invisible. The lower photo, on the other hand, clearly shows two footprints of a bobcat. Even if the shadow doesn't completely outline the pads of either the obvious one to the right or the fainter one to the left, there is enough in each depression to suggest a whole to our perceiving mind.

As I suggested above, the frame of the photo forces the reader to concentrate on a small space in which he is implicitly promised some evidence. However, without that frame the tracker in the field may easily overlook much evidence hidden by the absence of shadow. We cannot always keep the light on the opposite side of an area we are searching, but it helps to do so whenever we can. And the light may not necessarily be direct sunlight; sometimes simply sky-space in a broken canopy may suffice.

At mid-day with the summer sun directly overhead, I often shade a suspected site with my body and then search the area with a small, high-intensity flashlight held at a 45 degree angle. If track depressions are present, their contours will show in shaded relief.

Tracking Tip #2: Walk around It

For some reason, humans seem to pick tracks out of a disturbed background more readily if they are in a following orientation, that is,

Figure 7.3.

facing in the direction in which the animal was traveling. If at a presentation I show a slide of a track or trail running horizontally on the screen, people invariably tilt their heads to the side to view it. I suspect this tendency is atavistic, a throwback deep in our psyches to our primitive hunter-gatherer past where following animals was the means for life. Whatever the reason, for the tracker the simple tactic of circling a commotion in the dirt will often cause recognizable tracks to jump out to the mind's eye from the background at the moment in the circling when a following orientation is achieved. You may try this yourself in the photos. The one above (Figure 7.3) shows a confusing commotion of muddy disturbance without readily identifiable tracks. We may know something passed there and suspect animal tracks in the commotion. Indeed, in the left of the frame there appear to be pad impressions, but it is difficult to find anything solid on which to pin an identification.

Now look at the photo in Figure 7.4. Here the frame has simply been reoriented to a following point of view. From this perspective it is much easier to separate the tracks from the rest of the confusing background. Slightly left of center in the frame are the front (below) and hind (above) tracks of an eastern coyote in a displaced trot. In the

Figure 7.4.

field, the same effect of reorienting the photo can be achieved by simply stepping around a commotion to view it from all points of the compass. At the point where we are positioned behind the animal, looking in the direction of its movement, our mind will more readily dismiss the myriad insignificant details in the site and suddenly see the tracks.

Not only is this tactic useful when confronted by a surface roil, but it also helps to distinguish prints of different species that overlap one another. One might not readily confuse the print of a deer with that of a snowshoe hare where the sign is separate, but if the prints overlap and the animals had been moving in different directions, they can present a surprisingly puzzling appearance. By circling the site the distinction between the two is likely to become obvious at the point where you are positioned following each animal.

Tracking Tip #3: Look More Closely

A confusing background can obscure otherwise obvious evidence. The site in the photo (Figure 7.5) was on the shore of Quabbin Reservoir in central Massachusetts. Something had been scraping at the sand on the beach near the water's edge, but at first glance the site appears to be so chaotic as to yield little evidence of the identity of the scraper. However, if we pause and look a little more closely among the seemingly random scrapes, we will see the signature of the animal that did the work. At the center bottom is a clear front footprint of a river otter.

We can tell that this is a front print by the more symmetrical arrangement of the five toes compared to a hind print of this species, which shows the toes retarded on the medial side. Because raccoons also dig in mud and sand and can leave some surprisingly otter-like prints, some care must be taken to distinguish the two.

Spoil from a dig or scrape of any sort often holds the last footprint the animal made just before departure. This may be true of bear, coyote, fox, skunk and raccoon as well as otter. Here, on the shore of a lake, the animal was presumably "clamming" after fresh water

Figure 7.5.

mussels although no shells were found. It may also have been after sand worms or other subterranean creatures, prey that would leave no evidence behind.

A word about this clamming. Although fresh water mussels may have the taste and consistency of rubber bands to us, they are common prey for otters, minks, muskrats and raccoons. The well-developed canine teeth of otters and raccoons are used to puncture the mussel shell at its apex. The puncture hole apparently either kills the animal inside or cuts the muscle it uses to hold its protective shell closed. Sometimes minks, which have relatively small canines, will simple hook their claws in the seam of the shell and pry it open without puncturing it at all. Muskrats feeding on mussels use their sharp central incisors to pry open the shell in a similar way. Look for minute abrasions along the sharp edge of the shells for evidence of either method. Finally, gulls also feed on mollusks, but open them by flying up in the air and dropping them on a hard surface. The result will be smashed shells unlike any mammal-feeding.

Tracking Tip #4: Obvious Conceals the Subtle

We see with our minds, the eyes simply acting as a conduit of information to be translated into coded electrical impulses for our brains to decipher. One trait of the decoding mind is to simplify the data that travels through the eye, exaggerating what is obvious and deemphasizing or eliminating the background. This editing function of the mind, while useful in our civilized lives, may result in subtle information at a track site being overwhelmed in our awareness by more obvious details.

In Figure 7.6, the mind fastens on the obvious tracks of turkey hen and her young at the upper left as well as the deer indirect registration at the bottom left. We may also see the motorbike tire treads on the bottom right. These tracks in mud are so conspicuous that it is easy to miss the more subtle sand prints of a red fox in the same frame.

After the mud in which the turkey, deer and bike tracks were impressed had dried solid, a fox came along from the opposite direction with dew-wetted feet and stepped first in some sand and then onto the hardened mud. The tracks are so light that they are easily overlooked.

Another example of this sort of failure of perception occurs when an animal with large feet for its body size steps on the same surface

Figure 7.6.

as an animal with a closer ratio of foot breadth to weight. In this way a deeper squirrel's print next to or even within that of a snowshoe hare, for instance, may be noted while the larger animal's fainter print may be missed.

We need to remind our mind to see past or around the obvious, looking for other information in seemingly insignificant background. With a little practice this will become habitual when surveying a site.

Tracking Tip #5: Timing Is Important

Years ago in a respected nature magazine I saw a photo similar to this one with the trail of a small animal that disappeared between a pair of wing marks, and a caption describing the scene as the kill site of a kestrel on a vole. However, although a casual look at the photo showed the 4-print patterns of a small rodent, it was a bounding rodent. Since voles mostly scurry in a flat plane, the bounding rodent in question was clearly a mouse. Furthermore, since mice forage over the snow almost exclusively at night and kestrels, like all falcons and hawks, are night-blind, the predator in the photo was almost certainly an owl.

Careful examination of the prints in Figure 7.7 shows that here, as well, we are looking at a mouse rather than a vole as the prey.

Figure 7.7.

While the crust on the snow caused incomplete registrations of patterns, a few are clear; one in the lower left corner, for example, shows a convincing mouse pattern. Anyway, finding a vole trail of any length on the surface would be an exceptional event. This genus of little rodents, favorite prey of hawks, owls and other predators, tends as a result to hide under the surface of the snow as much as possible. There they construct elaborate tunnels at ground level to get around to their food, the rootlets and bark of young growth. Only on rare occasions, for instance when surface meltwater floods their tunnels and then refreezes or when they encounter an obstacle under the snow, do they emerge onto the surface. Because they live in tunnels with low overhead they don't use the bounding typical of other rodents with its high arc of stride. On the infrequent occasions when they emerge onto the surface, they follow their normal habit of flat scurrying. The trails they leave behind show an irregular pattern often back and forth over an obstacle under the snow, a trail that disappears back under the snow as soon as possible. Often, too, this runway will be positioned under some overhead cover like a low-lying branch that the animal senses may protect it from a swooping raptor.

Mice are also rodents, of course. But their food centers on nuts and seeds. As the overwintering structures of birches gradually disintegrate during the season, the seeds falling onto crusted snow are blown around and dispersed by the wind. This is the tree's strategy for spreading its kind. While lying on the snow surface, however, they are a ready source of nutrition that is seemingly irresistible to mice, which accept the danger of running around on the surface in order to gather them. The long series of 4-print bounding patterns found wherever wind-dispersed seeds are scattered on the snow are made by mice, not voles. As they wander on the surface, mice rely on their large eyes and ears to detect the approach of a raptor. The eyes and ears of voles, by contrast, are relatively small as neither sight nor hearing helps them in the darkness under the snow.

As the tracks on the snow in our illustration are those of a mouse, and mice are nocturnal wanderers, it stands to reason that the predator here, as was the case in the magazine, was an owl. The eight registering primary feathers supports this. Extrapolating from the width of the mouse patterns, the owl in this case was probably a barred owl whose calls I had heard in this stretch of woods on the evening before I took this picture. The photographer in the magazine article was misled by the common assumption that the event occurred in the same conditions as the observation of its evidence. Because he viewed the scene in daylight, he assumed the attack was in daylight as well.

Timing is important, not only for the predator and its prey, engaged in an adaptive race toward better defenses matched by better attacks, but also for the tracker following after the event. In reconstructing a scene from the track record, then, it is important to reimagine the event at the time and in the conditions in which it occurred. If these are uncertain, imagine the event at several different times and conditions of weather and surface until one scenario seems to fit. In this way you may be able to back into a sense for these things from the evidence at hand.

Tracking Tip #6: Counter-marking and Time Compressions

Our discussion in Tracking Tip #5 of the importance of considering timing can be extended to include simultaneity. When we find a site with more than one sign of wildlife presence (Figure 7.8), it is

406

Figure 7.8.

perhaps natural to assume that the sign was all made not only in the same conditions in which we view it, but also at one moment in time. Many years ago when I found my first otter haul-out, I discovered among the debris of fish scales and bones another scat, this one a well-defined tubular dropping with a pointed end. I assumed that this scat as well as the others was left by the otters when they had last used the site. Unfortunately, the illustration in the rather poor tracking book I was using at the time seemed to confirm this assumption. Not until I had acquired more experience did I realize that the drawing in that book was highly imaginative. Eventually, I also realized that the other scat at the haul-out had been left not at the time of the otters' visit but later by a passing coyote, perhaps asserting its property rights.

This story illustrates two points. First of all, it presents an example of the habit of counter-marking sign by many wild animals. Secondly, what we see at a tracking site may represent a compression of time onto a single page. While we see it instantly and in the present, it may contain evidence accumulated over a longer period.

Counter-marking is very common, especially among wide-ranging predators. A gray fox, for instance, may cover bobcat urine marks with its own. Coyotes often deposit their own scat over that of other animals that they encounter in their range, including those of its own species. When my friend, Kevin, arrived in April to open his cottage in western Maine, he detected the skunk-like odor of red fox markings around his property. Fearing for his cats, he set about counter-marking all the fox marks he could find with his own urine, hoping to warn off the potential cat-predator. A week later he returned to find each one of his own marks had been re-marked by the fox! After all, whose property was it anyway?

You may wish to review the incident reported in "The Inner Game: Seeing" (pages 56-58) about the snake, coyote and probable mink predation. In that case my friend had assumed, because he observed both the coyote tracks and the mangled snake at the same time, that the killing of the snake also occurred at the same time, whereas a more careful look at the site suggested that there may have been a hidden third party and a lapse in time. It is always risky to draw a straight cause and effect line between what we see in Nature without considering that the relationship among pieces of evidence may be more complex than it at first seems. Suspecting hidden causes and time interventions can suggest alternative interpretations.

Tracking Tip #7: Determining Direction

In which direction was the animal traveling? Examine the photo (Figure 7.9) closely. Note that each track has a small mount of powdery debris at the end closer to the camera. Novices often assume that this should appear behind the track as the animal accelerates, pushing backward and lifting up as its foot leaves the surface at the end of the stride. That may, in fact, be the case in high-energy gaits like loping (hind only) and galloping. However, the arrangement of the tracks in this trail indicates a walking animal, and walking imparts far less energy to the surface over which the animal is moving. As a result, in a walking or even trotting gait, debris should appear at the head of each print and not at its rear.

Reorienting our minds so that the animal is approaching the camera, we can see that each track has a rounded leading edge. This rounded print with a fair amount of straddle and a 15-inch step length identifies the animal as a bobcat. The mounding and powdery debris ahead of each track are caused by loose snow sticking to the fur on the leading edge of the paw and dropping off as the paw leaves the track and advances. Depending on the consistency of the snow, this debris may be mounding or clumps, or even a cone of fine snow diminishing ahead (spatially) of each print in the direction of travel.

In high-energy gaits such as loping, bounding and galloping there is much more push backward by the legs. This often results in mounding and debris both behind and in front of each track, the hind debris by the push-back, and the front debris by the lift-out. Once again depending on the surface, such gaits with their high arc of stride relative to a walk may also show impact debris in all directions around the track. This often occurs on soft, heavy surfaces such as mud and slushy snow.

In distinguishing loping from galloping you might look for lesser mounding behind the front prints in a lope, since less energy is imparted to the front legs in this gait than is the case with a true gallop where the front legs are actively engaged in supplying power to the gait. See pages 128-139 for an explanation of the differences between these two gaits.

As usual, there is a caveat. In mid- to late-winter as the sun climbs higher in the sky, it increasingly applies more radiation to the snow surface. The long shadows of January are replaced by intense

Figure 7.9.

sunlight. Irregularities in the snow that create shadows absorb more ambient light, and so they tend to melt differentially. A track is just such a disturbance, creating shadows and angles that absorb the intense light bouncing around off reflective surfaces. This can quickly create mounds and depressions in a print where there were none immediately after it was made. The careful tracker will take into account the season, the freshness of the track and the angle of the sun in interpreting mounding around a track. The "yeti effect" of sunlight on track impressions was described above (page 66).

In the case of the bobcat trail in the photo, the fineness of the debris forward of each track and the texture of the snow surface indicate a fresh trail after a recent snowfall, in which case the debris pattern was made at the time of the bobcat's passage.

Tracking Tip #8: Widen the Frame

In trying to identify animal sign it is easy to get so absorbed in its details that we miss the bigger picture. Sometimes this means we ignore the habitat context, or sometimes we may simply miss information in the immediate vicinity of the sign. In this case the scat in the photograph (Figure 7.10), found on a jeep trail through a woodland with dense understory in central Massachusetts, is clearly that of a fairly large carnivore. We can know this by the sweetish odor, which is distinctive for carnivore scat, and the general appearance including some fur at the upper part of the larger item. We may also notice that the scat lacks large bone chips on the surface. The scat of a coyote, the only other carnivore in this area that leaves a scat of similar size and quantity, often shows bone chips, but not always; scat resulting from a fresh kill, where the animal was consuming flesh or internal organs first, may not. Only when this meat is mostly gone does the coyote begin to gnaw bone, with chips then appearing on the surface of the resulting scat. Bobcats also have carnassial teeth that they use to cut through flesh and small bones but the muscles that operate them are weak compared to those of a coyote; thus the absence of large chips in a cat's scat.

To the absence of bone chips in the scat pictured here may be added other evidence for identification. An experienced tracker may recognize the ball-and-socket effect of one of the segmentations, a fairly good sign of a felid. (My friend and fellow tracker, Kevin Harding, suggests that this effect results from a pulsing of the anal muscles

411

Figure 7.10.

while the cat is excreting. On the other hand, Joe Choiniere suggests that it may simply be a function of eating style or digestion where fecal matter accumulates in the gut in small amounts that are pushed together as moisture is extracted and readied for excretion.) Obviously, experience helps in noticing such subtleties. However, a less experienced tracker may make the same identification just as reliably simply by widening the frame of his concentration to that of the photograph in Figure 7.11.

This wider-angle photo of the same event shows the telltale scrape that felids make in preparing a site for a deposit. Widening the frame even more, the novice tracker may note that this scene was on an unused jeep trail through a woodland with dense understory in central Massachusetts. This is a habitat favorable for stealth hunters. The stealth hunter in this part of the country that prowls through dense cover, prepares a scrape for scat deposits and leaves scat with ball-and-socket segmentations and without large bone chips is a bobcat.

Figure 7.11.

As you can see, we have involved three levels of concentration in this identification. First, we looked at the scat itself, took measurements of maximum width, examined its surface content, detected its odor and looked at its peculiar segmentation. Then we widened our narrow focus to include a few square feet around the scat, noting as we did the prepared scrape. Since dogs scrape with their back feet next to and after scat and urine deposits, we made sure that the scrape was made before rather than after excretion. Here the scat is directly in the scrape so, obviously, it had to have been deposited after its preparation, but bobcats sometimes prepare a scrape and then miss it when they position their hind end. We may also see whether there is debris on the surface of the scat from violent kicking beside it after deposition, the habit of dogs. One must be careful here to distinguish such accidental debris from the deliberate action of a cat in covering its scat to conceal it. Cats don't always do this, but may in the territory of a more dominant animal. Finally, we widen the frame even further to include the habitat in which the event took place. Did it occur in field, a mature forest with little underbrush or in an area with dense cover that is favored by predators that sneak around quietly looking for dozing or otherwise unwary prey? Collapsing the information from all three of these levels results in an accurate identification where information from just one level would not. (By the way, the butterfly is a duskywing absorbing nitrogen from the scat.)

Tracking Tip # 9: Things Fall Apart

This section reviews and elaborates upon information in "Species Notes: Black Bear" (pages 203-205), dealing with bear work on logs. As was explained in that section, when a fallen log disintegrates from simple decay, the clumps of wood and bark generally fall off and lay parallel to it. However, when a mammal works on the log, searching for ants, beetles and grubs, the pieces will be torn away and lay at an angle to the log (Figure 7.12). Three animals do this in the Northeast — bears, fishers and skunks. A couple of clues may identify which was the vandal. First, the distance the pieces lay from the log will suggest the strength of the animal: bears, of course, are much more powerful than fishers and skunks much weaker. Secondly, a

Figure 7.12.

414

careful search of the log may reveal claw marks as the animal raked its paw across it until the claws caught on a crack, giving the animal purchase to pull away a section. A search may show four or five claw marks of the bear about 4 inches across on the surface of the log and perpendicular to its lay. When a fisher does the same thing, the marks will be about 3 inches across and the pieces closer to the trunk. (With either animal, only four nail marks will normally be found, the fifth, medial toe being more vestigial.) With a skunk, the spread of the claw marks, if they are visible at all, will be less than 1½ inches across.

Since many things can cause marks in old wood, including fungal staining and wood-boring insects, some care needs to be taken in evaluating discovered marks. In shallow scratch marks look for relatively straight lines to distinguish mammal work from other causes. In cases where the claws may have caught immediately upon placement, there may be no scratches. In that case look along the edges of pulled-off sections for indentations and wood compression where the claws hooked into a crack in the wood.

Few things are straightforward in tracking, however. Some imaginative reconstruction may be needed to exclude the possibility that an already dead tree, well corrupted while standing in place, fell and shattered on impact. Look for other evidence, such as violent damage to other, adjacent trees from the falling timber, claw marks or the presence of ants or other invertebrates that may have served as a bear or skunk attractant. Backing off one's perspective may reveal nearby logs also damaged or rolled to uncover food hidden beneath. If the work appears to be fresh, the season in which it is found may also help. Bears do most of their log-demolishing after ants and grubs during periods when more plentiful and easily accessed food such as berries and nuts are not available. Early spring upon arousal from hibernation and early summer are the two starvation periods for bears when they may be driven to eating ants, grubs and beetles.

One must, of course, account for the possibility of mechanical damage by humans and our machinery. In areas frequented by people it is easy to overlook bear work as this sort of damage. The first bear mark tree I ever found, I and many others had passed by repeatedly, absently dismissing the wounds on a balsam fir as just this sort of mechanical damage. The tendency to do this is reinforced by the fact that bears often mistreat trees and sign boards along human walking paths that the bears appropriate from us after dark. These are the same locations where we might expect a trail-clearing crew or forestry machinery of some sort to scrape the bark off of trees. Finally, trees that fall across paths, human or animal, often get a frayed look as they disintegrate. This is not likely to be bear work, but rather the result of passers-by like deer, stepping repeatedly on the log or clipping it with the sharp leading edge of their hoof cloves as they move over it.

Tracking Tip #10: Rules to Be Broken

Look closely at these two scats. The photos are presented so that their relative sizes are true. (A penny is ¾ of an inch across). The scat in Figure 7.13 was placed in a prepared scrape on a snow-covered stump. The other (Figure 7.14) was also in a prepared scrape on the shore of a reservoir. Both were in central Massachusetts. Now, what species of wild animal left each deposit?

Trick question. They were both left by bobcats. The upper one was confirmed by tracks in the snow, and the lower one by the prepared scrape in which it lay. Note also the lack of contradictory evidence, such as no visible bone chips in either that might suggest a coyote.

The photos illustrate two points. First, no tracking guide (including my own, which contains several illustrations of bobcat scat) can cover all possibilities. Scat differs depending on size of the animal (which in turn may depend on sex and age), what it has been eating and how much it ate at one time. Although one often hears that bobcat scat is segmented, I occasionally have found coyote scat that is segmented as well. The visible presence or absence of large bone chips

Figure 7.13.

416

Figure 7.14.

on the surface is also given as a way of telling cat from canid. Bob-cats lack the teeth to crunch thick bones, but a deer carcass that has not yet been reduced by scavenging may yield coyote scat without bone chips as well. Thus rules that attempt to simplify and reduce possibilities of scat appearance, as well as other animal sign, cannot always be taken as valid. Identifications are based on the sum of the evidence, not on simplifying rules about particular aspects. The scat of any species, like its tracks, are usually variations on a theme. The trick is, with experience, to learn to recognize the theme.

Secondly, one should regard the size parameters presented in tracking guides, even trustworthy ones, as guidelines. At best, the sizes that are presented cover a normal range for an adult animal. But there can be exceptions outside the normal distribution. The scat shown in the bottom photo is unusually small, even for the female bobcat that left it. The one above it is very large. While its maximum width is within the normal range for bobcat, it is substantially longer than the theoretical maximum length that I have seen given as the rule for a bobcat. It might easily be mistaken for a cougar rather then the large male bobcat whose sign I find occasionally in that area. If you should

417

discover an out-sized scat or track, you have no way of knowing from its measurements whether you are dealing with the rule or with an exception. As always, look for other evidence.

Tracking Tip #11: Maybe it's more than one

An experienced tracker, when confronted by a seemingly patternless trail in the snow, immediately suspects that it was made by more than one animal. This is worth keeping it mind, especially when tracking social species. We have already seen how coyotes will follow exactly in one another's footsteps in soft snow as an energy saving tactic (Figure 7.15). Herd species like deer, elk and peccaries routinely follow one another out of habit, but rarely direct register in such a way and so the coincident trails of two of them may show an unrecognizable pattern as they overstep or understep each other.

In other cases you may find one animal following another of the same species that has trespassed on the owner's territory or hunting range. Furthermore, males often follow the scent trails of females during mating season. Females may follow males as well, although often in a coy, meandering fashion as if trying to suggest that they aren't really all that interested.

Figure 7.15.

A caveat here involves the trails of a skunk (Figure 7.16). These short-limbed animals often leave a loping trail where one 4-print pattern follows another with no defining space between and with variable order and spacing of prints. This may look like the trail of more than one skunk or perhaps a single animal tipsy on fermented berries. However, careful inspection will show the differing front and hind single registrations of 4X loping patterns that have been run together (Figure 7.17).

Not only will the trail that you are following have been made by animals of the same species but by animals of different species as well. A few winters ago a couple of my docents tracked what they felt sure was a gray fox. It had walked over the snow to a brookbed where, on the other side, its trail magically turned into what looked like that of a mink. They were mystified. Had they mistaken the identity of one or the other? A wider look at the scene and some imaginative reconstruction showed that the fox had taken a drink at the margin of the brook and then vaulted to the side up onto a fallen log where its prints were not at first seen. A mink, on the other hand, had swum down the brook and emerged from it on the other side at exactly the point where the fox had drunk, picking up and continuing the line of the fox's trail across the snow toward another drainage.

Bobcats often illustrate this tip, as well. They are notorious for using the trails of other animals to ease their passage over snow. Not only does the bobcat find it easier to follow a prey animal that has broken out the snow, but it may find the animal itself at the end of the trail and invite it to dinner.

Tracking Tip #12: Keep An Eye on Your Expectations

All trackers who have been at it for awhile have had the experience of looking at a piece of ground that seemed vacant of animal sign and then suddenly have a scat or animal hair or even a track seem to emerge like magic from the background. But a variation on this experience is the masquerading of wild animal sign as that of human activity. If you spent most of your woods-time in parks near human populations, it is easy to miss animal sign by assuming that what is in front of your eyes was made by people in one way or another. Such parks are laced with human "convenience trails," created by people who want to get somewhere more rapidly than the official trail system permits. You may get so used to seeing these trails that you tend to see them where

Figure 7.16.

Figure 7.17.

they are not. If your experience is mainly in a metropolitan park system, for instance, and then you go out to extensive forests in the countryside where you see a vague deer path through the woods, you may dismiss it without a thought as one of those convenience trails even though there are not enough people per square mile in the area to create one.

Any sign found along the human infrastructure of dirt roads and trails may have been made by either wild animals, humans or their pets. You may, in passing, notice a scar low on a tree that was made by porcupine gnawing and dismiss it as mechanical damage by the machinery of road crews. The same may be true of bear sign, as we have seen. Gnawing by bears on telephone poles along country roads may also be seen as mechanical damage by the machinery used to erect the pole. Furthermore, as we saw in the species notes, bears like to mouth trail signs, leaving tooth marks that are easy to mistake for the indentations of shotgun pellets as hunters sight in their weapons using the sign as a target. These latter examples illustrate a correlative tendency of the mind to see any mark on a human-made object as having been made by a human.

Conversely, some human activity may be seen as animal sign. A trail crew brushing back the margins of a hiking or other trail may make cuts on branches that look very much like ungulate browsing. Even away from roads and trails, however, it is a good idea to be circumspect in attributing sign to wild animals. While tracking bobcat with a private client deep in the woods of Quabbin, we came upon a large tree at the base of which were two parallel cuts in the duff of the forest floor. The cuts were less than a foot long, about 3-4 inches wide, about a foot apart and about three feet from the base of the tree toward which they were oriented. We puzzled over these for some time, suspecting pawing of the ground by deer after acorns or false truffles perhaps. However, when deer do that sort of thing, their scrapes tend to have a wide triangular shape rather than the narrow slits we were seeing. I pulled out my set of *Trackards for North American Mammals* and went down them one by one among all the mammals that might have made marks that size. One by one we eliminated bears, moose, snowshoe hares and so forth in descending size. Nothing fit. Then the dawn broke. We had neglected to consider the one other mammal that might be in these woods. The marks, we realized, had been made by the heels of a hunter's boots as he sat with his back against a tree waiting for a deer to show itself in the woods to his front. For most of my experience in these woods they had been closed to hunting. Only recently had an intense three-day hunt been permitted. Having developed an expectation from my past experience that I would not find sign of man or dog deep in the Quabbin woods, I allowed that experience to cloud my perception.

Tracking Tip #13: Several Lessons from a Bear and a Pot

While prowling around on the flood plain of the Megalloway River in northern New Hampshire in late June, I discovered the sign in the photographs (figures 7.18 and 7.19). A pot had been unearthed and, in the spoil, I found the hind print of a black bear. This behavior expressed both the bear's exquisite sense of smell and its curiosity. At first glance I thought perhaps this was an ant dig, given that this was very fresh sign in late June, the impoverished season where bears turn to feeding on ant nests and grubs in disintegrating logs. The pot with its holes I thought might have been ravaged by the bear out of frustration at not having found food in the dig.

Upon reflection, however, I realized that this analysis was wrong on a couple of counts. A closer look at the pot showed that the many

Figure 7.18.

Figure 7.19.

holes in it were not caused by a bear's canine teeth, but rather by bullets. Note the direction of the lips on several of the holes. If they had been punctures by teeth, the tags should have been inside the pot. Furthermore, the distribution of the holes appears random rather than expressing the symmetrical arrangement of a bear's teeth. Lastly, if the bear had mouthed the pot, it should have been flattened by its jaws. The fact that none of these clues was present showed that they had been made by bullets or shotgun pellets passing through the pot and exiting the other side.

424

Gradually a more likely scenario emerged. The pot had been used as target practice or for sighting-in by hunters. This targeting had not taken place on the site but probably somewhere upstream. And since most hunting in the North Country occurs in the fall, the targeting might have been done at least 10 months earlier. In the spring floods, the pot had been swept downstream and deposited in an eddy of the river where it was covered by silt settling out in late spring. Either the following June, or perhaps years later, it was sensed under the silt by the bear, which dug it up and, having satisfied its curiosity, left it. I had seen in the past instances where a bear had dug up an inedible object apparently out of curiosity as to a strange odor under the snow or earth. But in this case it was even possible that the pot had some lingering odor of food residue that attracted the bear's nose to its location.

This incident illustrates several tracking tips. First, look closely. The fresh spoil of a dig often shows the last print of the digger as it departed the scene (Figure 7.19). Secondly, a closer examination of the pot showed my error as to how the holes in it were made. Third, one must be aware of the tendency of the human mind to create time compressions. Because I had found the pot and the dig at the same time I assumed initially that the holes in the pot had been made at the same time as the dig. Fourth, the incident shows the value of reflection after the fact. Very often the errors in an initial analysis, or at least alternative explanations, occur later, perhaps on the drive home after a day's tracking or perhaps months or years later. Think about your tracking experiences and don't be afraid to revise your impressions or to admit mistakes. That is, after all, how one's knowledge progresses.

The last lesson that this incident illustrates is the need to document what you find with a camera. It was not until I thought about the event much later that I began to construct the alternative scenario. As I was doing so, I realized that what I needed was another look at the holes in the pot. If I had documented the incident only in writing, I could not get that additional look. After all, one cannot write down all the details of an event for later reference. To attempt to do so would require volumes for even a simple event. In a sense any event is fractal; the more closely you look, the more details you find — infinitely. One can never know until later what information you should have recorded at the time. While it is true that a photograph may not show a crucial detail which may be outside the frame or hidden by some object in the field of view, it can preserve many more details than can words. A picture, as we know, is worth a thousand of them, at least.

Tracking Tip #14: Consider the Substrate

We have already seen early in this book (page 78) that tracks and sign weather over time. Tracks in loose material like sand tend to fill to the angle of repose, and such tracks also can lengthen in the direction away from the wind, both effects greatly distorting them from their original appearance. Heavy wet snow has a plastic quality that can cause it to deform from its own water-sodden weight, even making a coyote trail look like that of a turkey! This effect can also occur with mud. The thin toe marks in the accompanying photo (Figure 7.20) suggest a heron, perhaps. However, the photo was taken at a puddle in a desert wash in Arizona, not a likely habitat for a fish- and frog-eating heron. Closer inspection showed that the mud on the sides of the track had folded in to create the impression of thin toes. In their original shape they would have showed the robust track of a raven.

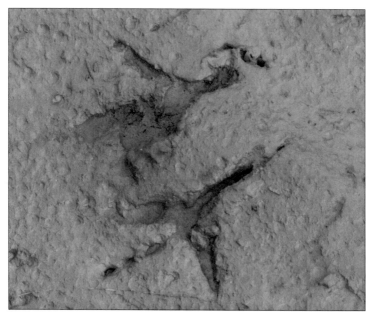

Figure 7.20.

Perhaps more broadly this tip might be titled "Consider the context" since the correct identification and interpretation here involved more than the substrate. It also required us mentally to back away from the details of the print and consider the region, habitat and species. (A black bird in the desert needs to drink a lot.).

Not only may the softness and deformability of mud distort the appearance of a track, but the relative hardness of the surface may, as well. The accompanying photos show two sets of peccary tracks as the animal moved from the firm edge of a wash to softer soil closer to water. The first photo (Figure 7.21) could easily be mistaken for the 4X bounding trail of a small desert rodent. Not until the animal moved over softer ground (Figure 7.22) do we recognize that the first set of tracks were merely the points of the cloven hooves of this pig-like ungulate in an understep walk.

Tracking Tip #15: Not All Sign Is Animal Sign

It is easy to let our mind get so dedicated to finding animal sign that we see it where it is not. Many anomalous marks may appear in soil or snow. A hemlock cone rolling across a slight dusting on a hard surface (either snow on crust or dust on hardpan) can leave a pretty good vole trail. Leaves and twigs blown by the wind can leave marks that may be misapprehended as tracks, as well, and a fallen branch buried just below the snow surface can look very much like a shrew tunnel. Water dripping from a branch overhead or from the rim of an overhanging cliff can imitate the back and forth, scurrying trail of a small mammal.

A falling tree or branch may skin the bark of an adjacent tree, leaving a mark that may look like a deer rub. The arrangement of pieces of wood and bark at an angle to a fallen log, which is often bear feeding sign, may have been caused, instead, by the impact of a dead tree as it crashed to the ground. Browse sign of an ungulate along a trail or road may turn out to be clipping by a brush hog or weed-whacker. (A good rule to follow is that if you find chewing on vegetation along a road or trail, look for the same sign a little deeper in where it would not have been made by a passing human.) Forestry equipment may leave scars on the trunks of trees along paths and dirt roads that look like anything from porcupine debarking to bear claw and tooth work on a mark tree. Although telephone poles are often used

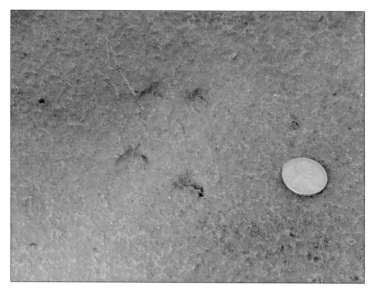

Figure 7.21.

Figure 7.22.

by bears for this purpose, injury to the base of a pole may also have been made during its shipment or erection by line crews.

A sub-title to this tip might read ". . . and not all animal sign is mammal sign." Faint tracks of juncos or other ground-feeding birds that hop rather than walk are often mistaken for bounding mice. Heron pellets regurgitated along the edge of a pond are easy to see as otter scat, and the smaller pellets of mergansers, lying on an off-shore rock or log where the birds have hauled out, can look a lot like muskrat scat, especially if one cannot wade or paddle out for a closer look.

Remember the lesson from "The Inner Game" (page 47) — "Never underestimate the power of self-suggestion!" Look more closely and think about what you are examining until you actually see it, remembering that seeing is done with your mind.

Tracking Problems

For the reader who has worked through the book sequentially this section serves as a sort of graduation exercise. Eighteen tracking problems are presented, each of which is intended to test your skills at track identification and interpretation. It is now time to take what you have learned in the past about finding and identifying and combine it with the techniques for interpretation that you have learned in this book and from your own experiences in the field.

Each problem begins with one or two photographs from actual tracking situations that I have recorded with camera and field note-book over the years. Then the circumstances in which the problem was found are presented, including all the details that the reader should need in order to solve the problem. These details may include the region of the country, season, weather and habitat in which the sign was found, whatever details would be apparent to the tracker on the scene and that are material to a solution. Following this description of the context for the problem, a series of questions are posed for the reader to ponder as he or she closely examines the photographs and digests the context.

The framing of the tracking event has been done for you. The next step in nearly all the problems, as it is in the field, is to identify the animal(s) involved. Since this is an advanced tracking guide, which was prepared for readers who are already conversant in identifica-tion, it is assumed that the reader will at least have access to one of the more reliable identification guides listed in the bibliography, either my own *Trackards for North American Mammals* and *The Companion Guide to Trackards for North American Mammals*, or one by Murie, Elbroch or Rezendes. The additional questions that are asked in each problem deal either with finding tracks or sign in the photos or with interpreting what is found for the presence or behavior of the animals.

There is a natural temptation to take a quick look at the problem, make a guess without actually thinking about the evidence and then go to the appendix for the solution. However, there are only a finite number of problems that can be contained in a book. Once they are gone, the learning opportunity that they represent will be over. Viewing the solutions before you have made a serious effort at solving the problems will short-circuit the exercise. To get the full benefit of the process, read the problem and examine the photograph at length. Then take as much time to think about it as you need in order to come to a firm conclusion on each question. If you feel stymied, sleep on it. Sometimes logical solutions to actual tracking problems I have encountered in the field come to me later when I can reflect on them with a fresh mind. Only after you feel fairly sure of the solution should you turn to the appendix.

I have tried to arrange these problems with the easier ones presented first. However, ease is relative to the previous experience of the reader, experience that differs with each of us. Certainly, all the problems are harder than I would impose on a beginner, and many, I expect, will require a review of material in this book or in one or another of the identification guides. So it should be. I think there is scarcely a tracker anywhere, however accomplished he or she may be, who can confidently answer off the top of his or her head all possible questions concerning the nearly infinite variety of animal evidence.

The solutions that I have provided in the appendix are my answers to my own questions. I do not insist that they are the only solutions. The more minds that are applied to the problems, the more likely other plausible explanations may arise. In this way the tracking art that we are trying to rediscover with our civilized minds may be enriched by the efforts of many. Good luck!

Tracking Problem #1

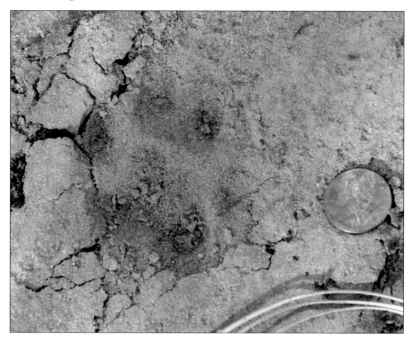

Figure 7.23.

Let's start with a relatively easy one. This photo was taken in central Arizona, but it might have been acquired almost anywhere in the temperate regions of the United States. Be careful: things are not always what they seem!

- Identify the animal that made this track.
- Is this the print of a front foot, a hind foot, or what?

Tracking Problem #2

Figure 7.24.   **Figure 7.25.**

Here is a problem that may send you to a good reference book with skull plates, such as those by Reid or Elbroch, both listed in the "Bibliography." The skull in the left photo was found in swampy, cutover woods along the Conway River in the White Mountains of northern New England. The skull on the right was collected by Kevin Harding in central Arizona but might very well have been found in the same location as the other. The relative size of the two is accurately presented.

- Identify the skulls. What features identify each?
- The current presence of both of these animals in the forests of northern New England relates to an interesting ecological story about the history of the Northeastern landscape referenced in both "Species Notes" and "Media: Forests" (pages 80-89, 322). Can you recount it?

Tracking Problem #3

Figure 7.26.

Much can be learned from the trails of wild animals. This photo was taken along the margin of a woodland pond in eastern Massachusetts. The snow over ice was fairly old, had been rained upon and refrozen. The diagonal trail from upper left to lower right of the frame had a step length of approximately 17 inches. Both the snow depth and track depth in this trail were about 1 inch with water-ice at the bottom causing the prints to look black against the white snow. The other diagonal trail had a track depth of about ½ inch and a straddle of about 4 inches.

- Identify the two animals whose trails appear in the photo.
- What gait was each using?
- In the trail of the larger prints, which direction was the animal moving?
- Were the two trails contemporaneous?
- Why might the larger animal have taken this route?

Tracking Problem #4

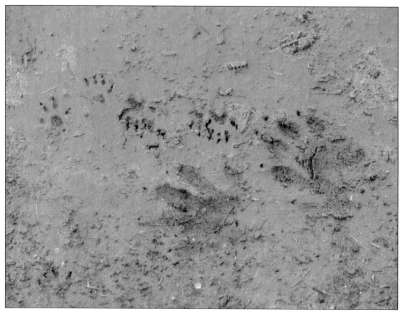

Figure 7.27.

This photo was taken in late May near the Connecticut River in northern Massachusetts. The sign of two animals registered in a dried mud puddle between a marshy backwater and the river. The width of the smaller animal's hind print is a little over ½ inch and the pattern width is about 2 ¼ inches. The trail of the larger animal consisted of pairings of front and hind prints with alternating slants, headed toward the river.

- Identify both animals.
- In the larger animal determine the gait. Was it slow walking, fast walking or trotting?
- Why might it have been headed toward the river?
- In the smaller animal which prints are front and hind? How can you tell the difference?
- Explain the behavior of the smaller animal?

Tracking Problem #5

Figure 7.28.

The photo was taken on the narrow beach of a reservoir in central Massachusetts. The shore is a few feet to the right of the frame and a higher bank with brushy edge is a few feet to the left. The tracks and trails of three animals (plus a careless tracker) are imprinted on damp sand.

- Using the size 10 human footprint for scale, identify all three animals and their gaits.
- As far as you are able, determine reasons for the presence and behavior of each animal.

435

Tracking Problem #6

Figure 7.29.

Figure 7.30.

This work was discovered in late spring along a trail in deep woods of a wildlife sanctuary in central Massachusetts. The surrounding forest was middle-aged second growth except for the tree in the photo. Some animal has pulled the bark off of about half the circumference of this ancient white oak to the height of a man. The tree itself, while still alive, sounded hollow when pounded. A search of the stripped surface revealed the marks in the close-up photo.

- What animal caused this de-barking?
- Was it sign of marking or feeding?
- At what season may it have occurred?
- What may account for the existence of this enormous tree in the middle of a much younger forest?

Tracking Problem #7

Figure 7.31.

The photo was taken along a Forest Service road above Jackson, New Hampshire, in soft snow near the boreal (spruce-fir) zone. About a foot of snow had fallen 48 hours earlier, after which had followed two days of dry, cold weather. The impressions were about 3 inches deep by 10 inches across.

- What species of wild animal made these impressions?
- What was the gait of the animal and what purpose did this manner of movement satisfy?
- Over time what will happen to the appearance of this trail?

Tracking Problem #8

Figure 7.32.

The trail in the photo was made by a red fox on the surface of a snow-covered field.

- Determine the gaits that the fox was using.
- In which direction was the animal moving?
- Where are the separations between patterns?
- Were these fast, medium or slow gaits?
- Was the animal traveling, attacking or escaping?

Tracking Problem #9

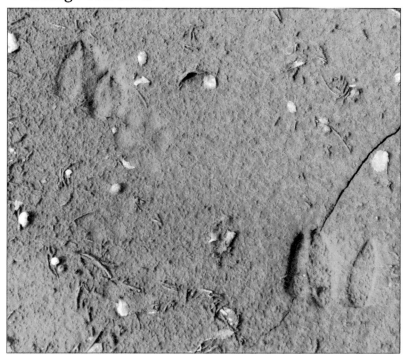

Figure 7.33.

The splayed print in the double registration on the lower right in this photo measures about 2 ½ inches across. The habitat was chaparral desert in central Arizona.

- Identify the species.
- What habitat characteristics and other evidence are involved in identification down to species level.

Tracking Problem #10

Figure 7.34.

This photograph was taken in the Red Rocks country of north-central Arizona, where the sandstone dust is as fine as talcum powder (note the fingertip impressions around the coin). However, it could well be in fine sand found nearly anywhere in the country. A narrow trail of these two-print patterns about 16 inches apart ran down a hiking trail. The patterns alternated sides of the trail with about a 3-inch straddle. In the deeper indentations of the prints sand granules have migrated downward with time to the angle of repose, softening the print definition. However, the shallower portions have retained fairly sharp edges.

- Which is the front print and which the hind? How do you know?
- Identify the mammal. (Carefully analyze what shows and what doesn't. Take into account the distortions of time and surface as well as the relative shapes of the two prints.)
- Determine the gait the animal had used, as well as its speed — fast, medium or slow.
- Given the gait and location, was the animal travel-hunting, pursuing or escaping?

Tracking Problem #11

Figure 7.35.

This photo was taken in west central Massachusetts in January after a thaw cycle. The snow depth was about 3 inches of frozen granular over bare ground. The trough running diagonally from lower right to upper left was about 6 inches across. The evidence of two animals appears in the picture.

- Identify the animals involved.
- Determine the gait each was using.
- What went on here?
- Discuss the timing of the two trails.

Tracking Problem #12

Figure 7.36.

Have mushrooms been falling from the sky, to be caught in the crook of pine branch? The location was in mixed pine/oak woods in central Massachusetts in mid-November. The fungus was about 6 feet above the ground.

- What animal species is most likely responsible?
- What is the reason for this behavior?

Tracking Problem #13

Figure 7.37.

The tracks in this frame were found on a firm dirt road in a Wildlife Management Area (a public hunting preserve) in north central Massachusetts. The pattern in the frame was repeated with a step length of approximately 18 inches. The patterns followed the course of the road through a forest that had frequent artificial openings created by game managers.

- Find the tracks in this frame.
- Identify the animal that left them.
- Determine the gait the animal was using and its relative speed.

443

Tracking Problem #14

Figure 7.38.

Here's a strange one. It seems some forest troll had been building pyramids on the spring snow! The location was central Massachusetts in mid-March. The habitat was a bottomland of hemlocks and white pine near a beaver flowage toward which the trail led. The snow was soft corn that had been subjected to the spring regimen of warm days and colder nights. Perpendicular to the direction of travel the straddle of the mounds (outside to outside) measured about 4 inches.

- Identify the mammal.
- Determine the pattern/gait.
- Explain the reason for such a gait and the surface conditions in which it occurred.
- Explain the strange appearance of the trail.

Tracking problem #15

Figure 7.39.

Figure 7.40.

The photos show the top (dorsal) and bottom (ventral) view of an item found in August on a hiking trail in central Massachusetts. The item was about 4 inches long. A few days later a similar item was found nearby along the same trail.

- What is it? (A little research in a reference book will get this down to species level.)
- What accounts for its presence in this location?-
- Speculate on which predator was responsible for its presence.

Tracking Problem #16

Figure 7.41.

Figure 7.42.

Figure 7.43.

This site was discovered in the woods on the side of Mount Grace in north central Massachusetts in April, ¼ mile from the village of Warwick. A faint skunk-like odor and a sparse trail of wind-dispersed feathers led to the discovery. Clockwise from upper left: feather pile, detail from the site, a scat measuring ½ inch in maximum diameter.

- Identify the predator.
- Identify the prey.
- Is this a kill site? What went on here?

Tracking Problem #17

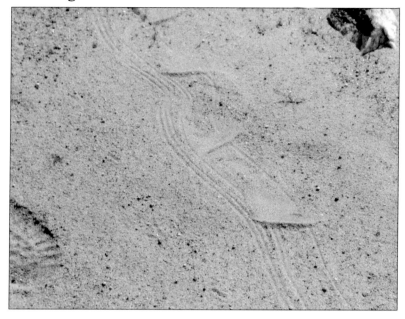

Figure 7.44.

This site was on the narrow beach of a reservoir in central New England at mid-summer. The water is several yards to the lower right and a brushy edge is the same distance to the upper left. My size 10 boot print may be used for scale.

- What animal did this?
- In which direction was it moving?
- What was it doing?

Tracking Problem #18

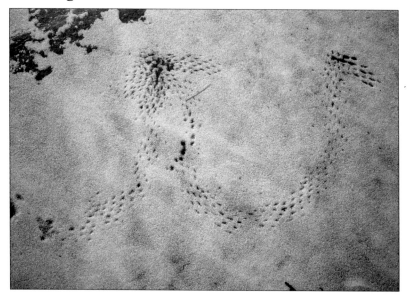

Figure 7.45.

The photograph was taken in central Massachusetts at the edge of a beaver pond surrounded by a mixed pine-hardwood forest. The time was mid-winter and the surface was a light film of snow over ice.

- Name the trail-maker at least down to genus level.
- Which Tracking Tip does this identification involve?

Mountain Lions

Cats of any sort have always fascinated me. If I'm on a tracking trip in New England and come upon a bobcat trail, other sign is neglected as I turn to tracking and learning about these expressive animals. Aside from the lynx the only other wild cat at least potentially in my area is the largest breeding felid in North America, variously named mountain lion, cougar, catamount, puma, panther and painter. While there have been many rumors of this animal being present in New England since its presumed extinction more than a century ago, there has been little hard evidence to support these rumors. Reports of mountain lion tracks usually turn out to be those of big dogs, bobcats or bears. The few that have been supported with some evidence are discounted by state game agencies as released exotic pets. In 2011, however, a young cougar, hit by a car in Connecticut, was determined by officials to have migrated from South Dakota via Wisconsin. How it got to the East Coast, however, is problematic. If it went across the upper peninsula of Michigan, it would have had to cross the Great Lakes at Sault Sainte Marie, then across the most populated and developed region of Canada, southern Ontario (which has an overabundance of deer and no competing predators, so why should it have kept going?), back across the lakes, then across New York State and finally across the Hudson River. This would have required either a number of long swims (not a developed skill of many cats) or unobserved bridge crossings. The alternative would be an end run southward around Lake Michigan and then across three farm states with little forest cover, once again ending with a crossing of the Hudson. You may detect that I am somewhat skeptical that this cat found its way under its own power to Connecticut.

In search of the real thing, I have made a half-dozen trips to Arizona in recent years to accompany my friend and fellow tracker, Kevin Harding, on many searches for evidence of these marvelous animals. Five of those years yielded success, but only after many days of searching. The larger the predator, the more thinly distributed they are over large areas, and central Arizona is large indeed. The following are three incidents on these trips that may yield some helpful information for the tracker intent on learning about these animals.

David Miller Trail — 2005

On the trail into the backcountry canyons of the Secret Mountain Wilderness, Kevin recounted the sad tale of David Miller. As I understand it, David was a young forest service ranger patrolling alone in the area where we were headed. As far as was known, he had hiked up out of the canyon where we presently were walking, going over a narrow pass and down into another similar canyon bottom. He disappeared without a trace. An intense search by the Forest Service turned up nothing. To this day his disappearance is unexplained, no doubt to the great pain of his family and friends. The prevailing opinion was that he had been taken by a mountain lion. But such an attack on a hiking trail should have left some evidence. None was ever found.

We arrived at a split in the trail where a bear-gnawed sign (page 215) pointed straight ahead to, ironically, the Bear Sign Trail and to the left for the David Miller Trail, renamed in memory of the young forest ranger. We chose the latter and climbed up rather steeply to a high pass where we stopped and ate lunch. The pass was quite constricted, so that anyone, animal or human, would have to use the trail through it in order to traverse from one canyon to the next. I noted that the sandstone walls on either side supported little vegetation for concealment either of man or wild animal.

A drought in central Arizona had turned the hiking trail we were on into talcum-powder dust roiled by a month of hiker traffic. It occurred to me that any cougar that wished to pass from the hunting ground of one canyon bottomland to the next as we were doing would have been forced to walk in the same dust. I began looking for tracks as we descended from the pass but distinguishing an animal track among the lug prints of dozens of hiking boots was difficult. About a hundred yards down from the pass something caught my eye (Figure 7.46). In the midst of the roil I made out two faint parallel ridges oriented to the trail direction. Could these be the ridges between the three posterior lobes of a cat's footprint? It was hard to tell at first because of the hiker commotion, but eventually the image of a cougar's print did indeed emerge from the background. Here before me was the first track of this species that I, an Easterner, had ever found. A search a little farther down showed more prints, clearer than the first (Figure 7.47). The cat had indeed used the same pass and same trail as we had in order to connect bottomlands. There was so much

Figure 7.46. The first cougar print, a front: vague but parallel ridges gave it away.

commotion in the trail that I couldn't get a lot of gait information, but I did get some good photos and measurements of individual prints.

- The front print width was 3 3/8 inches and secondary pad width was 2½ inches.
- The narrower hind print was about 3 inches wide by 3¼ inches long with a secondary pad width of 2 1/8 inches.
- It was difficult to age these tracks but the fact that we could only find them intermittently suggests that enough time had passed for hikers to erase most of them with their boot prints.

When we reached the canyon floor, the trail turned into deep sand and the cougar prints that we had picked up and lost intermittently disappeared altogether. The cat had probably had enough of the convenience of a hiker trail with its risks of human encounter and had dodged off into the prickly chaparral to hunt. The trail out to Dry

451

Figure 7.47. A clearer track found nearby. Flattened margins of the secondary pad and lack of splay in the toes show it to be a hind print.

Creek Road was long and tedious and so I let my mind wander to the tragedy of David Miller. In the course of the next hour or so while plodding along I constructed a scenario that would explain his disappearance without a trace.

With pack and radio he had done exactly what we had done, my theory went — he had climbed up and over the pass between the canyons. Perhaps, at some point up in the pass, two things had occurred. For one, he felt the need to answer the "call of nature." For another he unwittingly pushed off the trail ahead a cougar that had been using it, just as our cougar had, to traverse between its canyon-bottom hunting areas. The cat watched from cover as the ranger passed and then began to follow him at a safe distance down the trail. The ranger would not have wanted to squat with his trousers around his ankles in the pass where he could be observed by a passing hiker, so he determined to delay the call until he had descended low enough into the canyon to where it began to open out into a scrubland (Figure 7.48) and provide the chance to get far enough off the trail for privacy. There, after moving perhaps 100 feet or so from the trail and into the prickly bush (Figure 7.49), he removed his pack and radio and, fatally, squatted down in what a cougar would interpret as a submissive posture.

Cougar's are not known for courage. While it would be a bad idea for a human to confront a grizzly bear, accounts suggest that confronting a cougar usually results in the animal's rapid departure. Only if it can approach unobserved and find a vulnerable target will it attack. David Miller may have offered just such a target. Had the attack occurred on the trail, a scuffle would have been recorded in the sand and equipment would have fallen off, the radio perhaps. But back in the chaparral, with equipment laid neatly aside, no record of the attack would have been found. I looked at the dense, prickly scrub as we hiked along, realizing that one could scour it for hours until one's clothing was torn to shreds and find nothing so small as a couple of items of equipment. The body itself would have been dragged off by the cat, deeper into the canyon bottom away from potential disturbance.

Figure 7.48. The dense shrubland of this canyon bottom in the Secret Mountain Wilderness provides concealment and good hunting for any stealth-predator.

Such are the fantasies of a tracker on a long, tiring outmarch at the end of the day. Unfortunately, for the last few years the road into the trailhead has been in such disrepair that I have not been able to return to that canyon bottom to test my theory.

Turkey Creek — 2007

I had reason to reflect on the origin of the State's name as I arrived in the Red Rocks country of north central Arizona. After nine months of drought the trails were fine dust, roiled by months of hiker traffic. The evening before I left for home, however, the sky lit up with a spectacular sunset and the day after I departed the entire state was deluged. But during the two weeks of my annual November sojourn, the air was dry enough to match the colorful but gruesome place names that abound in the high desert such as "Dead Horse Ranch" and "Dead Man's Pass." There was no doubt in my mind what they had died from.

In past years, the tracking and birding had been straightforward. If you wish to see birds in Arizona, go to water; if you wish to see cougar tracks, go to water as well as to the passes between canyon bottoms where the big cats hunt. This year all bets were off. There was little impounded water anywhere; even the cattle "tanks" as the locals call man-made waterholes, were dry, and the dirt in the passes was so roiled by boot prints that picking tracks out was more than difficult. For days we searched for cougar sign among the ponderosa pines on the plateau south of Flagstaff and in the canyon trails of the Secret Mountain Wilderness and Sycamore Canyon with no result. We found bear sign in the form of torn apart logs on the plateau and gray fox tracks and scat in the Turkey Creek area south of Sedona. Raccoon and peccary tracks were everywhere, but no cougar.

Finally, on the fifth day of tracking, Kevin thought we should try the backcountry south of Red Rocks State Park. He had heard of an unrecorded water hole down there that we might be able to find. So, for a couple of hours we trekked down trails, up washes and over desert ridges through Arizona juniper and carefully around clumps of prickly pear. Eventually, from a lookout we spotted an escarpment perhaps a half mile off that looked as if it might be a creek edge. So we headed for it.

Once there, however, we found the wash at the base of the cliff to be bone dry like everything else. We decided to head up this broad

Figure 7.49. Kevin Harding on the long trail out of the wilderness. Prickly plants and shrubs like agave, cat's claw and manzanita testify to the long-term overgrazing of the area by cattle. Such growth makes off-trail penetration very difficult for humans, less so for cougars.

creekbed anyway on the chance that we might find pooled water. The tracks of free-ranging cattle were everywhere in this prickly, over-browsed rangeland. While Kevin went on ahead to look for the "lost" waterhole, I hung back to look at a few suspect prints.

One crumbled track at the edge of the wash looked somehow a little different from the cattle prints (Figure 7.50), which are round and about the same size as a cougar print. I oriented myself in the direction of the animal's travel and noted a small lateral ridge in the center. Even with the distortions of collapsing dust, a cow print should not show such a ridge. I looked more closely. The ridge seemed to be at the head of a large trapezoidal smooth area. Hmm. Any of this could have been the result of distortion by time or sliding sand. Best look around nearby in the line of travel for another. A couple of feet ahead I distinguished a faint depression on a plate of hard pan (Figure 7.51). It didn't look like much, but I continued to stare at it and began to experience the phenomenon familiar to any experienced tracker. A cougar track began to grow to my mind's eye out of the faint depression.

Figure 7.50. On the right in the frame is the first track to catch my eye among the cattle prints. Note the flattened leading edge of the secondary pad. The cat was moving right to left. A search ahead revealed a faint second print just above the penny marker.

At first I distinguished a general round shape with convolutions at the edge. Then I seemed to distinguish some lobing at the posterior, then tri-lobing. My heart skipped. Tri-lobing! The same feature of cougar prints that had caught my eye a couple of years earlier in the Secret Mountain Wilderness. Gradually I was able to make out nearly the entire outline of a large cat print. With two prints I now had a trail and with it a sense of the animal's direction. It was headed down the dry wash from the direction Kevin had taken. My friend was well up the creek by then, beyond calling distance, so I started to backtrack the cat in his direction. Once I knew the trail was there, its tracks began to jump out of the background all along its trail — two more twenty or so meters up the bed, then another (Figure 7.52), then two more, shallow but nearly perfect with the asymmetrical front print and the narrower, more regular hind clearly visible (Figure 7.53). Finally one more in some crumbling hard pan not far west of the unmarked water hole,

Figure 7.51. A close look at the faint track confirmed the identification. The wavy ridge at the bottom and the flattened top of the secondary pad were visible enough to mentally reconstruct the rest of the print.

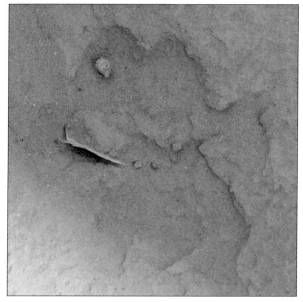

Figure 7.52. A hundred yards ahead a more obvious track showed that I hadn't been imagining what I wanted to see.

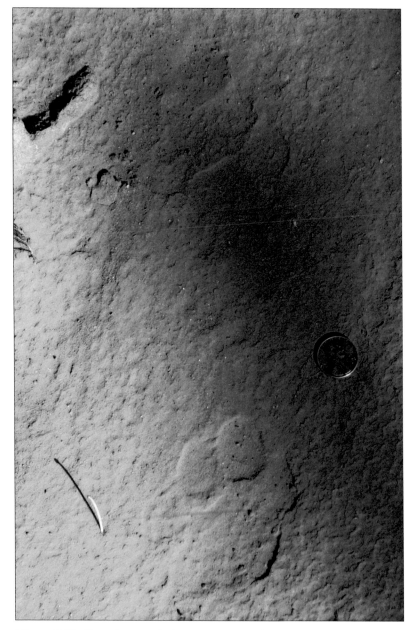

Figure 7.53. Near the waterhole an overstep walking pattern appeared. The front left print is below. The hind print above may be mentally reconstructed by locating the leading edge of the secondary pad.

a deep basin in the sandstone perhaps 5 feet across, next to which I soon found Kevin standing (Figure 7.54). No doubt this is where the cat had come from, both to drink in solitude and perhaps to ambush an unwary javelina, the pig-like ungulate whose bands roam all over the Arizona desert and form a major prey species for mountain lions in the area. After eight days of searching we finally had success. A great day!

Data collected from this event included:

- The track width was around 3 inches with a track depth that was generally very shallow on the hard pan of the creek bed. Most of the tracks except the first one were no more than 1 millimeter deep.
- Direct-register walking step length was 17-18 inches on the fairly flat surface of a dry wash. The cat only walked, avoiding the cobble in the center of the wash which it probably found as uncomfortable to walk on as did we.
- Straddle outside to outside was about 5 inches.
- At the waterhole, the track patterns dissolved where the cat apparently spent some time drinking and investigating prey scent. Here were the only single registration prints we found.

All the measurements were within the expected parameters for this species. Things (except the human skin apparently) tend to age slowly in the dry air of the Arizona desert. As a result, I do not have an accurate idea of the age of the tracks other than that they were made obviously since the last flooding of the wash that occurred months earlier. The condition of the first track suggested that the trail was fairly recent. It had been made in a slight slope of softer crusted sand where the animal's foot had slipped slightly sideways to the left, causing crumbling on the right side of the print. The print itself seemed to have had its fine features softened by time and what little wind might have found its way down the creekbed. If I had to guess, I would say that the trail was perhaps 2 days old, less likely a week. After that time I would expect the deterioration of this particular track in softer substrate to have proceeded to the point where the pad outline in the right (uphill) side of the track would have disappeared.

I have included here at the end of this piece a scenic shot of the Turkey Creek area (Figure 7.55) to give the reader a feel for the spectacular setting of our tracking adventures at this site. Grassy

Figure 7.54. The "lost" waterhole where the cat's trail had originated. In the dry desert such pools are magnets for wild animals along with their predators.

Figure 7.55. A "park" at Turkey Creek with the Nuns in the background.

"parks," chaparral and juniper, a natural seep, many dry washes that flood with the occasional rain — all have yielded more opportunities to track mountain lions since my initial visit recorded above.

Dry Creek — 2008

We had been searching for six days down at Turkey Creek where we had found cougar tracks last year, as well as up in the ponderosa pines of the Coconino Plateau and now in a dry creek bed northwest of Sedona in the red rocks country of north central Arizona. The bed was mostly cobble with a few puddles from recent rains. Only the sandy banks were a trackable surface, but try as we might, we found nothing but occasional peccary trails coming down to drink. After four hours of this we turned up a smaller tributary toward a remote cattle hole called Earl's tank.

Dead ahead in the distance was an impressive red and yellow buttress. Having few photos of myself on these Arizona treks, I asked Kevin to let me go on ahead. After fifty yards or so I would strike a suitable tracking pose and at my voice signal he could shoot a picture with the buttress in the background. Once in position I pointed randomly at the ground and looked back to see if he was ready to shoot, intending to look down in the direction I was pointing as if finding something on the surface. At that point, Kevin snapped the shutter prematurely and without my awareness (Figure 7.56). When I thought he was set, I looked down along my extended finger and, incredibly, that for which we had been searching futilely for the past week was at the end of my pointing finger! Two fresh prints of a cougar in an overstep walking pattern lay at my feet (Figure 7.57). Glancing right and left I could see more of the same. The cat had been descending from the direction of the tank down the side creek toward the main flow. I looked up at Kevin. "Mountain lion tracks," I shouted. "Yeah," he nodded, wrapping up the strap of the camera, as in "Yeah, right." "No really," I repeated. "Mountain lion tracks!" and pointed again at the ground at my feet. He came up and looked down in amazement. "I thought you were kidding!"

This incident resembled another a year or two earlier up on Munds Mesa. This plateau is shaped something like an hourglass, with a narrow constriction in its middle through which any transient animal would have to pass. Added to this was a cattle tank where a mountain lion

461

Figure 7.56. Pointing randomly at the ground, I waited for Kevin to get set to snap the picture.

could get a drink and perhaps capture prey that also relied on the tank for water. Up and down and across we searched fruitlessly, finding nothing near the constriction but the usual turmoil of hiking boots. At one point I wanted to test the surface for recording tracks, so I bent down and pressed the tips of my bunched fingers into the talcum dust. They made a perfect "fox" track. Satisfied that I could pick out a cat track from the commotion, I straightened up and started to walk away. Kevin stopped me. "Aren't you going to photograph it?" "Oh, no," I answered, "I was just testing the surface." "No," he insisted. "Look!" I glanced back at the impression I had made and perhaps two inches to the right of it was the faint but perfect impression of a cougar hind foot! If ever there was an example of the obvious obscuring the subtle in the mind of the tracker, here it was (pages 403-404). My mind was so focused on what I wanted to see, the deeper impressions of my fingertips, that I had completely missed the cat print.

Back at Earl's tank we spent the next two hours sorting out the trail of the cat. It had come down from the tank, perhaps two hundred yards distant, following the bed of the side creek. Kevin went on above

Figure 7.57. I looked down from the end of my finger at this: a double registration of the cat we had been searching for all morning.

and found many javelina tracks at the tank and more cat tracks above it. Using sticks poked into the soil we were able to reconstruct its route and gait. The cat had come down from above to drink at the tank and perhaps stalk a vulnerable javelina, then passed on downward in a constant overstep walk/trot, leaping down over several ledges to the last visible tracks at the place where we first found them. In all, we discovered perhaps twenty prints, enough to reconstruct the trail for perhaps two hundred yards. Choosing the best of the tracks (Figure 7.58), I made several casts from plaster that Kevin had brought along and then together we built covers for them to protect them from mountain bikers and their dogs while they dried. Although we had found cougar tracks before, I had never discovered a site where the soil conditions allowed casting. In the past, the soil had always been in

Figure 7.58. A nice right-front print of the Earl's Tank cougar. Note its wider spread and greater asymmetry compared to a hind.

fine powdered sandstone, far too delicate to hold the shape of a track under the flow of plaster.

Data from the event included:

- Front track width was 3¾ inches; hind print was 3¼ inches. Front secondary pad width was 2¼ inches. These suggest a medium-size cougar.
- In descending from the tank the cat used exclusively an over-step walk/trot. The step length for this gait varied from 14 inches where the overstep was so slight as to appear as a double registration to 20 inches where the overstep was about 7 inches. That would be a lot of variation for a walk, suggesting in a smaller animal a transition to a trot, but cougars have long legs, so one must scale up the variable.
- The straddle outside to outside was about 7 inches.
- Track depth was generally less than a centimeter but deep enough and in sufficiently firm substrate to produce several very good casts.
- Lack of pebbling within the tracks from a shower a day or two earlier indicated that these were fresh tracks.

We had no idea how fresh! Satisfied with the day's work, we retreated down the creek bed, retracing our original route. After a few hundred yards we followed our own tracks up onto a sandy bank and found ourselves staring in disbelief: several cougar tracks right between our own footprinted trails where we had passed side-by-side on the way up! "How could we have missed these?" we asked, and engaged in a bout of mutual recrimination: "You weren't looking hard enough" and "Well, you should have found them, they're closer to your footprints than mine," and so forth. Crestfallen, we continued our retreat. Only later did I realize that we hadn't missed the cougar prints in the sand at all. We had been searching diligently all the way up the creek, and these prints were not subtle — they were half and inch deep with plenty of contrast and directly in our sweep of vision. The logical answer was that they had been made *after* our initial passage. This could only mean that when we left the main wash to start up the bed to Earl's tank and found the first of the day's prints, we had just at that moment pushed the cat off the trail. It had been descending toward us, sensed our approach and dodged off into the juniper shrubs.

Perhaps it had crouched, watching and waiting for us to pass as we discovered its tracks. After we moved on up toward the tank, it re-emerged behind us and slipped silently down into the main wash where it left its tell-tale footprints in the sand between our own.

I close my accounts of tracking cougars in Arizona with a photograph I took south of Sedona in Carroll Canyon (Figure 7.59). It is intended to give the reader an idea of the incredibly beautiful landscape of the Red Rocks country in which I track mountain lions every year. Here the desert to the south butts up against the Coconino Plateau which has eroded over millions of years into fantastic rock formations with countless deep, narrow canyons to explore.

Figure 7.59. Buttes after rain — Sedona, Arizona.

What Does the Fox See?

I have before me a photograph of an aged red fox, chin resting on the ground between its front feet, ears alert and eyes directed slightly off-camera as if something there has caught its attention but not enough to get it to raise its head. I look at this for a long time, wondering what is in that observant gaze. This fox has had a lifetime of hunting, of raising offspring, of weathering winter storms. Does it remember all these experiences in such moments of rest? Does the animal reflect on them? We know that wild animals can remember, sometimes quite extraordinarily. A coyote, once fooled, remembers the location of the mistake for years after, carefully avoiding a repeat. Young animals with lengthy parental association are known to learn from and remember the lessons of the adults of their species. In fact, that long association apparently is necessary for them to learn the extra-instinctual lessons for survival.

Wild animals, then, can remember, sometimes perhaps better than we. Recall the legendary elephant. But what is the nature of that memory? We remember what we experience directly. But we also remember what we read in books, histories, fiction, textbooks, as well as what we have imagined on our own but never experienced. All of these have a distorting effect on memory. What we recall is seen through a thousand influencing and distorting lenses. The fox, on the other hand, has no histories that reevaluate with each generation the experiences of the past, telling us in effect how to remember them. And the fox probably has no distorting imagination to create events in its own recent past that may never actually have happened. Instead, its view into the past is direct, without the convolutions of history, rationalization or imagination. And this view goes farther back than its own experiences. Through instincts sharpened by century after century of natural selection, its view into the past reaches back many generations before the animal's physical existence. It is a view without frills, providing him an "instinctive intelligence" different from our own, an intelligence to which we are denied access, our instincts having been corrupted by the safeguards of civilization and our memories distorted by many complicating screens and lenses.

For those who have read through this book, we have come to the end of our journey together, to the point where my path turns one way and yours goes on. By way of farewell, I present the following account written many years ago of the day when I first experienced the insights described above. I hope that it and the previous chapters in our conversation together will be of some help to those for whom discovering the secrets of Nature hidden in the simple footsteps of an animal has become an absorbing adventure. Perhaps as well, this anecdote will help to cast our importance on our tiny, fragile planet in some reasonable and sustainable perspective.

Fox Wisdom

A revisionist theory about the fox has been circulating for a while that challenges the animal's legendary cunning, suggesting instead that the Reynard of fable doesn't actually rely on wiliness to catch its prey but more often stumbles on its dinners by accident. This theory insists that the fox, far from being wise to the ways of the forest, simply puts in a lot of aimless miles until it accidentally cuts the path of some potential prey, which it then captures by quick reflexes. There is an old saying, variously directed at wolves, coyotes and foxes, that they live by their feet. The revisionists suggest that the fox, at least, does so largely to the exclusion of an intelligent brain.

As susceptible as the next person to the intellectual charms of revisionism, I had this theory in mind one winter a few years back when I came upon the trail of a red fox as I started up the Champney Brook trail on Mount Chocorua and had occasion, since the animal and I were climbing the same mountain by the same route, to examine its trail at length and learn what I could from the record of its behavior in the snow.

Back in those days, I was still testing the idea that an animal leaves behind a diary from which we might with some practice read intention or accident, success or tragedy, even wisdom perhaps. With enough careful observation, I suspected, Thoreau himself might have discovered the reason for the hiccup in the fox's trail that he observed one winter morning across Walden Pond. For a fox does little for no reason, and one of the adjectives that well describes its winter behavior is "efficient." Unless engaged in the pan-species foolishness of courting rituals, the fox's movements are winnowed down to the necessary alone. So, in imitation of Thoreau, I aligned myself with the fox's trail so that I might align myself with its wisdom, if wisdom it expressed.

It was early on the coldest morning yet of the young winter. The temperature was in the teens and had been below zero in the valley only a couple of hours earlier when the tracks had been laid down. They were so fresh that the crystals of snow around their rim were still sharp and could easily be disturbed with a puff of breath. Here was a good size red fox, traditionally wise to the value of energy conservation, frittering away precious calories in the perfectly pointless act, even in practical human terms, of climbing a mountain.

For two miles and a couple of thousand feet I followed the perfect line of prints, the placement of which was as measured as a sewing machine's stitches (Figure 8.1). The impression of each hind foot landed entirely within the outline of each front foot on the same side, a process trackers call "direct registration." Foxes and other wild hunters do this as an economy, the front foot pre-packing the snow for the hind. Repeated thousands of times on a winter prowl, this small saving in energy can mean the difference between surviving or not.

There were coyotes in these woods, I knew, and a young coyote in December can have a step length and print-size that overlap that of the smaller fox. But a coyote is a big animal with small feet while a fox is a small animal with big feet. In the same conditions the fox's prints will look delicate compared to the robust impressions of a coyote. The pad outlines, too, said red fox, for these canids have fur-covered feet at all seasons, a useful vestige of their boreal heritage. In dry, loose snow this results in a diffuse track compared to the distinct impression of a coyote's naked pads.

For over an hour I followed the neat trail with its nearly perfect 14-inch step lengths and a straddle so narrow that the prints were arranged in a nearly straight line. I had certainly to grant the animal grace. The careless romping of a dog liberated from the monotony of the house leaves a coarse, haphazard trail. It can afford such carelessness since a can of Alpo and a hearth await it at the end of the day. All right, I had to give the fox dignity, even elegance perhaps, but intelligence?

As I hiked, I found the thin layer of fresh snow punctuated by the trapezoidal bounding patterns of white-footed mice (Figure 8.2) that had crossed the trail back and forth from one hole to another. Was it stupidity that I was witnessing, after the fact, that the fox didn't investigate each one as he came upon it? Instead, its measured stride never varied, not even an "indirect registration" to one side to suggest a turn of the head in the direction of a potential meal. Perhaps the revisionists were right: this was either stupidity or blindness. The

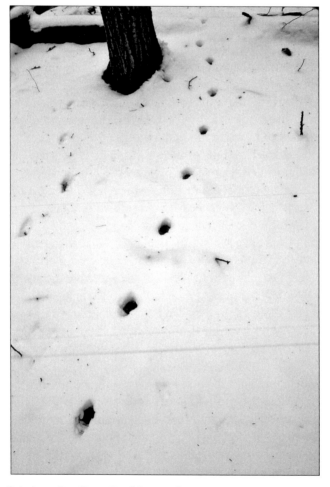

Figure 8.1. A perfect line of red fox tracks, as measured as sewing machine stitches.

unvarying perfection of its trail might have been the hobgoblin of a little mind, indeed. I began making mental excuses for this most beautiful species of wild canid. Perhaps this particular animal had, in fact, poor eyesight. Or wits. There must be half-witted foxes. Foxes . . . half-witted . . . foxes. . . . My own phrases were repeating themselves like a chant; the monotonous effort of the climb was making my mind whirr unmeshed. But stupid wild animals with poor senses don't survive, I insisted through the mental fog. And here clearly was an adult

Figure 8.2. A mouse trail in shadows.

471

with, by its size and the sobriety in its trail, at least a year of success behind it.

Occasionally, the animal would leave the trail at an angle, usually to give a quick sniff to a pillow of windthrow that might hide a rodent or hare, but then it angled back out to the trail without so much as disturbing any of the fresh snow on the branches to discover if there was something under the pile. How does such an animal live? By luck? If so, in these temperatures it had better get lucky soon.

Finally, three-quarters of the way to the top of the shoulder of the mountain the tracks stopped dead. The animal's trail then led off to the left for a few yards and ended in a patch of disturbed snow. Another yard, angling back toward the trail, showed some blood and the gut sack of a small animal, neatly excised and left on the snow. I poked at it with my glove; the entrails had frozen solid in the morning cold. I had seen this sort of surgery before by foxes and had always marveled at how an animal without hands could manage it, like peeling a grape with your teeth and elbows. Canids can barely articulate their toes and so their digits would seem to be useless for any task more delicate than digging out rodents. This rigidity shows in the lovely symmetry of the fox's track, with no mobile toes adjusting to the surface from step to step as with a cat or a weasel. But to my frustration, such symmetry conceals the workings of the heart, as beauty will. One track alone is remarkably cryptic; at least a whole pattern of them, if not a lengthy trail, is needed to imagine more accurately what it was experiencing. I felt fortunate, then, to have a couple of miles of continuous trail in good condition in which to read this animal's mind.

I felt better about my fox now. Here, in the snow before me, was success at least, if not intelligence. But how had it been achieved and why here after the animal had ignored so many possibilities lower on the mountain? I inspected the footprints to the kill site closely. After the fox had stopped on the hiking trail, it moved toward the prey with the tiny steps of a caution that testified to the urgency of its need. First it had moved a little to the left and then back and to the right. There was no mark of its brush, nor were the tracks elongated to show any part of the animal's heel — it was not crouched to take advantage of the pillows and undulations of the snow between it and its quarry. Instead, the animal appeared to have been moving upright.

It was only at this point that insight struck; I had been viewing the fox's behavior through human eyes. We are a sight-dependent species; the fox is not. It was not seeing its prey, but rather hearing it. The meander of the trail was an effort to triangulate with its ears some sound beneath the surface, perhaps a squeak or the muffled sound of tiny footfalls or the sound of chewing as a vole gnawed the bark at the base of a hobblebush. I could "see" the fox now, in my imagination as a human will, upright, ears perked forward to catch every sound. Slowly it had moved forward, lifting and placing each foot deliberately and carefully ahead of the other by no more than a couple of inches. Finally, there was a short space clear of tracks. The fox, having gauged the location of the vole, leaped into the air, ears perked forward, in a pounce that to humans looks playful but that is deadly earnest to the fox. The high arcing pounce is an effort to get both its big front feet together in the air so that they come down at the same time, covering as much as eight square inches, punching through any crust in the hope of pinning some part of the prey long enough to get its own sharp muzzle into the snow to grasp the creature with its teeth.

I sat on some blowdown and opened my pack for a candy bar, pondering what was in front of me. Now it was clear why the fox had ignored the trails of mice. They could have been made at any time during the previous night, mice with their big eyes being nocturnal wanderers after birch seeds scattered over the snow. To hunt down every mouse trail would be a fool's errand; it's what I would have done, having lost my instinctive intelligence many generations ago and replaced it in the puny span of one life with the intelligence that comes from books. I had long ago resigned myself to the fact that in a primitive state I would have become an hors d'oevre for the saber-toothed tiger very early on. I was not the end product of thousands of years of ruthless natural selection as was my red fox. Civilization had been invented to protect me from that. Reynard, on a remorseless energy budget, can afford no such human mistakes for which he will be indulged and forgiven. Unless the fox's incredibly jazzed-up senses tell it through some lingering whiff or some errant sound, that one of those trails will lead with fair certainty to warm-blooded prey beneath the snow, then the fox will, in fact must, ignore it.

That was it, of course. I had phrased it without realizing: "instinctive intelligence." The fox doesn't think in the human sense of sequential reasoning. It is a creature of action. In some mysterious evolutionary way of trial and fatal error, the thinking had already been done for it, the wisdom accomplished and encapsulated in that seeming oxymoron. What was left for the fox was to act upon it.

Somewhat heartened by my rehabilitation of the fox, I finished the candy bar and, renewed by the energy of the sugar, prepared to move on up the mountain. And so did the fox, itself refreshed not by a Milky Way but by the transmuted energy of one of its nearer stars contained in the meager flesh of a vole. Within a hundred yards, the shivers that had set in while I had sat were calmed by the effort. There it was again, another flash from the mental front: that was why the fox was climbing out of the valley on the coldest morning of the year. It must descend into the pool of cold on the valley floor every evening for the abundant prey that live there. But then it warms itself, as I myself was doing, by the exertion of the climb back up to some high, sheltered spot with a sunny southern exposure.

As I half expected, when I reached the top of the shoulder of the mountain west of the main ridge, the fox's trail led off from the hiking path toward the south side of that shoulder. Let humans continue on to the heat-robbing winds of the summit. Whatever their reasons may be, athletic or aesthetic, they are foolish compared with spending the day curled up with nose and feet under a bushy tail, absorbing the energy of the sun directly while digesting it indirectly.

I didn't follow the fox over the shoulder to discover the animal itself, perhaps to inform it in a patronizing way that humans had deduced from textbooks of meteorology, which the animal couldn't read, that the day would turn cloudy with a chance of snow rather than yielding the sunshine its instinct predicted. It would be cruel to disturb the animal in its doze and make it run for no other reason than the pleasure of seeing it when it had been so perfectly and necessarily careful about such energy expenditures. Or perhaps it was simply that I had set another goal, a mountain peak more explainable to the achievement-minded back in the lowlands than the pursuit of fox wisdom. It was a failing perhaps that I was unwilling to be put off from accomplishing that summit, even to follow farther and perhaps suffer more enlightenment at the paws of the fox.

Glossary

In an effort to reduce the amount of jargon in this guide, I have tried to use common words as they are generally understood rather than assigning them specialized meanings. Nonetheless, in any technical field there will be some terms not used in everyday speech. Below is a glossary of tracking terms as they are used in this book.

General Terms

Alternating: a zigzag trail of tracks representing a walking or aligned trotting gait.

Alternating Slant: tracks arranged on a diagonal relative to the direction of travel but with the diagonals reversed from pattern to successive pattern. See *Repeating slant* below.

Concave: curved inward; think of a cave. Opposite of *Convex.*

Convex: curved outward like the surface of a ball or the dome of a hilltop. Opposite of *Concave.*

Crepuscular: active at dawn and dusk.

Digitigrade: moving either on the toes alone or on the toes and *secondary pad(s).* See *Plantigrade* and *Secondary pads* below.

Direct Registration: See *Registration* below.

Diurnal: active during the daytime.

Dorsal: viewed from above the animal. See *Ventral* below.

Double Registration: See *Registration* below.

Even: tracks or pads arranged side-by-side so that a line across the leading edge of both is perpendicular to the direction of travel.

Gestalt: here, the fundamental form of a track or trail that survives variations and may be used to identify the animal. See *Signature* below.

Indirect Registration: See *Registration* below.

Lateral: toward the side, rather than the centerline of an animal's body when viewed from above. See *Medial* below.

Medial: toward the centerline of an animal's body when viewed from above. See *Lateral* above.

Pattern Width: the distance across a track pattern from the lateral edge of the leftmost print to the lateral edge of the rightmost print, measured perpendicular to the direction of travel. Note the difference between this term, *Straddle,* and *Trail width.*

Plantigrade: moving on the entire sole of the foot including the *tertiary pad* or area, as with bears, skunks and humans. See *Digitigrade* above.

Print: in this book, the impression made by a single foot. Compare to *Track.* In other tracking guides these terms are synonymous.

Quadruped: literally four-footed as opposed to "bipedal," or two-footed. This is the Latin form of the Greek-derived "tetrapod" used in some books.

Registration: the process of creating an impression with the foot. "Single registration" (SR) occurs when the print of each foot appears without overlap or contact with any other print. "Direct registration" (DR) occurs when a hind print coincides perfectly with a front. "Indirect registration" (IR) means that the hind foot landed slightly off-center from the front print, leaving the appearance of a single distorted print. "Double registration" (DblR) occurs when the hind print overlaps the front but so widely off-center that the presence of two prints is clear. Note that some other tracking guides use "double registration" to refer to two single registrations that are simply close together in a pattern.

Repeating Slant: arrangement of tracks on a diagonal relative to the direction of travel and paralleling that diagonal from pattern to successive pattern rather than reversing it. See *Alternating slant* above.

Retarded: located toward the rear of a print or pattern; the opposite of *advanced*. In this guide "retarded" often refers to a print feature that is located toward the rear relative to the same feature on the opposite side of the print, giving the print an asymmetrical appearance. It may also refer to a print located rearward in a pattern relative to another on the opposite side.

Rotary: a four-print (4X) rotary pattern that has an order of placement: right-left-left-right or left-right-right-left from the beginning of the pattern to its end. Because wider rotary patterns have the shape of the letter C or a comma, they are sometimes referred to as "C-patterns". See *Transverse* below.

Scat: the excrement of a wild animal. The most helpful measurement of scat is at the point of maximum diameter.

Secondary Pad(s): the pad or pads that register in a print immediately to the rear of the toe prints. In other guides this pad may be variously referred to as the "plantar pad" or the "palm pad."

Sign: any evidence for the former presence of an animal. Although tracks may be considered sign, the term is often reserved for everything except tracks, such as scat, digs, rubs, chews, bones, hair, scent marks and so forth. The British often use the term "spoor" to refer to sign.

Signature: a mark or outline appearing in part of a print that, while not always present, definitely identifies a species when it does appear. The smooth leading arc on the *secondary pad* of a raccoon is an example.

Step Length: in this guide, a measurement from any point in the print of one front foot to the same point next to the print of the other front foot (or hind foot to hind foot) in the course of a single step. Step length is measured parallel to the direction of travel. In slow gaits, like walking and trotting, step lengths are more or less constant. However, in rolling gaits, like loping and galloping, step lengths (unlike "strides") usually alternate short and long. See *Stride* below.

Straddle: in this guide, the distance across the two front (or hind) prints measured from the *lateral* side of each and perpendicular to the direction of travel. In *displaced* gaits, the straddles of both front and hind prints are narrower than the *pattern width*. Note that other tracking guides may use these terms synonymously. See *Displaced* below; see *Lateral* and *Pattern width* above.

Stride: the distance between two consecutive prints of the same foot. A stride amounts to two consecutive step lengths and is more or less constant along a trail in which the animal does not vary its gait. Note that other guides often use "stride" and "step length" synonymously.

Tertiary Pad(s): generally any pads in a print that register to the rear of the *secondary pad(s)*.

Track: in this book, a single entity that may be the print of only one foot or the partially or perfectly overlain impressions of more than one foot. In a coyote pack's winter trail a single track may be made by the successive overlain placement of multiple prints of several animals. Other books use the terms synonymously.

Track Width: the width of a track measured perpendicular to the direction of travel. Usually this is from the medial edge of the most *medial* toe pad to the *lateral* edge of the most lateral toe pad. If a large secondary pad exceeds the width of the toe spread, its width becomes the track width. See *Lateral* and *Medial* above.

Trail Width: a measurement of the widest part of the impression of an animal's body perpendicular to the direction of travel. In most cases this is the same as *pattern width,* but in some short-legged animals in snow, such as porcupines and otters, body or fur impressions may make the trail width wider.

Transverse: a four-print (4X) "transverse" pattern has the order of placement: right-left-right-left or left-right-left-right. Wide and long patterns have a shape more or less like the letter T, thus "T-pattern." Short patterns are often called "1-2-1." See *Rotary* for comparison.

Ventral: viewed from below the animal. See *Dorsal* above.

Vestigial: describes a toe pad or other foot part that is gradually disappearing in the course of evolution. Many wild animals, like bears, have a vestigial fifth medial toe that registers only slightly or not at all in a print. The human "little toe" is becoming vestigial but is located on the *lateral* side of the foot, the opposite of other mammals.

Withdrawn: held above the general plane of a print. A withdrawn pad is deemphasized in a track, registering lightly or not at all. The fifth *medial* toe of a black bear is often withdrawn from prints.

X (as in 2X, 3X and 4X): indicates the number of *tracks* evident in a pattern. Note that these may range from direct to single registrations; thus, a 2X pattern shows only two tracks but these may be the impressions (prints) of 2 or 4 feet, or more in the case of two animals following in each other's footprints. In a 3X pattern, one hind foot lands in the impression of one of the front fcct, producing a single track to go along with the prints of the other two feet and giving the impression of a 3-legged animal.

Gait Terms

Aligned: a gait in which the hind foot on each side moves directly over the print of the front foot on that side. In an aligned gait the animal's spine is arranged parallel to the direction of travel. See *Displaced* below.

Arc of Stride: the arc described by following a point on an animal's torso, viewed from the side, through one cycle of gait motion (a full *stride*). Walking has a flat arc of stride while bounding has a high arc and may be referred to as having a lot of *verticality*. See *Verticality* above.

Bound: a high-energy *rolling* gait similar to a *lope* in its mechanics but with more *verticality* resulting in a higher arc of stride. It is most often used by an animal to clear *resistance*, that is, some obstruction in its path such as soft snow or forest debris. See *Verticality* below.

Coincident: in this guide, any gait in which the front and hind of an animal rise and fall together through the cycle of a stride. *Trotting* and *stotting* are coincident gaits. Compare to *Rolling*.

Displaced: in this guide, a gait in which the hind foot moves past rather than over the print of the front foot of the same side. To do this the animal displaces its hind end to the right or left, holding its spine at a slant relative to the direction of travel. When this posture is used in a trotting gait, it is sometimes referred to as a "side-trot." See *Aligned* above.

Flat: having a low *arc of stride*, that is, a gait having insignificant vertical motion.

Gallop: a high-energy rolling gait that resembles a *lope* in its basic mechanics. It differs from a lope in that the center of gravity is lower and power is imparted to the gait by the front legs as well as the hind.

Hop: a bounding gait by a rabbit. "Hop-bound" is a gait with a high *arc of stride* used by many rodents *bounding* through soft snow. The front feet land and prepack the snow for the hind feet that follow in and at least partially cover the front. There is usually a slight pause while the animal extricates front from under hind and launches itself with its powerful hind legs into another short, high arc.

Lope: an easy run in which two feet (two front or two hind) are in contact with ground while the other two are moving forward. This is a *rolling* gait: when the shoulders are down, the haunches are up and when the haunches are down the shoulders are up. There may be brief moments of both gathered and extended *suspension*. Most propulsion comes mainly from the hind legs, the front being used mostly as props.

Lope-bound: an easy run with a higher *arc of stride* than a pure *lope* and a tighter grouping of prints. Lope-bounding is used, often by long legged animals, for deep snow or low debris, alternating with other gaits such as pure loping or bounding as changes in the surface are encountered.

Rolling: any gait with a rocking horse motion in which the front end of the animal is up when the hind is down and vice versa. Loping, galloping and bounding are rolling gaits. Compare to *Coincident*.

Rotary: a gait in which the order of footfall is right-left-left-right or left-right-right-left. Compare to *Transverse*.

Scurry: a very fast walking or aligned trotting gait by a small animal such as a mouse, vole or shrew. It is a *flat* gait with *coincident* rise and fall of both ends of the animal.

Suspension: any period in the stride cycle where all four of a mammal's feet are off the ground. In "extended suspension" the animal's body is stretched out in mid-leap; in "gathered suspension" the animal's hind feet have moved forward past its front feet so that front and hind form an X under the body when viewed from the side.

Stot: a *coincident* bounding gait peculiar in North America to mule deer and one species of flying squirrel in which all four feet of the animal leave the ground at the same time in a sort of pogo-stick rise and fall.

Transverse: a gait in which the order of footfall is right-left-right-left or left-right-left-right. Compare to *Rotary*.

Trot: a slow run in which two feet, one front and one hind, are in contact with the ground while the other two are in synchronized motion. The pairings may be on the same side of the body or, more usually with wild animals, diagonally opposite. A trot always involves at least a brief moment of *suspension*. Trotting is *coincident*, that is, the shoulders and haunches rise and fall together.

Verticality: the upward as opposed to horizontal movement in a gait.

Walk: in a quadruped, a gait in which one foot moves forward at a time leaving the other three in contact with the ground. Through most of the stride cycle tripod stability is maintained. There is no *suspension* in a walking gait.

Pad Nomenclature

Several competing systems have been used by different authors to designate the pads on various animals' feet. For one reason or another, none of these systems seems adequate, being too complex, vague, illogical or inconsistent from animal to animal. At the risk of making an already confusing situation even more so, a simple nomenclature, which is used in this guide, is advanced below. Whatever it lacks in anatomical precision, it has proven adequately descriptive for fieldwork and is much simpler and more consistent than other methods in publication.

Primary pads are the most forward grouping of pads on an animal's front or hind foot. Also called "digital pads" or "toe pads."

Secondary pad(s) refers to the second pad or group of pads which are located immediately to the rear of the primary pads. When there is more than one of them, they are usually contiguous to one another but separate from the primary and tertiary pads. The ball of a human foot is a secondary pad.

Tertiary pad(s) refers to any third pad or grouping of pads that shows on the heel area of a print. Some animals have them and some don't. The front foot of a red squirrel, illustrated below, usually shows these pads in most prints, but the hind prints of the same animal do not. Thus this squirrel, like many other sciurids and rodents, is normally *digitigrade* on its hind feet and *plantigrade* on its front.

Phalangials are toe bone impressions, linear marks that often appear in prints between the primary and secondary pads.

In mammals with fused secondary and tertiary pads, such as bears, skunks and humans, the whole area of the sole is often called the "plantar pad". In this book I may refer to the "secondary area" or "tertiary area" of this fused pad as its ball-of-foot or heel area, respectively.

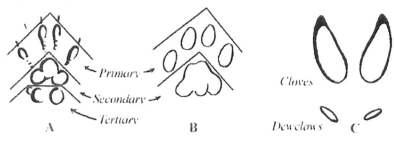

Diagram G-1

This nomenclature works best for soft padded animals (Diagram G-1A and B). The prints of hoofed animals (ungulates) (Diagram G-1C) may show dewclaws where the secondary pads should be. But dewclaws are really the *retarded* first and fourth toes of the animal's foot, which have moved back up the limb in the course of evolution. In the case of ungulates, then, the usual nomenclature is suspended in favor of the traditional "cloves" and "dewclaws."

It should be noted that animals which normally register four toe pads in a print, like dogs and cats, have a vestigial fifth medial toe located farther up the foot in the tertiary area. However, this toe seldom shows in a print and so the problem it poses in this nomenclature will be ignored. It should also be noted that the hind foot of such animals actually extends up to the angular joint that we intuitively think of as the animal's elbow. But the heel ("elbow") of this foot has no fleshy pad, since it is seldom in contact with the ground and thus needs no cushioning. The heel of the front foot is high up and close to the animal's trunk.

Solutions
to the Tracking Problems

Solution to Tracking Problem #1

At first glance, one may note the absence of nails, the oval toe prints and the relatively large secondary pad with flattening along its anterior edge, all characteristics of a felid. Using the penny (3/4 of an inch in diameter) for scale we see that the track is about 2 inches across, making it a bobcat, right?

However, look again at the secondary pad. Is that really what it is? With a careful re-examination you may see that this "pad" is actually the four toes of a narrow hind foot that has slightly under-stepped the front, creating an indirect registration. Once we "see" the reality here, we may recognize this as a gray fox track. The combination of a wide, round front print with a narrow hind, you may remember, is a common profile for this species. Also, gray foxes tend to toe-down radically with the hind foot so that the hind secondary pad registers very lightly or, as here, not at all. The round, cat-like toe pads and lack of nails complete the deception. The track width suggests that this was a large male spreading its front toes. See a similar illustration in "Species Notes: Gray Fox" (page 323).

Solution to Tracking Problem #2

The long, narrow snout and various other details identify both skulls as members of family canidae, or dog-like mammals. Measurement of length as well as details of shape exclude coyote. The two remaining candidates are red and gray fox. Although the gray fox in the Northeast is slightly smaller than the red, the difference is too close to be useful. Instead, final identification can be made by details on the cranium. In this dorsal view, red foxes show a pair of temporal ridges in a narrow V-shape down the center while a gray fox skull shows these as a broad U. This detail alone safely distinguishes the red fox skull on the left from that of its distantly related cousin.

Additional details on the gray fox's mandible (lower jaw) also serve to distinguish it from the red. Note the obvious "step" or indentation at the proximal end of the jaw line (on the left in the right-hand photo). Grays have this but reds do not. Also, note the flatness of the mandible's ventral (lower) outline. A red fox's mandible has a distinct rocker, that is, it is distinctly convex on its lower edge, from front to rear.

Although each of these foxes represents its own genus and is only distantly related to the other, their skull structure and other features have come to resemble one another through convergent evolution. This is the process in which, over time, two unlike species begin to develop similar physical appearance because they are filling a similar niche. Thus a vulture, descended from storks, comes to look like a hawk because soaring is a useful way to find carrion and a hooked beak is useful for opening dead carcasses. By the same process, red foxes and gray foxes have come to resemble one another because large ears and a long snout are useful in capturing small rodents.

Red foxes are traditionally considered animals of pastoral landscapes while the gray is the arboreal fox, better suited to forests because it can climb trees. With the arrival of European colonists who, by cutting and burning, quickly converted the gloomy forests of New England to the open pastoral landscape of old England, the gray fox declined and the red fox proliferated. The last gray fox in the Concord, Massachusetts, area was killed for bounty in the early 1800s.

As agriculture declined after the mid-1800s and the cut-over forests of the Northeast re-grew, the pastoral red fox that had done well in the colonial and post-colonial landscape had to accustom itself to the new conditions. In the absence of competition from the eradicated gray fox, it managed not only to survive but to do quite well. I have often found their trails deep in the mountains around clearcuts and old log yards grown back to brush. The regrowth of the forest, however, has resulted in the gradual recovery by grays of their ancestral ranges due to the competitive advantage of climbing ability. In some even-age forests they seem to be replacing reds. Such was the case on the Conway River in northern New Hampshire where I had tracked red foxes on both sides every winter for several years. Then one winter the trails were gone on the east side, replaced by the un-mistakable tracks of a gray fox, the first whose sign I had seen that far north. I was witnessing, at a microscale and real-time, the return of a native predator to its ancient range and to a habitat in which it could compete successfully with the default resident fox. The presence of the red fox skull in a forested habitat invaded by newly arrived grays may suggest this transition.

While grays may be able to out-compete reds in even-age forests, in older-growth woodlands, such as official wilderness areas embedded in the national forests, where there is increasing vertical complexity, there is habitat for both these animals.

Solution to Tracking Problem #3

In this photo, the trails of red fox and a gray squirrel crossed.

The trail of the gray squirrel moves from upper right to lower left in a bounding gait. Each set of impressions is two larger hind feet advanced and wide apart with the two smaller front feet behind and closer together. The condition of the snow blends the front and hind prints on each side together into one long track. The paired hind feet show a splay that should immediately distinguish this pattern from that of a rabbit, the bounding pattern of which normally shows parallel hind prints.

The other trail is a slow displaced trot, a gait used by canids. Its elegant economy points to a wild canid that must be very careful about expenditures of energy in the winter. The very narrow straddle and 17-inch step length in this gait identifies the animal as a fox, and finally the oblong rather than round front prints distinguish it as a red fox rather than a gray.

The fox was trotting from lower right, away from the camera. In a displaced trot, the smaller hind foot passes by the front foot of the same side and lands advanced forward of the front. The difference in foot size is apparent in the photo.

The two trails were not made at the same time as neither trail patterns shows the least reaction to the other's presence. The appearance of the two trails also suggests that they were made at different times, the fox's first and probably in the evening since foxes tend to be crepuscular, and the diurnal squirrel's later.

Red foxes often use the margins of ponds in winter for three reasons. First the snow there is level and without obstruction, allowing for displaced trotting. This gait is very efficient for traveling as it has a very low arc of stride and thus little wasted vertical motion (pages 113-114). This is important for an animal that must cover many miles on its hunting routes. A second, related reason is that the snow on a pond edge is often packed by wind and sun, providing firm and therefore efficient footing. Finally ponds embedded in a forest usually have a dense, brushy border used as cover by small animals that the fox is hunting.

Solution to Tracking Problem #4

The smaller animal was a chipmunk, the larger a raccoon. Note the raccoon's club-like toes extended for support and traction on mud. Also note the shape of the secondary pad, with a smoothly rounded rather than lumpy anterior edge. This latter is a feature unique to raccoons and serves to distinguish them reliably from other animals with which this "great fooler" might be confused.

The prints of the raccoon in the photo show a slow walk. The arrangement of the feet in the pairings of this animal's trail shows its speed. If the hind print is "retarded," that is, slanted to the rear of the front, the animal is walking slowly. If the pairing is even, it is walking faster. If the hind is advanced ahead of the front, the raccoon is either walking very fast or trotting. With each of these increases in speed the step length will increase as well.

Raccoons in the wild spend a lot of time looking for food along the edges of water, searching for organisms in the mud. The route from the marshy backwater to the river was probably part of its foraging circuit.

All squirrel-like animals have similar feet — four registering toes on the front foot and five on the hind. Furthermore, on chipmunks and squirrels the three central toes on the hind are arranged evenly rather than in an arc as is the case with the smaller weasels with which these prints can be confused. Thus, in the print pattern on the left of the frame, the rounded track with four registering toes is a front print and the one just to its right is a hind. One might almost be able to lay a straight-edge along the three central toes in this print. Another feature of the three central toes can be helpful in distinguishing right hind from left: the lateral two of the central three often register slightly closer together. This characteristic is a little clearer in the hind print in the center of the frame, showing it to be a left hind print in a double registration over the left front.

Squirrels, chipmunks and mice normally move in a four-print bounding pattern, with two hind prints placed widely apart at the head of the pattern and the two front prints farther back and closer together (Diagram A-1).

Above is a sequence with decreasing speed from left to right. The pattern on the right represents a foraging gait where the animal is levering itself slowly forward, searching for food under its nose. This pattern is evident in the right hand set of prints in the photo with animal headed toward the bottom of the frame. While the registration of the central two front prints is slight, their presence can still be detected. In this case, given the month, the chipmunk may have been nibbling at tree-pollen grains or other edible debris blown onto the sticky mud.

Solution to Tracking Problem #5

On the left of the frame, moving from top to bottom, is the trail of a bobcat in an overstep aligned walk or trot, with the amount of over-step suggesting the latter. Note the lack of nails, the oval pads and the large secondary pad as well as the roundness of the prints.

Moving diagonally from top center to lower right is the trail of an eastern coyote in a direct-register walking gait. Note the arrowhead shape of the three visible tracks and the lack of lateral and medial nail marks as well as the narrow straddle and perfection of the direct registrations — all coyote characteristics.

Diagonally from lower left to upper right is the trail of a river otter. You can see the five asymmetrically arranged toes in the hind prints and the lightly registering secondary pads. The three tight groups of four prints per pattern indicate a lope-bounding gait.

Bobcats are wary of open spaces, preferring to move unobserved through cover. Although the temptation of easy transit on the beach overcame most of its concern, the cat kept to the upper part of the beach, close to the neighboring brush in case a quick disappearance were needed.

Coyotes are bold by comparison. Since we are their only predator, they watch us, learn our habits, including when we are likely to be around and when not, adjusting their own routes to avoid our own. This one has marched down the center of the beach, confident in its own intelligence.

The otter had been ashore at a rolling site or scat station somewhere within the brushy border and was making its way back to its preferred habitat, water. The open shore is a danger area to be crossed quickly to the safety of the water and thus the use of a speedy lope-bounding gait.

Solution to Tracking Problem #6

The native forest in New England was cleared by newly-arrived European colonists very early on. While much of the exposed, stony soil that resulted was unsuitable for crops, it could be used for pasture, once appropriate English grasses had been imported. Because livestock raising was the only use for much of the land, farmers left a few trees standing in their pastures as shade for their animals. One of the species of tree that the farmers left in place was the white oak, perhaps because, when standing alone, this tree develops a broad, shady crown. As an additional benefit, it also periodically produces a large mast crop of acorns. Unlike acorns of other oaks, white oak acorns contain little bitter tannin and so were useful for fattening the livestock sheltered beneath them. After the decline of New England agriculture in the mid-19th Century the forest grew back in the old fields and pastures, surrounding these old trees with a new young forest. Today these ancient trees can still be found deep in woods that, at an earlier time, had been pasture. Foresters, who often advise woodlot owners to remove them, call them "wolf trees" because they deny sunlight to other marketable trees that might grow in their space. Nevertheless, in wildlife sanctuaries and other preserved habitats they are allowed to grow old and die in peace and in place, at which point they begin their second life, providing food as well as denning and nesting cavities in their hollow interior.

When black bears emerge from hibernation in the early spring, there is little for them to eat. At this time they seek out seeps where skunk cabbage has emerged, sometimes so early that the new growth stands in snow. They may also avail themselves of the various creatures that live in root tunnels under dead trees or stumps. Thirdly, they may seek beetles and grubs that have over-wintered bcneath the bark of dead and dying trees. This apparently was what happened here. A hungry bear encountered this tree, sensed food under the surface and stripped the tree of loose bark as high as it could reach.

Bears often mark trees, apparently to advertise their presence or ownership of a range. But that kind of work is quite different from what we see here. At a mark tree, which is often a sappy variety such as red pine or balsam fir, the bear reaches up and claws downward creating a wound, runny with sap. Then it turns and rubs its back on the roughened surface. While scratching, the bear may turn its head and bite off a chunk, leaving behind two parallel slanted grooves from its canine teeth.

The two seasons when there is little to eat for a bear are mid-summer and early spring. In June and July the tender plants in the wet spring meadows harden up, becoming woody and difficult to digest, and the berry crop won't ripen until August.

However, in this case lack of leaf fall covering the scattered slabs of bark at the base of the tree suggests recent work of that early spring rather than the previous summer.

In both seasons, bears not only go after beetles and grubs under the bark of old standing wolf trees, but they also dig out anthills as well as rolling and tearing apart old rotting logs on the forest floor. Unfortunately, mid-summer is also the peak of the human camping season, and bears have learned what red and white plastic boxes are for.

Solution to Tracking Problem #7

A snowshoe hare created these lumpy ovals in the fresh snow by spreading its toes to maximum width and packing it down into "hare islands."

Winter-adapted animals have a variety of ways of coping with deep, soft snow. Some, like bobcats, coyotes and deer, follow other animals, allowing their predecessors to do the work of breaking out the trail. Others have adapted physically: both lynxes and fishers, for example, have grown very big feet to distribute their weight over a larger areas and so stay "afloat" on the surface. No animal has developed this adaptation as much as the snowshoe hare. Each of the hare islands was made by repeatedly hop-bounding from one spot to the next.

By following the same route over time and landing each time slightly ahead or behind the previous prints, the hare packs down the snow in longer and longer ovals until eventually they connect in one long route. Over this packed trail, the animal can get around to feeding sites as well as escape all but the fastest predators.

In the Great North Woods, the lynx is the hare's main predator. It too has grown very large feet for flotation on the snow and so can chase a hare around its circuit. However, unless the hare is caught napping, is weakened by disease or starvation, or is forced out of its packed trail, the cat is likely to come up wanting.

Solution to Tracking Problem #8

At the lower right are two displaced trot patterns from which the fox transitioned to a 1-2-1 lope for the remainder of the trail. The lope patterns are run together without any defining inter-pattern spacing.

The fox was moving from lower right to upper left. You can tell this by the order of single registrations with the larger front registering first in each pattern, that is, closer to the camera.

On the lower right are two patterns of 2 (2X) single registrations with the smaller hind advanced. These are followed by pattern in which the right-front registered first, followed by the left-front and then next to it the right-hind. At the top of the pattern is the left-hind. The order of footfalls makes this loping gait transverse. The first lope pattern is run together with the next, beginning also with the right-front and with no defining inter-pattern space.

That the two displaced trotting patterns are slow is indicated by the minimal space between front and hind prints. In a faster trot there would be more, stretching the pattern out. That the following transverse lopes are also fairly slow is indicated by the lack of overstep in the central two prints in each pattern. In a very slow lope, the hind print would be to the rear of the front; in a fast lope it would be advanced ahead of the front.

This fox was in no particular hurry, probably traveling in search of targets of opportunity along its route. The high center of gravity in both these gaits conserves energy, and so either is what a red fox would use for traveling on a firm, even surface.

490

Solution to Tracking Problem #9

The ungulate tracks in the frame were made by deer, of which only two species range into central Arizona, the white-tail and the mule deer. White-tails prefer the dense cover of wooded streamsides, while mule deer also forage away from water and out in the more open chaparral desert where these tracks were found. The track measurement also suggests a mule deer since, at this latitude, they are smaller than white-tails. Furthermore, in the double registration at the lower right, the closed print shows a flattening to almost a slight concavity along its left edge, a feature common in mule deer hind prints but seldom found in white-tails.

This was a bit of a trick question to illustrate Tracking Tip #4: The Obvious Conceals the Subtle (pages 403-404). There are actually two species of animals in the frame. Did you notice the faint canid prints just to the right and below the deer tracks on the left? There are only two wild canids in central Arizona, the small gray fox and the larger western coyote. Comparing the print-size of the canid to the mule deer tracks makes them about 2 inches across and identifies them as the latter.

Solution to Tracking Problem #10

The front print is the lower one in the picture shown with the problem. Because the heaviest part of a mammal's body is its head, the toes of the front, bearing more weight, will often splay more widely than is the case with the hind, an effect easily seen in this pair of tracks.

As is the case here, most identifications are made on the sum of the evidence rather than on one or two details. First of all, prints with a symmetrical arrangement of toe pads and a central pyramid of negative space generally denote a canid. With this in mind, notice next the relative shapes of the two prints; among canids, a combination of a round front print and a narrow hind is a fairly reliable tip-off for gray fox. This identification is supported by the near absence of nail marks (gray foxes have retractable claws like a cat) and by the shape of the secondary pad on the hind print. The posterior edge of this pad is scalloped, again like that of a cat. This scalloping is the posterior edge of the winged ball effect that also denotes this species. Although this effect appears on both front and hind feet, it is most pronounced in this pair of prints on the hind, the effect on the front being disguised by distortions of weight, gravity and time. The entire secondary pad would look like Diagram A-2 if it registered clearly.

Diagram A-2

The art of tracking could be said to be the art of seeing what almost isn't there, but if you look very closely you will find the slight but definite impression of the "wings" on either side of the main pad.

On the front print, the suggestion of a bar across the secondary pad might suggest red fox since a naked callous on the pad of that fox often registers as a chevron mark in this location. However, this is the only bit that says "red fox;" all the other evidence — lack of nail marks, roundness of the front print with a narrow hind, and the shape of the secondary pad — indicates its gray cousin (see pages 70-71 for the probable reason for this distortion). Remember, we are basing identification on the preponderance of evidence rather than a single detail. Besides, a check of a field guide will show that red foxes don't range into this part of Arizona.

The normal maximum step length for a walking gray fox is about 13 inches when the animal is direct registering. Here an overstep of only 2 inches or so could suggest a walk, but the 16-inch step length favors an aligned overstep trot, a gait favored by gray foxes more than reds. On the same surface, with shallow track indentation, a red fox would most certainly use a displaced "side-trot." Both gaits are used when the animals are travel-hunting.

Solution to Tracking Problem #11

A wavy trough of appropriate size in January snow is almost always evidence of the passage of a porcupine. These animals have relatively short legs and rotund bodies, the combination of which results in a trough where the animal was sinking down more than a few inches.

Less obvious are the footprints of a bobcat in the porcupine's trail. Look carefully at the photograph and you will see a series of darker direct registrations zigzagging in the trough.

Both animals were walking. The porcupine, which carries its mobile defense on its back, has little need for speed. The bobcat would prefer to walk, a gait with a low arc of stride, as an energy saving measure in cold weather.

Bobcats seek out the trails of other animals in the winter to ease their passage around their hunting circuit. Better to walk in the already-packed trail of another animal than to waste energy breaking out one's own. What's more, there is always the possibility of encountering the trail-maker himself in a vulnerable situation away from its den. Extrapolating the trough width shown here puts the step length of the larger animal at about 12 inches, short for a bobcat but within the expected parameters, especially if the footing was poor. Each of the cat's prints in this walking pattern is a direct registration of front and hind. Although the one at the lower right of the photo looks long enough to be a fox track, the rest are quite round, suggesting that the oval appearance of the questionable one was due to an under/overstep of the left hind foot. Thus, this track can be excluded from consideration. The pronounced zigzag of the predator's walking trail suggests against fox, as well, an animal that usually leaves a very straight trail. Here once again the identification, lacking print detail, is based on the sum of the evidence rather than on any single trait.

The porcupine's track depth indicates that it was made before the thaw when the snow was deeper and softer. If the trail had been made at the current reduced depth, the passing animal would have made little or no trough. The softening of snow detail also suggests that this is an old trail. Obviously, the cat used the trail sometime after the porcupine's passage, but its use, too, occurred during the thaw or shortly after its end. The larger animal's footprints are black because there was still enough insulated moisture under the snow to melt the bottom of the prints before both trails froze up again into the condition in the photo.

Solution to Tracking Problem #12

Squirrels like mushrooms as much as we do. This one has been harvested and placed for drying before being included in the animal's winter larder underground. Since gray squirrels do not establish concentrated winter caches as red squirrels do and since the southern flying squirrel found in this range isn't know to store mushrooms, this is probably the work of the red squirrel.

By placing the fungus in the crook of dead branches on the lower part of a pine tree, the squirrel accomplishes two things. First, it removes it above the scent-plane of other foragers prowling the forest floor. Secondly, it protects it from contamination by bacteria in the soil that would quickly turn an unharvested mushroom into a slimy black mass. It is possible that chemical constituents in the pine retard the growth of bacteria as well. Eventually, the fungus, after thorough drying, will be included in a winter cache underground where cold temperatures will refrigerate it until it is eaten.

Solution to Tracking Problem #13

Here is another example of tracking as a matter of seeing what almost isn't there. The animal sign in this frame consists of two faint prints, one on the lower left and the other on the upper right. The lower left track is a front print and the upper one a hind.

The only clear feature in the front print on the lower left is a lateral bar that many trackers will recognize immediately as the bar that traverses the secondary pad on the front foot of a red fox.

The pattern of two single registrations arranged on a slant and repeating from pattern to pattern is a displaced trot pattern common to red foxes and coyotes. As we have seen before, all the front prints are lined up on one side of the trail and retarded in each two-print group, while the hind prints are advanced and all on the other side of the trail. The wide separation of the two tracks in the photo indicates a very fast trot.

Red foxes and coyotes hunt on their feet, putting in a lot of miles in a given evening in order to visit favorite hunting sites where they have had success in the past. Forest openings with dense regrowth are a favorite habitat for many of the fox's prey animals such as ruffed grouse and cottontail rabbits. This red fox was probably using the human infrastructure of a dirt road to make its rounds to these sites. Because such roads are level, they allow the red fox to use its displaced trotting gait, which is the most efficient gait in its repertoire due to its relative lack of energy-wasting verticality.

Solution to Tracking Problem #14

This problem illustrates three principles:

- The trail or other evidence that the tracker finds may not have been produced in the same conditions in which it is viewed.

- It is always important to consider the effects of recent weather on the evidence.

- Sometimes the usual sequence of identification to interpretation must be reversed.

Here we have a trail of mounds arranged in a repeating slant with pattern widths of about 4 inches. We know from our study of patterns and gaits that such a trail can be left by either a canid in a displaced trot ("side-trot") or by a bounding member of the weasel family. In the latter case, the straddle measurement here suggests either a mink or a long-tailed weasel.

If a canid in a displaced trot made the trail, it would be a succession of pairs of single registrations, one front and one hind. If it were a member of the weasel family, then the pairings would be direct registrations of two feet, one front and one hind in each track. Given that there is no detail left in the mounds for track morphology, how can we know which?

The mounds were created by the differential melting of soft versus packed snow. Disturbed snow becomes dense as it absorbs the energy of any compression force. As the animal landed on each foot, it compressed the snow underneath, hardening it and making it resistant to melting. These compressions radiated out and downward from the original outline of the prints into cones below the animal's feet. As a thaw set in, the surrounding undisturbed and therefore softer snow melted, leaving behind the hardened cones, now raised above the surface in a negative image of the original trail. Snow can be magical stuff!

Canids use a displaced trot on smooth, firm, even surfaces for its efficiency in getting around. Mustelids use a 2X bounding gait in deep, unconsolidated snow where they are sinking down more than the length of their short legs. The height of the cones shows that the latter conditions existed when the trail was laid down, and this, of course, points to a mustelid (weasel-family member) as the creator of the trail of pyramids. Because there has been so much distortion of the trail by

495

weather, it is difficult to tell from the straddle measurement whether it was made by a mink or long-tailed weasel. The fact that the trail headed in the direction of the beaver flow is circumstantial for the more aquatic mink, however. And its regularity also suggests mink since, as weasel species decline in size, their behavior becomes more antic, showing irregular spacing and direction. Thus, in this case, I would give the identification as the larger mink an 85 percent certainty.

Solution to Tracking Problem #15

The item in the photographs is the tail of flying squirrel. Unlike the tails of other squirrels, the fur is arranged on a single plane on either side of the tail shaft. To my eye, these tails always looked perfectly coifed, as if the animal had just returned from the beauty parlor. Because flying squirrels use their tails for a rudder as they glide from one tree to another, they are kept in perfect condition. Both northern and southern flying squirrels range into that part of New England. By size and lack of darker tip, this one was more likely the smaller southern species.

This little animal was the victim of predation. The predator had nipped off the tail before carrying off the carcass. The presence of two examples of this on the trail suggests that it was a favored hunting lane used by either an avian or terrestrial predator. Because the tail of most animals holds little nutrition, it is a common practice for predators to remove it to lighten the load.

Figuring which predator was responsible requires a close examination of the tail and a consideration of the habits both of the squirrel and local predators. Flying squirrels are nocturnal animals that glide from tree trunk to tree trunk gathering nuts or other edibles. During the day, they usually den up in a hollow. Since hawks are night blind, they can be eliminated from consideration. The fact that flying squirrels spend a lot of time up in the trees rather than on the ground marginally prejudices against terrestrial predators like red foxes. Both bobcats and gray foxes can climb trees, but neither is particularly nimble while doing so. A gray fox may chase a gray or red squirrel up a tree and then corner it as the crown of the tree narrows. But a flying squirrel could simply glide away from such an attempt. Bobcats use trees as hunting perches from which to survey the ground below rather than as actual hunting sites. A fisher might trap a squirrel in a tree cavity den while the little animal was sleeping during the day, but this is where a close examination of the evidence is needed.

In order for any mammal to nip off the tail of a squirrel, it must mouth it in such a way that there should be disturbance farther out on the tail. If the predator uses its front incisors, the tail has to be in its mouth. Such a process would cause saliva to mat the fur. If the mammal chewed off the tail with its side teeth (premolars or carnassials) then the marks of the teeth on the other side of its mouth should also show along with saliva matting. The tail in question, however, is in perfect condition, showing no matting or crimping anywhere along its length. The predator, then, must be able to nip off the tail with a narrow instrument such as the tip of a beak. Since night-blind hawks have already been eliminated, this leaves only owls as the most likely predator: they hunt at night and have narrow beaks. Like other predators, they also use hiking trails and dirt roads as hunting lanes. In this area there are two common owls that might take a flying squirrel-sized animal — great horned owls and barred owls. Because the latter is by far the commoner in these woods, it was the most likely predator.

Solution to Tracking Problem #16

The scat pictured — a single, constricted segment — is classic fur-scat of red fox, the result of its last meal, not this one. It apparently was left by the fox as a property marker. The identification is supported by the odor in the woods around. Red fox urine scent marks smell like sweet skunk, but are distinguishable from the skunk by their relative lack of strength. The urine of other local predators such as bobcat and gray fox does not have this odor. Coyote urine, to my nose at least, smells like skunk mixed with burnt hair, but the odor is so weak that it is rarely detectable at any distance.

The detail photo of the site reveals the prey. The glossy purple feather is from the neck area of the common feral pigeon, a species that proliferates near human structures such as the nearby village. The larger gray flight feathers confirm this identification.

This was not a kill site; no blood is evident in the debris. Instead, it is a plucking site and perhaps a feeding site as well. After it makes a kill, a red fox will flee the scene with its prey, stopping only when it reaches concealment. There it will lighten the load by ridding the carcass of excess, inedible parts such as feathers. Then it may feed on the remains or transport them to a den to feed kits. In this case, the kill was made most likely on the grass of the village common, and the pigeon was carried into nearby woods for plucking and feeding.

It should be noted that we cannot be absolutely certain the scat and site odor were left by the predator of the pigeon. A red fox, discovering the plucking site of another predator, either of its own or a different species, may leave scat and urine marks as an assertion of its own territoriality (page 406).

Solution to Tracking Problem #17

Not all animal sign is mammal sign. The sinuous marks in the sand are those of a snake that has been hunting the edge of the reservoir.

The swept sand and pushed up ridges show that it was moving from lower right to upper left.

Its hunting foray into the open met with success as it appears to have caught a frog that it carried in its jaws back to the concealing shrubbery for consumption. The long striations are probable the unfortunate frog's toes.

Solution to Tracking Problem #18

And, not all sign is animal sign. The species was not, as it may appear, a vole or shrew that had left back-and-forth trails while foraging on the snow. Actually, the species was a white pine. The "trail" was made by a fallen cone that had been blown over the surface of the snow by the wind. The "prints" are actually the contact points of the opened scales of the dried cone. As it was blown along, the tapered shape of the cone caused it to rotate around its base faster than its tip, creating the curved trail of marks.

The Tracking Tip involved in the solution to this problem is #15 — not all sign is animal sign (pages 427-429).

Preparations

Everyone who practices the art and science of tracking wildlife and interpreting the evidence assembled begins early on to accumulate a kit containing odds and ends that are helpful in the field. Included in the first section of this appendix are the contents of my field kit as an example of what might be carried by other trackers without adding too much weight. While digital cameras have become ubiquitous nowadays, a few suggestions that I have found helpful in the use of this tool are also included along with a lengthy description of a data card on which may be recorded information that is not easily captured by the camera's lens. I have in several places in the body of the book recommended the creation of plaster casts of tracks as an aid in their analysis. Many other tracking books contain directions for these, but I present my own method here because it includes a few tips not found in the works of others.

The second section of this appendix deals with processes that, while clearly tangential to tracking, nonetheless may be very valuable to anyone who travels far from civilization, especially in colder latitudes and altitudes. "How to Stay Warm Outdoors in the Winter" contains information that I make available to my own clients in the form of a handout before we go out into the woods. The simple traditional advice of "dressing in layers" isn't nearly enough to keep one comfortable and out of danger in the woods. Although most of us feel we know how to accomplish the challenge of staying warm, some additional advice may be called for since much of the processes of heat generation and conservation are not intuitive. Two brief, final sections provide basic information on land navigation and emergency shelter-building. Neither of these subjects is treated exhaustively, but I do provide information that may be of critical assistance to the tracker unsure of his present location or finding himself far from a heated shelter.

Tracking Tools

The Tracking Kit

Every tracker keeps a small kit (Figure B.1) containing the odds and ends of equipment he may need while searching and following wild animal sign. I use a commercially available book-sized pouch suspended from a shoulder strap to hold mine. It has multiple pockets for the following, recommended contents.

- *Trackards for North American Mammals* and *The Companion Guide to Trackards for North American Mammals* contain life-size images of animal tracks as well as information on step length, straddle, and other details essential for identification. The cards and the book slip nicely into an open slot on the back of the pouch.

- Tape measure and (optional) caliper measure for fine work.

- Flashlight. A LED light with the newer technology that produces a very bright beam that will illuminate the dark interior of dens even on bright days. Holding it low across a faint track will accentuate shadows and bring the track into relief.

- Magnifying glass. I have removed a rectangular reading lens from its holder and keep it in a soft sack in one of the pouch pockets. You may also wish to carry a loupe for very close work.

- Pre-printed data cards or 3X5 cards and two pencils with soft lead (one always runs out just when you need it).

- Rubber gloves and a baggie for handling and saving specimens.

- An old penny. For photos I use a penny for scale. It is almost exactly ¾ of an inch in diameter and a size familiar to every one since we all handle them. A tape measure in a photo is hard to read and few people have a good intuitive sense of an inch. I prefer an old penny because light reflecting on a bright new one may fool the auto exposure on my camera.

- Colored plastic tape for marking sign location. Remember to remove it after it is no longer useful.

Figure B.1. A typical tracking kit.

- GPS, paper map and compass for both finding your way and marking the location of sign.

The Camera

Before the digital revolution I used a heavy Nikon FM3, all-manual body with a Nikon 105mm micro lens. It took beautiful crystal-clear slides, free of any aberrations and so sharp that you could blow up the resulting images and examine the facets on a grain of sand. In addition, I carried a Nikon 25mm wide-angle lens for shots of trails and habitat. I mounted the camera on a Benbo tripod and head, the sort with a boom that allows the camera to be positioned anywhere at any angle. Although this was a heavy rig, the results were worth it. Now, however, slide film is gone from the market and the rig is obsolete.

Nowadays the camera that I carry is a light point and shoot, a Canon S3IS. It is a 6 mega-pixel camera with a zoom lens. Although it is equipped with image stabilization, that gizmo works only if I hold the camera horizontally. For most track photos I am pointing downward and so I still have to carry a tripod in order to get very sharp images. For this purpose I use a little pack tripod with pull-out legs and an adjustable head. Rather than carry it in or on my pack, I find I am more likely to use it for every shot if I secure it inside my waist belt by one folded leg so that it is ready at hand.

Camera Technique

For close-ups of tracks I use a penny in the frame for scale and try to take the image both in sunlight, if it is available, as well as another frame in the open shade provided by my shadow. In the winter the low sun of December and January gives the best results. As the sun climbs into February and March I use open shade more as the direct sunlight on snow is harsh and too contrasty. In the summer I avoid shooting at mid-day with sun overhead. Details that you may be able to see will be invisible in the resulting two-dimensional image. Overcast days are also less than ideal; the flat light tends to wash out images.

I rarely use the camera flash for track shots as the light source is too close to the lens and the results are the same as with the sun overhead. On my old Nikon I could detach the flash and hold it low to one side of the print to get shaded relief. On my current camera, unfortunately, the flash is integral to the body.

My new camera doesn't have a cable release, an unfortunate oversight by the manufacturer. Instead, I must use the time-delay feature in order to get a rock-steady shot. The combination of a tripod and the timer provides very sharp images, many of which appear in this book.

As far as exposure is concerned, with a digital camera it is better to underexpose an image than overexpose it. In an underexposed digital image the dark areas can be lightened in your photo editing program so that you can see details invisible in the original image. This is just the opposite of slide film where a dark area is simply a black hole in the image, but light areas, when darkened in your image-editing program, will show detail. I was distressed after my first trip to Arizona that my 35mm slides of the long-sought cougar tracks turned out to be badly overexposed and washed out. Scanning them into my editing program, however, allowed me to darken them, resulting in excellent detail in the overexposed areas of the shots.

Most of the rest of the techniques for photographing tracks and sign are common sense and improve with a little practice. The great thing about digital cameras is that you can see the results immediately rather than having to wait a week for the slides to come back from the lab. If the digital image is unsatisfactory you will see it immediately and can adjust on the spot for a better shot.

Finally, Joe Choiniere has a simple and novel way of investigating dens such as hollows in a standing tree. Rather than trying to maneuver his head and a flashlight into the tree hole, he just puts his camera in and shoots a frame with a flash. The instant imaging of his digital camera provides a quick way to see if a porcupine or other critter is in residence.

Track Traps

In snowless months it is notoriously hard to find readable tracks, at least in the Northeast where I do most of my own work. Mud and sand can record tracks but the mud is usually in the wrong place or is so deep that it is hard for a human to go out on it to examine tracks made by a light animal for which weight was no problem. Also annual leaf fall usually covers the mud with a layer of dead vegetation that resists good track impression. On the other hand, sand tracks, unless the sand is wet, quickly fill in from the sides and get windswept.

Rather than search endlessly for the perfect surface in the perfect place, many trackers resort to track traps. First, one finds a natural funnel that concentrates animal passage, such as a gap in a stone wall or a berm across a bog. Then, he takes a hand cultivator and roughs up the surface, coming back in a day or two to see what animals have walked over the softened ground. When I made a few of these simple traps 20 years ago or so, I was disappointed in the results since, although I could see vague impressions in the lumpy surface, I often found it hard to get enough information out of them to identify the animal. This drove me to certain improvements in the technique. For quite a while I would carry a mesh lingerie bag and a wooden shingle with a handle glued to one side. With a trowel I would scoop up the softened earth and sift it trough the mesh of the bag back onto the trap site. Finally, I would use the shingle to smooth and tamp down the surface. The results were a little better; at least I could identify what passed over the trap. However, I then realized that the trap

503

provided little more than the identity of the animal. The surface was too low-contrast to allow a decent photograph, and the limited area of the trap revealed little about the behavior of the animal other than its identity. Only small animals left enough information to determine gait. Furthermore, the traps were quite short-lived. Although freshly roughened earth is dark and moist, it soon dries out into a fragile surface, an unusable medium for a good plaster cast. Then, with the first rain, the entire trap, with the natural structure of the soil sifted out by the mesh bag, would pack so hard that no tracks would record in it.

Paul Rezendes used to improve his roughened-earth traps with bait in the form of a Q-tip moistened with some noxious fluid from a small vial. This bait, he claimed, attracted animals that would scratch all around it. An entire day visiting these sites with him one time yielded exactly no tracks. And anyway, if the animals were so attracted to the baited Q-tip that they walked and scratched all around in it, I imagined the likelihood would be that one would get a commotion of prints overstepping one another in an artificially induced and unnatural behavior. I never bothered to try it myself.

Some researchers advocate the use of track plates. These are white metal squares placed as with the track trap and preceded by lamp black. The animal walks down the funnel, steps in the carbon and then steps onto the white plate, leaving a chiaroscuro record of its passage. This technique, once again, provides little more than identification and in the case of small mammals, their gait. The plates are too small to collect the prints of larger animals, which, sensing the presence of a strange object, are unlikely to step on it. Furthermore, the tracks, although very clear black on white, do not greatly resemble the tracks that one would find on natural surfaces. The hardness of the plate compresses and distorts the shape of the soft pads of mammals, yielding little that is photographically useful other than, once again, the animal's identity.

With these caveats in mind, track traps and plates are useful for researchers who are doing censuses, since they care little about accurate detail or about the behavior of animals and only want to determine their presence or absence in a habitat. Track traps of the sort I described above with an improved surface can also be useful as a teaching tool. The eyes of students, who see in three dimensions and without a fixed perspective, will be able to perceive more in an actual track than they will in a photo taken through a camera's lens with its fixed frame and perspective as well as its two-dimensional result.

Remote Cameras

These have become much more sophisticated than the old, low-resolution systems that hunters would mount on a tree near a lick to find out what was coming to it. Now the sky is the limit on the technology available to record the passage of animals. A group in the Boston area has been researching man-made wildlife passages under busy highways to see if animals will actually use them. At first they relied on tracks in the soil of these passages to identify the animals. But later they installed motion-detection still cameras equipped with infra-red capability so that they could get ghostly images of the animals themselves as they passed through in the dark. I'm sure that somewhere, someone is working on a motion-detecting, infra-red video camera as well. All one seems to need for this is a wallet a lot fatter than mine. In the end, I'm not sure I see that these expensive devices yield much more than the tracks of the animals do. However, I suppose the researchers who employ these cameras would argue that a picture of an animal using an underpass carries more weight with a highway department or conservation commission than the mere testimony of a tracker.

Making Tracks

Plaster Casts

A cast is a great tool for becoming intimate with the details and nuances of a track, the foot that made it and the living animal above that foot. It is amazing how much detail in an animal's track is missed in the field. A print in mud, let's say, is an impression that recedes from the eye and is often viewed in less than ideal light. Even a photograph of a track can conceal as much as it reveals since it is only a two-dimensional representation with a fixed perspective. You cannot vary the angle of the light or its quality once the photo is taken; you are stuck with the conditions that existed when the shutter was depressed. I am often surprised with how much more I can learn about an animal's track while sitting in my office holding a cast, examining it with eye and fingertip, in comfort, without the whine of mosquitoes in my ears or cold rain trickling down my neck. In my study I can hold a cast of a bobcat track, for instance, and examine it closely, moving it in and out, holding it at different angles to my desk light to reveal subtle, hidden features. In this way I can see and touch the cat-like depths of the toe pads of a gray fox that are invisible in a two-dimensional photo. And I can feel with my fingertips the depth of those pads, where in the field such an action would distort the details themselves. In casts of bear tracks I can see immediately why the little toe on the medial side of the foot often fails to register, since the casts, held on edge, reveal that this toe is often held above the plane of the others, withdrawing it from shallow prints. Better than in the field, I can reach a level of intimacy with the animal, almost feel its body above the track and absorb the raw information, information that, once internalized, draws one into a closer intimacy with the animal that in turn can result in new intuitions about its elusive life. In the field the inconsistency of light, the awkwardness and discomfort of one's posture and the time limitations imposed by other responsibilities make for hurried and imprecise inspection. Errors are inevitable in the difficult circumstances of field work, but a well-made plaster cast, studied at leisure with the eye and the hand, tells the truth.

Another use for casting can be simple identification in the field. A few years ago I conducted a tracking walk for a local land trust. In search of readable prints we went down to the muddy outlet of a bog where we found a trail of three tracks of an animal that had moved up the bed. That was all, just those three tracks, after which the trail disappeared onto more solid ground. The individual prints were at the bottom of dark holes in the mud, each a couple inches below the surface. The light under the hemlock canopy was poor and the tracks were low-contrast dark brown on dark brown. The step length of the animal was about 12 inches, the straddle was fairly narrow and, at least at the surface, the track width was less than 2 inches. By habitat, straddle and step length, I guessed at a walking gray fox, but I could sense the group doubted my guess since the evidence was so scant and obscure. I had brought along a small casting kit intending to use a demonstration of cast-making to fill out the walk should the route prove to be unproductive. Here was a perfect opportunity, although one with some risk to my credibility should I be proven wrong.

The first cast (Figure B.2 - left) that I made and cleaned on the spot would assure an experienced eye that the gray fox identification

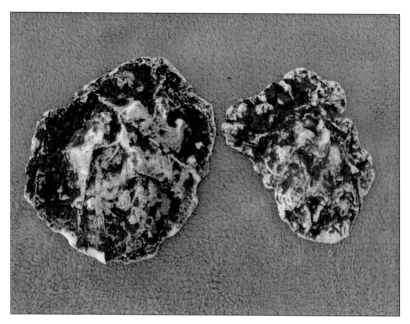

Figure B.2. Although the double registration on the left may be satisfactory as that of a gray fox to an experienced eye, the cast on the right is more conclusive. It shows the familiar narrow hind print with small secondary dot and a wide space between it and the toe pads.

was correct, but the impression was so rough that it failed to satisfy the group, so I made another (Figure B.2 - right). This one I left to dry until the end of the walk. Upon our return the cast had dried just enough for me to scrape off most of the mud, revealing a more recognizable gray fox print to the satisfaction of the skeptics.

Making the Cast

Plaster casts are cheaply and easily made (Figure B.3A through I). Although a variety of other materials have been tried, the results in terms of expense, ease and detail of the result from plaster have not been equaled. Go to the hardware store and buy a tub of it. Decant it into various appropriate containers. At my age, I use an empty plastic Citruscel jar; it holds enough for the largest casts, and the full container holds just about as much as I want to carry long distances. If you find a track that cannot be fully covered by this much plaster, an

507

Figure B.3A. The kit.

elephant perhaps, you can fill the track with the contents of the jar and allow it to harden while you go back to the car for a refill. There is no reason not to reinforce a cast later after the original has set.

In addition to the jar of powdered plaster in your casting kit (Figure B.3A), you should carry a narrow container for mixing. I cut off the top of a half-pint plastic milk container and carry it empty in my pack. A narrow container like this is better for mixing than a broad bowl because you will be using a stick from the forest litter lying about as a stirrer. Such a stick would mix the plaster in a bowl too slowly and unevenly, but in a narrow container the mixing is very quick and uniform. Furthermore, the flexibility of such a bottle allows for easy and precise pouring close to the track.

Tracks in mud often have a little water pooled in them. This must be removed before attempting to cast in order to avoid a soft, chalky surface on the finished product. The easiest way to do this is to take a paper towel, fold it over to a point and dip the point into each of the pools. The water will be drawn by capillary action up into the absorbent paper.

Figure B.3B. Drawing the boundary.

I usually draw a circle around the track in the earth or mud so that once I begin to pour the plaster and the track is mostly covered, I have an idea of where the now hidden track is located and how broadly I need to cover the area around it (Figure B.3B).

Pour as much of the powdered plaster into the mixing container as you estimate will be needed (Figure B.3C). In a very large track don't worry about not having prepared enough at this point. You can always mix more and reinforce the cast with a second batch. Clean a stick for mixing and then pour about 1/3 as much water as plaster into the narrow container. The less water you use, the more durable the cast will be. If you use too much water, the cast will have a chalky texture that will cause detail to be erased during cleaning. Thoroughly stir the plaster into a smooth consistency. You will notice some bubbles appear in the mix from the stirring motion. My friend, Joe, whose hands appear in the accompanying photos, encourages these bubbles to rise and dissipate by tapping the bottom of the container on the ground. Wear eye protection if you do this because droplets of the abrasive mix may shoot upward from the narrow container into your eyes.

509

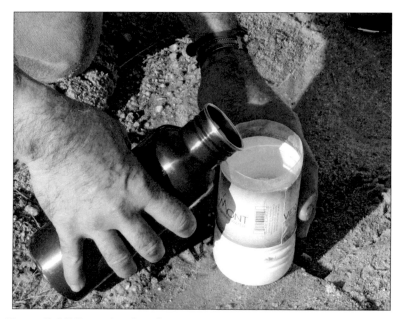

Figure B.3C. Preparing the mix.

Figure B.3D. Applying a thin mix to the tracks with a drip stick.

510

If the track is on a level surface, you may begin to pour the plaster into the track as soon as it is mixed and while it is still quite thin (Figure B.3D). By not waiting until the mix thickens, you will preserve more of the fine detail in the track, including the nails. If you wait until the mix is the consistency of pancake batter, the plaster may not penetrate into the nail holes. Instead, a bubble may form deep in the holes and result in flaws in the finished cast.

It is important to pour carefully. The flexible milk container allows me to hold its rim close to the ground to lessen the impact of the poured liquid. In very delicate substrate I may use a drip stick, as shown in the accompanying photo, allowing the poured fluid to run down its length to the tip, which is held just above the surface. Or, in damp sand, I may pour the mix off the edge of the track and allow it to flow into the depression. Different substrates will require different approaches. After the thin mix is poured enough to fill the track or tracks, wait a few moments until the mix begins to thicken and then pour the rest to fill out the outline you drew around them (Figure B.3E and F). If you are casting a set of tracks, a single application over a large area may be too thin to create a sturdy cast. Simply mix another batch and apply it as a reinforcement layer over the first. Occasionally, I may bury long, thin green sticks in the plaster to increase the sturdiness.

The hardest part then is to wait. Go away and come back in an hour or two rather than sitting and staring impatiently at it. After your return, try to indent the plaster with a fingernail. If it seems solid, use a sharp stick to begin poking around the cast, penetrating the mud or earth at a 45-degree angle to undermine the track (Figure B.3G). Keep poking around the edge until you can feel the cast and the underlying soil release from the background. At that point you can pick the cast out of the soil, invert it and pick off the loose surface earth, being careful not to expose the plaster underneath. The second hardest thing to do is to wait again, until the next day. Once the back of the cast feels dry to the touch, take a pencil with the lead broken off or use the rounded end of toothbrush and pick away the dried earth over the track impression, gradually exposing the vague outline of the track. Do not try to remove all the mud, but only down to where anything loose has been picked off and you can just see the track outline through the remaining soil (Figure B.3H). I usually wait until the next day when the cast is fully hardened and then take a soft toothbrush (not

Figure B.3E. Pouring the cover layer.

Figure B.3F. Smoothing the reinforcing layer.

Figure B.3G. Freeing the cast.

Figure B.3H. Rough-cleaning the cast.

513

Figure B.3I. Detailed cleaning the tracks.

your current one!) and begin lightly brushing off the dirt where it covers the features of the track, leaving the soil around the periphery of the track and between the pads (Figure B.3I). Gradually the pads of the track will come into relief as white areas against the darker soil of the background. Once the track is exposed, stop. The dirt remaining will serve as background, throwing the track into relief and making it easier to see its details. If you brush this material off, the track will be white against white with detail more difficult to discern. Also, don't be too scrupulous about cleaning the mud in the various grooves and seams on the pad impressions themselves. The track of a red fox, for instance, will show the fur covering its foot better if there is still a little dirt between the hair impressions.

If at some point in the process the cast should break, all is not lost. Simply place the broken cast face down on the ground with its parts firmly in place. You may wish to bind them with an elastic band that can be cut off later. Then mix and pour a reinforcing layer of plaster over them.

I have seen in books where one is advised to fill in the negative spaces between the various pads of a print with a brush and dark paint. I find this process often alters the gestalt of the original to the point where it reduces is usefulness for later study. Where and where not to paint involves a human judgment that was not in the original. It is better, it seems to me, to clean the track as I have suggested, so that its pad outlines emerge rather softly out of the background just as they do in real life. After the cleaning is accomplished, you may find that there are irrelevant bumps and creases in the periphery of the cast that also stand out. These can distract the eye from the central impression, and so to get rid of them I sometimes color them in with an earth-tone carbon pencil available in any art supply store.

Some people go to the bother of forming a frame around the track with a strip of cardboard and a paper clip. I have found this extra step to be unnecessary for most casts. If you time the pouring of the thickened mix properly, it will not spread very far from the point of contact. I keep it as simple as possible; including extra steps and extra gear with which to make the frame complicates the process to the point where you may find yourself sometimes unwilling to make a cast in the first place.

I have never had much success with casting snow tracks, and the results I have seen from other trackers' collections have been less than satisfactory. The warmth of the mixed plaster and the heat of the reaction that takes place in the poured liquid inevitably melt out all the fine detail in the track. True, one can make out the vague form of an animal track, but not much can be learned about its morphology by studying it afterwards.

To date, I have collected about 200 of these casts and refer to them often by way of absorbing the "gestalt" of the tracks of various species. Because they are a negative image of the original track, the features are raised to the eye rather than recessed from it. This feature also allows one better to examine its details with a light source where shadows might have obscured some detail in the original recessed track.

Drawing and Tracing

Another way of absorbing and internalizing the details of the track or sign that one encounters is to draw it in the field or from a photograph in the ease of one's home study. Some people have more facility with this than others, but with a soft pencil, an eraser and willingness to endure the trial and error of initial attempts, one can

515

discover interesting aspects that may have escaped one's initial notice. I mention an eraser because, in my case at least, I find I must redraw portions of a track over and over until I arrive at an image that captures its fundamental character.

Since I lack the skill of an artist, I am more likely to trace the outline of a track from a photograph I have taken, filling in the interior details by shading. Even this process is instructive and a good deal more accurate than a sketch made in the field. While tracing and shading, I am able to absorb the gestalt of the print, to internalize the shapes in close detail and add them to my catalogue of mental images, ready for use later on.

Recording Information: Data Cards

To paraphrase Darwin, any information not gathered to prove or disprove a theory is useless. Another way of saying this is that unless the gathering of information is guided by a goal, something to prove, the randomly collected information will usually be missing some key element needed later to use for or against a theory latterly arrived at. Darwin's argument was intended to emphasize that one must first come up with the theory and then collect the data rather than the other way around.

For our purposes, Darwin's warning applies in this way: it is possible to measure the secondary pad alone or any other pad of an animal's print in an almost infinite number of ways. Without the goal of identifying the animal or interpreting its behavior, many of these measurements will be wasted effort. The key to identifying and interpreting animal sign, then, is to know from experience what sorts of information to collect toward the goals of identification as well as interpretation for gait and relationship to habitat.

Here is the information that I encourage tracker trainees to collect on their data cards and to use for the identification of tracks and trails.

1. Date:
2. Location (region and UTM coordinates):
3. Habitat:
4. Surface conditions (have space for details as necessary):
5. Track depth:
6. Track appearances: front/hind (have space to draw):
7. Circle - SR DR IR DblR:

8. Track width:

9. Pattern (have space to draw):

10. Pattern width (front or hind straddle):

11. Step length:

This is certainly a manageable amount of information to collect, and it will easily fit on a 3X5 card in the tracker's kit. While it would be possible to collect pages of information on a single track, the above information is the sort that is most likely to lead to successful identification and interpretation.

Let's go through these items in order and consider why each can be important.

1. Date: This is mostly of administrative value to order the cards that you collect. However, it can be useful for identification as well. If you think you may have a jumping mouse track in January in New England, something is wrong. Jumping mice are profound hibernators.

2. Location: Identify the general area, for example "Federated Women's Clubs State Forest" or "Petersham, MA." Then provide the UTM coordinates for the precise location. Use a GPS for this, including the datum and the block series. I use UTM for land use over short distances (as opposed to crossing the Atlantic, for instance) because it grids off the surface into neat 1000 meter squares. USGS topo maps, which are usually gridded, show lines in UTM at 1:25,000. A plastic template using this grid size, available at any outdoor shop, allows you to locate the spot with great precision on the paper map. The alternate system is the "Lat/lon" grid. Although there are tic marks on the borders of the paper maps to allow use of this, the actual grid is not ruled onto the map, and so one would have to use a very long ruler to superimpose manually such a grid on the map. And then the blocks would be shaped like parallelograms rather than squares, changing with each degree of latitude to a different shape, preventing the use of any template for locating the spot. Finally, lat/long is based on a scale of 60 minutes/seconds, further confusing the location to those of us who have mentally accommodated the decimal system.

Don't forget to consider the datum. USGS paper maps use NAD 27, that is, North American Datum 27. Many GPS units are set by default to NAD 83. Change this in your GPS set up program so that your coordinates will match those on the paper maps. Otherwise your readings, when transferred to the map, will be about 40 meters off from their actual location.

Finally, include the map series — such as "19T." Any map system is faced with the problem of representing the curved surface of the Earth on the flat surface of a map. The UTM system blocks off the surface of the planet into rectangles approximately 100 miles across. In an area this small, the north-south grid lines are reasonably close to true north over the whole map series within the block. However, at the edges of each block there is an overlapping wedge with the next map block. In this wedge it is possible to have two coordinates for the same spot, one each for the two overlapping map series blocks. Thus, in Massachusetts, for instance, if you are in the central area of the state where the overlap occurs, it is important to know whether the coordinates correspond with the 18T block series or the 19T series.

A typical GPS entry on a data card might be: "UTM NAD27 19T 3042082E 6250232N." In practice, I drop off the first several terms and start with the "19T…" because I always use UTM NAD27.

Of the 8-digit coordinates, the last 5 are the important ones. The first two of these represent one each of the vertical and horizontal grid lines on the map; the last three are a measurement in hundreds of meters from those lines. Thus, for the above coordinates I would first locate the vertical "42" line by reading along the bottom or top of the map sheet. Then measure 82 meters east (right) from that line and draw (or imagine) a vertical line at that point. Next go to the "northing" coordinates and find the "50" line along the right or left border of the map. Once that line is found, measure up 232 meters northward from that line and draw (or imagine) a horizontal line. The intersection of the two drawn (or imagined) lines is the location of the spot where the tracks or sign was found.

This may seem like a lot of work to locate wildlife sign, but in practice it is done quite quickly, simply by writing down the map series and coordinates that show on your GPS screen. Actually plotting the spot on a paper map is something that is usually done, if it is done at all, later on at leisure. Writing down the coordinates, or at least saving them in the GPS as "waypoints," is important because it make the location impervious to the fading of memory or changing of light.

3. Habitat: Look around. What sort of wildlife habitat are you in — forest, field, field edge, beaver pond edge, etc.? This is significant because different kinds of animals live or hunt in different habitats. It also may give you a clue as to what the target animal was doing. Here is a typical entry: "Medium-age NHA w/dense young pine regrowth" — that is, medium-sized northern hardwood association trees with a pine undergrowth.

A number of things can follow from this. First, the area was probably cut a few decades ago and has now grown up to birch, beech, ash, hemlock with an admixture of pines. The pines have sprouted at ground level to dominate the understory. Technically this is not supposed to happen, since pines like sunlight and there is less sunlight under a closed canopy of hardwoods. The reason may be that deer have browsed back the competing hardwood regrowth without touching the pine, giving the latter a competitive advantage. Finally, the density of the pine regrowth provides cover for snowshoe hares. The tracker, investigating this habitat, may adjust his mental search images to expect bobcat and fisher, both of which typically hunt this sort of dense habitat in search of their principal winter prey. Now, instead of looking randomly and missing much, he targets his mind to look for hare scat and tracks along with the patterns of tracks and other sign of their predators.

4. Surface: Under this category I place a series of lines representing the various layers of snow or ground that may affect the appearance of the track. A typical winter entry may look like this:

2" soft sn

Brkbl cr

1' sn

These entries describe 2 inches of soft snow over a breakable crust that was covering 1 foot of older snow over ground. Only include those layers that actually affect the movement of the animal or the appearance of its tracks. Wild animals adjust their gaits to conditions as well as intent. A running record of this sort of information will eventually reveal to the tracker the patterns he can expect various species of mammals to use on different surfaces. In the above conditions fishers, for instance, are more likely to use a transverse or rotary lope rather than a weasel bound. This expectation, once again, allows the tracker to hone more sharply his search image for appropriate sign.

5. Track Depth: This is related to the previous entry. How deeply is the animal sinking into the snow with each cycle of its gait?

In snow where the consistency does not provide pad details but simply an outline of the foot, this can be significant to identification. A series of slant-2X repeating patterns, for instance, can be two entirely different gaits by two different species. If they are direct registrations set deep in the snow, then the tracker is looking at a bounding gait of one or another of the weasels. If the patterns are of shallow single registrations on a firm, even surface, on the other hand, he is most likely looking at displaced trotting trail of a canid such as a red fox or eastern coyote. Superficially, the two patterns look similar, at least in dry snow or other conditions where neither pad details nor evidence of single or direct registration appear in the prints.

6. Track Appearance: Draw what you see. Drawing the print forces you to look closely at its details. Don't worry that you have no talent for art. Experience will begin to emphasize to your mind details that are significant to identification, and your drawings will improve. At first you may only represent the bare outline of the track. Later you may begin to shade the tracks to provide the illusion of depth. A series of horizontal lines in the deeper parts of the track will provide this valuable illusion. I suggest drawing the track so that it is oriented with the anterior of the paw print toward the top of the card. Most people are uncomfortable viewing a track sideways and so this orientation will obviate the need to turn the card to get the track with the anterior side up.

The advent of digital cameras might seem to have obviated the need to draw anything, and photos of tracks can certainly be useful (I have about 5000 of them), but a photo will usually be brown on brown or gray on white, often lacking the definition that may have been apparent to the eye on the scene. Furthermore, a drawing will force you to pay the kind of attention to detail that will improve your skills as a tracker. A photo is too easily filed and ignored.

7. Circle SR DR IR DblR: With reference to the track depicted in item 6 above, circle whether it is a single registration, a direct registration, an indirect registration or a double registration. Obviously, if the track contains the overlay of more than one print, this overlay will distort its appearance away from the expected appearance in tracking guides as well as in the mind's eye of the tracker.

8. Track Width: An overlay of prints will also affect its measurement, resulting in, for instance, an IR print measuring wider than an SR of the same species. Of all the measurements that one could make of a track, width has always seemed to me to be the one that is most helpful to identification. Some trackers have recommended measuring track length as well, but I have found this to be quantification for quantification's sake and usually a waste of time since sometimes the tertiary area of a track of a single species may register or partially register or register not at all. Combined with an accurate drawing of track appearance in 6 above, track width is more consistently useful in differentiating species. In weasels, for instance, many of which have nearly identical track morphology, a width of 1 3/8 inches in a clear, shallow SR print will quickly distinguish mink from the smaller long-tailed weasel and the larger marten or fisher.

These measurements are not scientifically precise and so cannot be relied upon alone for a positive identification. After all, an animal can spread or close its toes for various reasons, altering its track width in the process. A bobcat in dry, firm mid-winter snow will usually leave a print no wider than 2 inches. However, in damp, soft, collapsible spring snow it can spread its toes out for support to nearly 4 inches! This is why we collect additional information on snow surface and time of year as well as make accurate drawings of the prints. Identifications as well as interpretations are usually founded on cumulative evidence.

9. Pattern: Next draw the pattern in which the track depicted in 6 above appears. I use a series of circles or ovals, shading in the oval for the front prints with my pencil to distinguish them from the hind. Since various animals use typical gaits for given conditions, even where track details are absent, the pattern alone may be enough for an identification when combined with data on the surface conditions. As an example of this, where the registration of padding is vague, a gray squirrel track and bounding pattern can look similar to that of a cottontail rabbit. However, a careful drawing of the pattern will reveal that the front feet of the squirrel are usually arranged on a shallow slant with the prints separated while the front prints of the rabbit are either one directly in front of the other or joined side-by-side.

10. Pattern Width (Front or Hind Straddle): Measure whichever pattern width — front or hind straddle — is useful, and do so from outside edge to outside edge for no other reason than consistency. In aligned gaits with species that have wider pelvises than chests, the hind straddle is more useful — as, for example, in distinguishing among sciurids. In species with about equal chests and pelvises, the pattern width, front straddle and hind straddle will all be the same width, at least in aligned gaits. In displaced gaits, the pattern width will be wider than either front or hind straddle. It will also be less useful than the straddle measurements since the straddles are more constant through the various gaits.

11. Step Length: Measure step length of successive prints from front to front or hind to hind, parallel to the direction of travel. In vague prints, such as in snowed-in or melted-out snow, measure center to center, again parallel to the direction of travel. Some trackers measure from one track to the next at a diagonal rather than parallel to the direction of travel. The problem there is that in some snow conditions, spring collapsible snow for example, animals will greatly widen the straddle of their walking gait. If you measure such a pattern from print to print on a diagonal, you will get a somewhat longer step length than in firmer conditions where the straddle is narrower.

The gait for which this measurement is most useful is walking since walking has little or no vertical component. In higher energy gaits with an arc of stride, verticality is the invisible third dimension of a gait that does not show easily in the two-dimensional pattern on the ground. We can't easily tell how much upward motion the animal is introducing into a lope, for instance, and so a measurement of step length for a lope will reveal nothing more than that it is a lope. In walking and some trotting gaits, the step length remains more or less constant, whereas in a lope, gallop, and many bounding patterns it alternates short-long-short-long etc.

Take several measurements of the step length from sections of the trail that are level since animals will adjust their gaits when moving up or down a hill. Sometimes variability in step length itself can be evidence that may help identify the animal. Red foxes walking on a uniform surface will give amazingly consistent step lengths of 15 inches with a very narrow distribution of no more than an inch. Bobcats often walk in a 15-inch step as well, but with more variability due to longer legs and the greater flexibility of a tree-climber.

This completes the entries on a typical data card, a process that takes much longer to explain than to actually do. Quantifications that are unlikely to lead to identification or interpretation have been excluded, but enough information about both the track and its pattern as well as the background that may affect their appearance and measurement has been included to allow for review and re-creation later when memory of the actual event has dimmed. Frequent review of these cards as the tracker's experience progresses will itself reveal to him factors and patterns that are personal to him. Education, we may remember, is a matter of re-expressing learned information in our own terms, a personalizing that imbeds it permanently in our consciousness.

How to Stay Warm Outdoors in the Winter

Wildlife tracking requires concentration. And since concentration is best achieved in physical comfort, it may be well, especially for readers in northern climates, to spend some time here discussing how to stay warm on cold weather tracking excursions.

This section contains information that I have gleaned from books of tested wisdom as well as from eighty winter ascents in the White Mountains of northern New Hampshire, many of which included overnight camps. No single selection of clothing and equipment is proposed in this chapter. Different selections will "cover the seams" both literally and figuratively in different ways. You will have to use your own judgment. Informing that judgment is the purpose of this chapter. Since tracking wild animals routinely takes us off trail and far from heated shelter, a thorough understanding of this information could conceivably save you a good deal more than just a few hours of discomfort.

I begin with a handful of sayings that have served as useful reminders to myself on various adventures.

- "Man is a weak, hairless creature capable of survival only between narrow limits of heat and cold." — roughly from Jack London.
- "To stay warm in the winter, take care not to get cold." — Eskimo saying.
- "Caveat emptor." — Roman warning.
- "Cowardice in the face of natural forces is a virtue." — The author.

The Physics of Cold

Heat Loss

Warm air (on the surface of your skin, for instance) always moves toward cold air (the winter air surrounding you). Heat travels from a warm body to cold air in four ways: conduction, convection, evaporation and radiation. Radiational heat loss is difficult to control and so will be ignored in this simplified discussion. For our purposes as winter travelers most heat is lost by the first three.

Conduction

All substances, whether solid, liquid or gas, will conduct heat from warm to cold by contact. The more poorly they conduct, the better their insulating ability. Dry air is a good insulator, while damp air is less so (water vapor conducts) and free water is poorest of all. Obviously, then, it is important to stay dry, or rather to keep our insulating layers as dry as possible so that they will conduct heat away from our bodies at the slowest possible rate.

Convection

For our purposes this means the removal of warmth by moving air. When we are standing in still air, a bubble of warmth of our own making surrounds us, rising in a column over our heads. If a breeze comes along, it blows this bubble away causing us to feel chilled. TV weathermen often dramatize their broadcasts by talking about how cold the wind chill makes it feel. The good news is that most convective heat loss can easily be defeated by wearing a simple hooded windshell of tightly woven, uncoated material.

Evaporation

In dry air, free water evaporates, that is, it turns into water vapor, carried invisibly in the air. Warm air can hold more water as vapor than cold air. Evaporation, the changing of free water to vapor, causes the damp surface that is the moisture source to cool. If you are the source of both the heat and the moisture (sweat) and evaporation is allowed to occur on the surface of your skin, your skin will cool. If, however, the moisture is transported away to the outer surface of your garments before it evaporates, this cooling takes place relatively harmlessly at some distance from the skin.

Physiology

Sweat

Our bodies have a thermostat that tries to keep our core temperature — that of our brain and internal organs — at an even 98.6 degrees whatever the outside temperature or level of exertion. When

Figure B.4. During heavy exercise you can wear very little despite the cold, keeping the rest of your clothes dry in your pack for when you stop — as this tracker did at Eagle Pass, Mount Lafayette, New Hampshire.

that core temperature starts to rise, our pores open and tiny pools of water form in them. We sweat. The opening of pores to emit moisture is often detected as a prickly feeling on our skin. As this sweat collects, it evaporates, at least in dry air, and cools us, maintaining the delicate equilibrium in our body. The implications for cold weather activities are obvious; if you don't want to get cold, don't overheat. Remove clothing during periods of exertion (Figure B.4), and wear clothing that will either transport sweat away from your skin rapidly or clothing that will continue to insulate even when wet. It is fair to say that management of cold in the winter is mostly the management of moisture. In that line, it should also be noted that even if free water is not collecting in our pores, we are always sweating. In dry air typical of a winter day, we are still losing body moisture through insensible perspiration, the escape of water from the pores as vapor rather than as free (and therefore "sensible") water. Furthermore, if we are exerting ourselves, and in snow it is hard not to, we also increase the loss of water in the form of moisture in our breath. In terms of our moisture demands, a dry winter day is like a day in the desert — so increase your intake of water. If you do not, your cells will become dehydrated and fail to function properly. This can affect your brain, causing you not to think straight. It may also affect the coordinated firing of muscle cells and the efficiency with which they burn fuel. A good way to check if you're properly hydrated is to note the color of your urine; if it's pale, you are OK, but if it's bright yellow, increase your fluid intake. As for fluids, plain water is best. Anything added to water, for flavor, energy or electrolyte replacement for instance, delays the uptake of water by our system.

Vasoconstriction and Core Temperature

Our bodies come equipped with primitive survival mechanisms over which we have little conscious control. One of these is vasoconstriction. As warm blood flows from the heart through arteries out to various using destinations, it warms the tissues of vital organs and supplies them with oxygen, allowing them to continue to function. When it reaches its destination, it flows through tiny capillaries where oxygen and waste products are exchanged. When the destination is the extremities, the warm blood also exchanges heat to the tissues, allowing muscles to continue to fire in a coordinated way. This is what allows us, in air cooler than our bodies, to manipulate things like knots and zippers at least for a while before our fingers numb. After these exchanges, the venous blood returns to the trunk area of our bodies somewhat cooled.

Trouble begins to occur when we chill. If we do not take steps to conserve heat, such as by putting on clothes, or to generate heat to replace the loss by an increase in muscular activity, such as by moving faster, our body will do the latter for us by involuntary muscular activity: shivering. If our body temperature continues to drop, however, the body takes additional defensive measures all by itself. From trial and error over the millennia of human evolution, the body has learned what it can survive without. It can survive without extremities like fingers, toes, a nose and so forth, but it cannot survive without a brain, heart, liver, lungs and kidneys, and it cannot reproduce without genitalia. These, then, are considered core functions, and significantly they are all concentrated in the core area of our body: head and torso, close to the warming flow of blood from the heart. As we chill to a dangerous level, the body begins to shut down blood flow to our extremities in favor of the vital organs in our core area: our hands and feet become painfully cold.

The mechanism for shutting down blood flow — and thus heat loss — to the extremities is called vasoconstriction. Responding to the sensation of cold on our skin, the brain directs the release of a chemical that causes the walls of the blood vessels in our hands and feet to constrict. Our extremities stop radiating heat so that our core temperature can stay up and we stay alive.

Fear

Vasoconstriction occurs not only in response to cold but also to fear. Reduction of blood flow to the extremities in fearful situations seems to be a primitive defense of the vitals against blood loss through injury to arms or legs that might be incurred on the hunt or in battle. If you were to get lost in the woods in the winter, you would experience fear; the primitive physiological protection would kick in, constriction would occur, and blood flow to your limbs would be reduced, leaving you with cold and potentially frostbitten hands and feet. Furthermore, fear would cause you to breathe heavily. Cold air passing over the moist surfaces of mouth, throat and lungs would cause evaporative cooling and more loss of heat to the atmosphere. The bottom line message is simple: fear can kill. If you know yourself, and know that you cannot subdue such fear, always carry a small thermos of hot

527

fluid. Drinking hot fluid delivers warmth to the interior of your body's core to make up for the loss. Purposeful activity, not the spastic activity of panic, will also help elevate your body temperature and open the blood flow to the extremities.

Physical Condition

A person who is in good cardiovascular condition is able easily to mobilize stored energy for burning and furthermore has a heart and circulatory system that are capable of delivering blood efficiently to the using organs including the muscles and the surface of the skin. Such a person warms up quickly under exertion and burns energy efficiently. The benefits of good condition, then, are obvious. One of the ways for keeping warm is through purposeful exertion, elevating the rate with which one burns body fuel. To stay warm, move faster or put on more clothes. If you get too warm and begin to generate sweat as free water in your clothes, slow down or take off some clothes. Obviously, the ability to continue to move quickly and efficiently requires good cardiovascular condition.

Nutrition

Winter is not the best time to go on a diet if you are planning long excursions outdoors. The national obsession with eliminating fat from our diet seems to follow from the mistaken notion that fat that we eat turns into fat on our bodies. In fact, everything we eat turns into fat on our bodies, if we eat too much. The best thing about ingested fat is that, gram for gram, it contains more calories than other food components. Furthermore, being slow to digest, it tends to meter out its calories in body heat over a long period of time. My traditional supper in winter camps on the mountain is Spam or Canadian bacon in a pocket of Syrian bread (it survives squeezing in a pack). This fatty meal keeps my sleeping bag warm all night. For vegetarians, nuts and seeds will do much the same thing. I am told that at the National Outdoor Leadership School, winter students learn to put a dollop of butter in their cocoa in the morning for the same reason. On the trail, chocolate or other candy will give an immediate energy boost, but only if it is taken while under heavy exertion; at rest, it tends to get stored as fat rather than burned. Once stored, this fat is resistant to mobilization as energy. The body, in another primitive tropical reflex against the next famine, is reluctant to give up those calories to something as unanticipated as the need to keep warm. If you wish to avoid unpleasant candy-mouth, you might follow the practice of a friend of mine who pre-bakes tiny red potatoes for use as trail snacks.

How the Red Fox Does It

As I have suggested, humans originated in the tropics and are rather badly adapted to life in the cold, at least compared to wild animals with which we share northern latitudes. Over the years I have tracked many wild animals in the winter each of which has developed its strategies for survival in this season to a high degree. Tracking them, one sees immediately that their behavior is very purposeful, with little wasted motion. (In fact, this efficiency is the easiest way to distinguish their trails from those of domestic animals that rely on man for their survival.) From among these winter-adapted animals I choose the red fox as an exemple from which we can learn a lot about life in the cold.

Physically, the red fox is a very slim animal, its seeming bulk in the winter a result of a thick layer of dense white underfur. This fur is marvelous in design and function. The individual hairs, like all white fur, are hollow, providing dead air as insulation. The surface guard hairs, on the other hand, are solid, plugged with melanin granules that give them their darker color. These dark granules absorb sunlight, so the fox seeks southern exposures for its rest spots, especially ones out of the wind and under evergreen boughs. Here he can curl up into a ball to present as little radiational and convective surface to the air as possible, wrap his long, thick brush (a "sleeping bag" he carries behind him wherever he goes) over his feet and nose, and take a nap, absorbing the short-wave radiation of the sun in his dark guard hairs. When the fox is exposed to moving air, while crossing an open field for instance, his fur acts like a portable forest, slowing down the passage of air over his skin through friction and reducing convective heat loss accordingly, just as if he were in a deep forest where the trees overhead would slow the wind. Turning a fleece-lined coat inside out accomplishes much the same thing. Patagonia tried marketing a jacket with a synthetic fleece outer a few years ago, but it wasn't sleek enough to sell to the fashion-minded and so the idea was abandoned. My own warmest, although not lightest, coat is a hooded one with wool fleece on the outside and a breathable liner.

When the fox walks in snow, he packs it with his front paws and then places each back paw in the print of the front, an efficiency that saves a tiny amount of energy with each of his thousands of daily steps. Finally, unlike most other mammals in North America, the red fox's feet are covered almost completely with long stiff hairs that insulate his feet from the cold. Animals that lack this adaptation often find a dark object such as a stump and stand or curl up on it. The darkness of the object absorbs solar energy that is then given back to the animal by conduction.

Human behavior in the cold has not been ruled by millennia of ruthless selection, as has that of the fox. We are after all "weak, hairless creatures . . ." But the advantage of the human race has come from our large brain, not from our physical powers or instincts. And since we are tropical creatures under the skin rather than boreal ones, we cannot rely on our instincts to tell us what to do. Instead, we must use our reasoning ability and imagination to help us adapt to the cold. Jack London lived long in the arctic and knew cold well. The man in his story, "To Build a Fire," died in the cold because he was neither imaginative enough to foresee trouble nor smart enough to get himself out of it. He lacked the human brain-power to compensate for the loss over the millennia of instinct and fur possessed by his dog, which trotted off down the trail to new fire and food providers, leaving his "master" sitting frozen in the snow.

Clothing and Equipment

No equipment list or set of rules will impart good judgment; equipment is useless, even dangerous, unless it is used wisely. Understanding the design of our equipment and its intended use is important. Comparing that design to what you learn from this chapter and from your own experience is also important. Knowledge can protect you from purchasing equipment that was designed more for profit than for use, and there is much of the latter on the market.

Marketing Techniques

It is a lot cheaper to produce clothing and equipment made out of oil than out of other natural materials. When gear for outdoor adventures became big business back in the 1970s, marketers began by campaigning against natural materials in favor of oil-based synthetics such as nylon and polypropylene, for no other reason, it seems, than to increase their profit. Instead of being designed for real use, as had previously been the case, many new items were then designed to appeal to the buyer in the store — they had to look good and feel good at once, and the merits of their design had to be instantly apparent. From a marketing viewpoint, any hidden merit that could only be discovered or understood by experience outside the store was no merit at all. Furthermore, many companies eliminated field testing as part of their design process, preferring to go straight from the drawing board to production in order to beat the competition in the direction of novelty. Finally, they prevailed upon magazines in which they advertised to express enthusiasm for their novel products in equipment reviews. Editors, fully aware of where the real money comes from, wrote valentine pieces extolling the virtues of these products. Some even went so far as to insert paragraphs into articles submitted to them by writers, promoting the newest wave of equipment or un-selling the old. The bottom line for the consumer — regard with skepticism anything you read in a magazine about the virtues of a product. "Caveat emptor."

Waterproofing

"Waterproof" is seen as a good word in the marketing lexicon. Marketers know that the public would rather buy something that is waterproof than and item that is not. The advantage for the designer is that nylon can be loosely woven (fewer threads per inch mean less production cost), then coated with a cheap water-repellant compound to give the garment substance to the hand. The problem with such garments is that they aren't "breathable," that is, they will not allow water vapor to pass out of the underlying garments to evaporate (and cool) harmlessly away from the skin. Instead, this moisture builds up in the under-layers, reducing their insulating capacity by increasing conductive heat loss. Waterproof gaiters are a good example. Much of the moisture put out by your feet is exhausted out the top of the boot by foot flexion (a good reason to avoid very stiff boots with wrap-around cuffs). If the leggings you are wearing are made of cotton or uncoated nylon, this moisture will pass through to the outside air. If the leggings are coated, as nearly all are to some degree these days, the moisture is trapped and builds up in your socks, eventually conducting heat away from your feet. The manufacturers would counter, I am sure, with the fact that people won't buy unwaterproofed gaiters because few understand the complexities of moisture transport, and so the gear makers refuse to produce them. Either you need to find gaiters made by a foreign company that hasn't caught on to sophisticated marketing and is still trying to make stuff that really works (I am nursing a pair made in Finland into very old age) or make your own.

Fabrics

There are many fabrics on the market today from which outdoor clothing is made, ranging from traditional wool and cotton both pure and in various blends to synthetic fabrics with a bewildering array of exotic high-tech names. I will try below to make some sense out of the situation.

Cotton: As I have indicated, any dry non-conductive material will keep you pretty much as warm as any other of the same thickness as long as it is dry. Cotton is no exception. The problem with garments made of this fiber is that, once they become damp, either from insensible perspiration, liquid sweat, rain or melting snow, the fibers soak through and lose their insulating ability. Once the garment becomes saturated enough to contain free water, it will actively conduct body heat out to the surrounding air. This is an advantage on a hot summer day but not on a cold winter one.

Cotton does have its winter uses, however. Very tightly woven cotton makes a very good windbreaker material, especially in damp or wet conditions. It is breathable and windproof when dry; when damp, the fibers swell to make it water repellant as well. Although it is more delicate than an uncoated nylon shell, it is more resistant to water penetration. Unfortunately, only a few manufacturers produce cotton windbreakers today, and those do so at premium prices, mainly for the well-heeled yachting crowd.

Besides use as a windbreaker, cotton has few other cold-weather functions. Don't try to use it as insulation in situations where you are likely to sweat. Cheap waffle-pattern cotton long underwear that feels so good at home has bundled the bones of dead hikers in the mountains for the several generations since long woolies went out of fashion. Ditto cotton blue jeans.

Wool: This is my fabric of choice for insulation. Although it is hard to care for and expensive, it has certain advantages over the competition, the most important of which is that it will continue to keep you warm even if it gets wet. Totally saturated, with free water held in between its fibers, wool still retains at least 35 percent of its insulating capacity. This is the result of the fact that wool has oily, hollow fibers. The oil (lanolin) in effect waterproofs the individual fibers. While cotton fibers soak through and become conductive, and synthetic fibers are dry but solid inside, the dead air inside each wool fiber remains dry and insulative. This is the reason the synthetic fabric designers keep trying to un-sell it in their magazine ads while trying to imitate it in the lab. Countless generations of Norwegian fishermen can't be wrong.

On the down side, one reasonable criticism of wool is that it dries slowly, holding onto moisture rather than transporting it away. On the other hand, although wearing it damp will make you feel damp, you will feel warm and damp. On camps I bring two sets, wearing one while the other dries. I also find that wool long johns make better pajamas on a winter camp than synthetics. Once the humidity inside my sleeping bag reaches 100 percent due to insensible perspiration, synthetic long underwear will feel clammy while wool will not. Another complaint is that wool long underwear is itchy. This may have been the case once upon a time, but the extremely fine merino wool used in modern underclothing is comfortable to all but a very few people. (I find that machine washing in a side-loading rotary, not agitator, washing machine in cold water will keep the fibers from abrading and breaking up into a fuzzy, irritating surface.) Finally, wool long underwear is certainly expensive and hard to find. We have the synthetic fabric industry to thank for this. Their marketers were so successful in un-selling wool that many of those sheep pastures are now occupied by strip malls and condo developments, unlikely ever to return to their former use. In recent years, however, wool has been making a comeback. Very good merino wool long underwear and socks can now be found in outdoor shops and catalogues, albeit at premium prices.

Synthetic Materials: These come in a bewildering array of names and claimed virtues, and every year a host of new ones appears in the catalogues. Most of this appears to be marketing hype since they all appear to work about the same. Impermeable fibers are formed from petroleum molecules and spun into fabric. Differences among them seem to be mostly in the spinning, as designers try to make them act more like wool or to feel soft against the skin. Their real virtue is that, because the fibers are solid and will not admit moisture, they cannot soak through and conduct heat away from the body as do cotton fibers. Furthermore, garments made of synthetics were, at least initially, cheaper than wool. However, once wool was largely un-sold and banished from the shelves, the price of synthetic garments began mysteriously creeping upward. Funny how that happens.

While synthetic long underwear is appropriate for winter wear, it is important that the user understand that these fabrics are mainly intended for transporting conductive moisture away from the skin out to the surface of one's clothes where the cooling associated with its evaporation can occur harmlessly. They are not good insulators themselves. This is important in use. If you are wearing an impervious outer garment such as a coated nylon shell, your sweat can't go anywhere, eventually building up inside the fabric as free conductive water. The more you sweat, the less insulative your undergarments become. If you get soaked from without, by melting snowfall for instance, the same holds true. I find that the best conditions for synthetic long underwear — and I own several sets — is in very cold, dry conditions of heavy exercise with breathable top garments such as a wool or pile sweater and a truly breathable wind shell. For damp conditions, I stick to my wool long johns alone.

The last knock on synthetics has to do with how long you can wear them between launderings. The reason that Vermont farmers could wear their wool long johns all winter without changing them, without developing fungus infections and without alienating their families, was that wool absorbs body oils into the interior of the fibers where they are unavailable for aerobic bacteria to feed upon. The smell you detect when you wear a synthetic for more than one day is the odor of dead bacteria that multiplied and fed on your body oils as they collected on the surface of the impermeable fibers. Having a short life span, the bacteria died in place, creating the odor. Repeatedly wearing the same synthetic garment next to your skin for several days (on a wet early spring camp, for instance) may result in fungal growth as well. I learned this the hard way.

Gore-tex: This was the first and best known of numerous materials that purport to be both waterproof and breathable. Essentially a membrane is sandwiched between two layers of uncoated fabric. This micro-porous membrane allows water vapor to pass through but blocks free water. Since most free water is generated from without, in the form of rain or melting snow, it can't get in. And since body moisture, at least at a moderate rate of generation, is usually in the form of water vapor, it is drawn from the warm moist surface of your skin through the micro-pores toward the colder outside air, theoretically allowing your insulating layers to dry. The "new and improved" version, I am told, works better than the original (which was touted at the time like the Second Coming). Still, the reports from the field suggest that membrane fabrics are neither as breathable as an uncoated wind shell nor as waterproof as a coated fabric with sealed seams. Consider several things. First, the membrane is delicate so that it must be protected inside and out by nylon fabric. The water vapor, then, must pass through three layers to get out. If it condenses on one of the inner layers, as often happens in cold conditions, it either turns into free water or remains as frost, essentially blocking further transport. Furthermore, Gore-tex will not transmit free water outward any more than it does inward. If your level of exertion is intense enough and you have not removed inner layers, then moisture will build up inside your clothes, reducing their insulation value. (In a damp, howling fog or driving cold rain where you are climbing hard, take off your middle layers and keep them dry in your pack. The moment of discomfort involved in the change is worth the result.) Finally, no membrane will transmit water vapor once the outside layer of fabric becomes saturated with free water, nor if its inside layer becomes soaked with sweat, nor if the micro-pores become clogged with body salts or oils. In all these situations it becomes essentially a very expensive coated fabric.

With those caveats in mind, Gore-tex type garments are both attractive, if expensive, and fulfill their promises to a degree. You can end run the expense by purchasing or making a good breathable wind parka and a coated storm parka, carrying each as the weather dictates. Whichever way you go, remember that any equipment, Gore-tex or otherwise, will only work for you if you understand its limitations, maintain it carefully and use it appropriately.

Down: Goose down, as well as eider down, duck down, etc., is the layer of very fine, light plumules under the coarse outer feathers on the breast of a bird. This light, non-conductive material has been recognized for centuries as superior insulation for bed covering in unheated rooms before the development of central heating. Moreover, it has the great advantage for mountaineers in that garments and sleeping bags made with it are extremely light and compressible, important traits for a climber with a lot of gear and a small pack. In the 1960s, down gear became all the rage in alpine ski clothing as the general public suddenly wanted to look like rugged alpinists, despite the fact that such use on the slopes or at the mall required neither lightness nor compressibility. This created a run on the world's goose down market, driving down the supply and driving up the price. Today the supply has been depleted to the point that quality down is hard to find and outrageously expensive. Much of what now is claimed as "down" is actually heavily adulterated with coarser outer feathers, a practice that is legal up to a certain percentage.

The good news for the budget conscious is that unless extreme lightness or compressibility is important to you, you can save a lot of money by purchasing synthetic fill garments. These can't be stuffed in the corner of a pack and are a little heavier, but they will keep you just as warm. Remember that any dry, non-conductive material of equal loft will keep you pretty much as warm as any other; even dead leaves or crushed newspaper will do the job. Furthermore when synthetic fill garments get wet they maintain their insulating loft while wet down disappears like cotton candy.

Layering: Dressing in layers allows you to regulate body heat and the production of sweat by adding and removing them. By far the most important layer is the one next to your skin. As has been explained, your long underwear should be made of wool or a synthetic fabric. Fit is important: it should be snug enough to prevent drafts from moving over the surface of your damp skin, causing evaporative cooling and triggering the release of the chemical that spasms capillaries. Yet it should not be so tight as to restrict circulation.

The middle layers may be composed of any non-conductive material. Keep them dry by removing them under exertion. Donning a windshell to frustrate convective cooling will usually work better under exertion than adding mid-layer insulation in order to reduce conductive cooling, with the obvious advantage that the inner layers remain dry in your pack for when you stop.

On a still day you will not need an outer layer of windproof material. But when the wind is blowing, put on an overlayer of uncoated nylon taffeta or tightly woven cotton and regulate your temperature

536

with the hood and a hat (Figure B.5). (Incidentally, you can often tell whether a nylon wind parka is coated or not by holding it at an angle to the light. If the finish on one side is shinier than the other it was probably treated for water repellency. Wrap the garment around your hand; if your hand immediately feels sweaty, the fabric has been coated. Finally, look at the armpit of the parka; if it has a small metal grommet

Figure B.5. Warm as toast despite appearances. A few well-chosen layers were sufficient for the wind-blasted summit of Mount Eisenhower, New Hampshire.

or two, the holes they surround are meant to try to ventilate a water-proofed garment.)

A Final Word on Fabrics: Down or synthetic garments will only conserve your body heat; they don't generate warmth themselves. If you are not warm to start with, insulation will not make you warm. Reconsider the wisdom of the Eskimo's deceptively simple statement.

The Extremities

Socks: Wear wool only; accept no substitute no matter how glowingly it is touted. Feet inside a boot under exercise always generate moisture, and even in the best prepared boot, much of that moisture gets trapped inside. Synthetics are mostly a transport medium rather than an insulating one; saturated, they lose most if not all their ability to insulate. Since foot moisture has nowhere to get transported to, it will collect in the socks. Wool maintains 35 percent of its insulating capacity even if saturated — with sweat or water from outside the boot. A good trick on day trips is to wear different socks and shoes in the car on the way to the trailhead, changing to dry socks and boots upon arrival — for some reason, feet tend to get sweaty in the morning.

You may have read about the trick of putting an impermeable layer between your foot and your socks. Plastic wrap or baggies supposedly have been used for this purpose. Somehow I doubt that the people who promote this have ever actually used it under exertion. The physical principle of a vapor barrier to keep insulation dry is indisputable. However, it works better in principle than in practice. What actually happens is that the baggie concentrates sweat on the skin of your foot, softening it over time and promoting the formation of sheet blisters, the sort that will incapacitate you for the rest of the winter.

Boots: Properly treated leather is best. Leather has pores like the skin that it is made from. While not strictly "breathable," the pores in leather allow some moisture to migrate out of the boot to evaporate harmlessly on its exterior. It will do this, however, only if it is *not* waterproofed. Waterproofing compounds contain wax, which clogs the pores. The waxed boot will not allow water in, but it also will not allow water out. Many very good European boots come pre-waxed, but you should realize that they are intended for summer use on slushy

Swiss glaciers at warm temperatures, not for winter use on North American mountains. (In winter, the Swiss ski.)

The oil in saddle soap should be enough to make the leather on the outside of your boots water repellant with perhaps some light boot grease added to the critical areas. Leather lining is best for the same reasons as for the outside. Unfortunately, leather-lined boots are increasingly hard to find. Instead, a cheap fabric lining hides a layer of open cell foam that acts like a sponge, soaking up foot moisture, presumably to be dried out later by the stove. Whether any of the moisture that accumulates manages to pass to the outer leather while you're still in the field probably depends on the treatment of that leather. Avoid boots whose leather has been painted! Incredibly, some manufacturers actually do this to hide the low quality or thickness of the leather in the boot, promoting it in ads once again with the magic phrase "100% Waterproof." Read "sweaty."

Sorel type boots with rubber bottoms, leather uppers and wool felt liners are quite good for day use. Even though the foot section is impervious, moisture is collected in the wool felt, which continues to insulate even when damp. The only problem with this type of boot is that it tends to be soft: snowshoe and crampon straps can constrict blood flow to the foot, and the soles tend to be so flexible that control in delicate situations is difficult. Another problem is that the wool felt liners of early models have more recently been replaced with cheaper synthetic liners. These feel the same in the store but are worse than useless in the field.

Fit is important. Boots should be roomy enough to accommodate two pairs of socks without constriction, especially in the forefoot. Your foot should almost rattle inside the boot. Sometimes less sock means more warmth.

Mitts: I use a couple of combinations of mitts. Wool or poly liner gloves inside wool fingerless mitts inside breathable shells are almost always enough. The thin liners inside the fingerless "millarmitts" delay numbing long enough to get knots tied and buckles fastened. With any exertion at all the addition of overmitt shells is enough for the rest of the time, that is, assuming one's core temperature is being maintained. Liner mittens that fit inside the shells should be kept dry in the pack or outside pocket for emergencies or when stopped for lunch. Another setup is to use the kind of mittens that have a fold-over flap for the fingers and a hole for the thumb; I have a pair of Helly-Hansens of this design that I use for winter hiking. When I'm on skis, I just use ski gloves and trust to muscular activity to keep my hands warm. For backcountry skiing, however, I always carry the mitt/overmitt combination in my pack.

Hats: The old saw "If your feet are cold, put on a hat" is true, and, if you have absorbed the information on physiology at the beginning of this section, it is easy to see why. Since, in order to protect the brain, the capillaries in the head area do not vasoconstrict when you begin to chill as do those in your extremities, your head acts as a radiator of heat from the core area of your body. Unless you interrupt the loss of heat with some sort of insulation, your core temperature will drop, you will feel chilled and, unless steps are taken to restore that temperature, you will lapse into exposure. This is the reason that, although humans have lost primitive fur from most parts of their body, we retain hair on our head. But the remaining hair on our neotropical heads is not enough in cold conditions and must be supplemented with a hat. Wool is best but tends to irritate some people's foreheads. My warmest combination for the slow movement of winter tracking is a synthetic inner hat with a wool knit over the top. I wear the synthetic alone when moving fast and add the wool when I slow down or stop.

As far as design is concerned, the balaclava style works best. The hat can be folded up to the top of the head to expose the face and neck during heavy exercise when you want to dissipate heat and then pulled down over your entire head when you stop. When I hike mountains in the winter in still air, I normally wear only a wool undershirt on my upper body, regulating my comfort by putting on, folding up or taking off a wool balaclava. If the wind starts to blow, I simply pull the balaclava down over my neck and don a wind shell with a hood. Other more chic styles of hats of synthetic materials with fold up earflaps approximate this effect but none do it as well as the old reliable wool balaclava. Only if your head, neck and trunk are warm will the tiny blood vessels in your extremities open up and flood them with warming blood. "If your feet are cold, put on a hat."

Once your fingers and toes are cold, even increasing the warmth in your core area will not bring immediate relief. The release of the chemical that was triggered by a threat to your core temperature and that in turn constricted your capillaries takes a while to wear off. That's why those painfully cold feet I got as a child skating on the ponds, didn't warm up right away when I came in and sat by the stove. "To stay warm in the winter, take care not to get cold."

Preventing the Freezing Pack Syndrome: One more question, and a tip. What do you do about the moisture that inevitably collects in your undershirt where it is under your pack, especially when you have been exerting yourself heavily? Most people have had the chilling experience of swinging their pack back on after a stop on the trail and feeling the compression of their moist garments under its pressure as a sudden and very unpleasant chill. My solution, used on many winter ascents, is to fashion a cotton towel as a liner for the back of the pack, cutting and sewing it in such a way that tabs extend under the shoulder straps as well. While exerting yourself, the sweat generated gets blotted up by the towel instead of collecting in your long underwear under your pack. I fasten the towel liner to the pack with sew-on snaps so that it can be removed and laundered between excursions. Try it; it works.

Recommended Reading

Washburn, Bradford. *Frostbite.* Boston, MA. Museum of Science, Boston, 1963. A pamphlet containing a detailed discussion of human physiology in the cold by a scientist and old man of the mountains. Quite readable but now out of print. Library copies should still be available in some networks.

Marchand, Peter. *Life in the Cold.* Hanover, NH. University Press of New England, 1987. You may wish to skip the chapters on plant adaptation, but study the ones on snow structure, cell physiology, and wild animals and humans in the cold.

Manning, Harvey, et al. *Mountaineering, the Freedom of the Hills.* Seattle, WA: Seattle Mountaineers, 1967. Mainly a book on mountain climbing, but my old edition contains much wisdom about how you and your equipment should work in the winter.

London, Jack. "To Build a Fire." *Youth's Companion* 76. An often-anthologized short story by a man who knew men in the cold. "Cowardice in the face of natural forces . . ."

I provide the following section to registrants for my winter tracking programs. You may wish to reproduce it for yourself.

Summary Points for Keeping Warm in Winter

- Remember that it is easier to stay warm than it is to get warm once you have chilled.
- It is most important to keep your core temperature high, and this in turn will keep blood flowing to your feet and hands. Putting on three pairs of mittens will do no good if your trunk area is chilled.
- Regular exercise improves circulation in feet and hands.
- Eat fatty foods in the winter before long excursions in the cold.
- Keep yourself well hydrated during long spells in the cold. Get used to peeing in the woods.
- Dress in layers that can be added or removed to maintain your comfort.
- Remove outer layers of clothing *before* you sweat into them.
- Do not wear cotton next to your skin, especially if you expect to sweat. Buy a set of synthetic or wool long underwear, sized to fit closely but not tight.
- Wear wool socks only. Don't stuff your boots so much that you constrict circulation.
- Regulate your level of comfort by putting on, adjusting or removing a hat.
- Buy a breathable wind shell with a hood. Use the hood to add to the warmth of the hat.
- A scarf may seem old fashioned but it works well for keeping out drafts and is easy to adjust.
- Except in really wet conditions, avoid wearing waterproof garments.
- If you expect to spend a long time in the cold, change to wool socks and comfortable boots just before you set out. A long drive to a trailhead in a warm car while wearing the footgear you intend to hike in will make your feet sweat and compromise some of their insulation.

Finding Your Way

It is easy to get so absorbed in following an animal's trail that the tracker might lose his sense of where he is, and since animals often stray from human trails, it is easy to get at least temporarily lost. Several books listed in the bibliography deal with land navigation, and I do not intend to duplicate their efforts here. Rather, I would like to underline some points that I have found useful over a lifetime of travel in forests, jungles and mountains in all seasons.

Disorientation

Every backcountry tracker should experience this at least once, preferably in controlled circumstances of low risk. Disorientation occurs when the tracker's head insists that the correct direction of travel is one way when actually it is another, often opposite to the expected one. I recall once years ago when I was birding in a dense spruce forest out in the Passaconaway region of New Hampshire. Unable to see much of anything but the trees around me, my mind adopted a wrong-headed sense of direction. To get out, it told me to go in one direction, south, when the compass told me to go in the opposite direction. South would have taken me deep into the mountain valleys and hills below Mount Passaconaway, while northward, so close that I could hear auto traffic, was the Kancamagus Highway. Even with that road noise, I had a hard time turning my head around to the truth of the lay of the land. This incident and others has led me to a sort of self-knowledge that has little to do with the condition of my soul, but rather simply to the knowledge that I have a poor sense of direction. Knowing this about myself has conditioned me to the habit of subconsciously orienting myself whenever I am in unfamiliar terrain. I don't take out a compass on the train platform, but rather seek out the sun according to the time of day to determine roughly where south is and where invisible landmarks around me should be. As hard as it may be to reorient one's mind to the truth, trust the compass.

GPS Notes

The GPS — global position system — device is a marvelous instrument. While it has taken much of the adventure out of backcountry adventures, it nonetheless will tell you where you are and in what direction you want to go. However, this device may contribute to a false sense of security in those who are not thoroughly familiar with its use and who do not have backup skills for when it fails. A GPS and a cell phone (for rescue) may get people into a lot of trouble. Both instruments rely on batteries that tend to fail under extreme weather conditions. Furthermore, cell phone coverage varies by customer base and so tends to be unreliable exactly in those areas where it is more likely to be needed to call for rescue. As for the GPS function, the user must take the time and effort to set the instrument up properly and then become skilled in the use of its many functions. There is, indeed, a learning curve that demands close attention to the manual that comes with each GPS unit. The skills described within it should be practiced in low-risk and low-stress situations *before* depending on the GPS in the wild.

In addition to the information in the manual as well as in the guides to land navigation cited in the bibliography, there are a few other things you should know. The first involves reliability. A friend has a relatively inexpensive GPS made for hikers by a well-known company. This unit often shows a large error in guiding him back to a recorded waypoint: as much as 200 meters on several occasions. In bad weather with low visibility this obviously can be a critical problem that only prior use and practice will reveal.

Another problem involves distance traveled. While your trail guidebook may tell you that a trail junction is 2 miles away, the "two-miles" of the map-maker may not be the same as the two miles shown on your GPS. Depending on how close together you have set your data points to record, the distance to that trail junction may be off from the expected distance by hundreds of yards. For instance, if your data points are set to the minimum distance, they will record every small deviation you make to the right or left to avoid a log or a boulder, giving you a much longer distance than may have been the case with the map-maker or trail sign designer who rolled a map-measurer over a topo sheet to get that distance. As a result, if you rely on the distance covered shown on your device and don't learn to read the ground and relate it to contours on a paper (or GPS) map, you may miss the junction. Arriving at the point where the device's trip computer tells you the junction should be and finding nothing, you may not know which way to go, backward or forward. Let's say you assume you haven't gotten there yet and press on. After a quarter of a mile, you decide, "No, it must have been in the other direction," so you backtrack a half mile. Still not finding the junction, you decide well, you just didn't go forward far enough the first time, and so you return, this time a half mile beyond the GPS junction position. This backwards and forwards can go on with mounting anxiety until you are overcome by exhaustion or nightfall. If you are lucky, you may find the actual junction before dark and learn a lesson from your experience. If you don't, you may need the information in the following, and last, section of this chapter.

A second caution deals with batteries. The common alkaline batteries you can buy in any drugstore do not function well in cold weather. Even though such batteries are well charged, if they are allowed to get cold, their power will drop so that your GPS will show low battery on its meter and a message will pop up telling you to change batteries. Doing so will not help, however, if the alkaline batteries you replace them with are also cold. Lithium batteries are the power supply of choice in cold weather since they will put out the same power level throughout their life regardless of temperature. Of course, always have a set of backups in your GPS case, because one danger with this sort of battery is that the meter on the screen may show a fully charged battery until suddenly it drops to nothing. If you are traveling above treeline in blowing fog and depending on the track-back feature of your GPS to guide your return to treeline, then you may have an unpleasant surprise when, having reached your destination, you pull it out of your pocket to consult it for the way back and discover that the lithium batteries had failed somewhere along your route and you have no recorded track to follow down to safety.

While on the subject of tracking above treeline, weather often changes with amazing speed. A clear day in which you can see the horizon and your own trail back to safety can suddenly change to a fog that can hide even nearby land features. Appalachian mountains in my region have a fringe of dense ground spruce that grows in a tonsure around bare summits. If you have not marked your exit from this treeline, a sudden fog may cause you trouble finding it again. An all too common scenario in our eastern mountains is that a traveler above treeline tries to get back down out of the wind and fog into the forest below and mistakes his entrance/exit point. An instinct that safety is below causes him to continue plunging downward anyway. He flounders in the snow-glutted ground spruce, breaks through crevices between hidden boulders, perhaps injuring himself but certainly exhausting himself fighting the snow. Cotton denim jeans and cotton-waffle "insulated" long underwear become saturated with sweat. Exhaustion and conductive heat-loss finish the job. Whenever I leave treeline, even on a clear, sunny day, I mark my exit point with colored streamers and record a waypoint on my GPS. This has helped me out a number of times when sudden weather changes bring blowing fog.

A word about "spruce traps" — in deep snow country that harmless seedling sticking out of the surface may actually be the tip of a much larger tree hidden under the snow. Step on it and you will get a surprise. The snow will collapse and you will fall downward. In very deep snow you may sink so low that snow falling in after you may bury you. The snowshoes on your feet that supported you over the snow now act as "deadboys," that is, as anchors to which you are connected that are difficult to pull upwards in the snow that accumulates above them. Climbing up the tree to the surface is also hindered by the snowshoes. Even if the snow is not that deep, you may have difficulty extricating yourself without help. One of my early winter climbing experiences when I was in my 20s had me falling into a spruce trap up to my armpits. Only the help of my climbing companion saved me as he was able to lift the heavy overnight pack off my shoulders while I extricated myself. Obviously then, having a companion in a remote location in bad weather is important. And, of course, avoid stepping on conifer "seedlings."

Map and Compass

Facility with map and compass until very recently was a basic skill among back-country travelers. Among the unwise, the advent of GPS technology has led to such skills being marginalized. In situations where you must depend on yourself in remote locations, redundant systems are the best. If one chooses to wander far from heated shelter, one's GPS should be backed up by skill at the traditional methods of land navigation, methods that do not depend on battery life. One is often cautioned to be sure to take a map and compass with you for excursions off-trail. However, these tools will do you little good unless you are skilled at using them.

The fundamental skill of traditional land navigation has less to do with manipulating compass dials and map grids than it does with "map inspection." This is the ability to relate the contours on a map to features on the ground. I suggest an exercise that is best done at first in the winter when there are no leaves to obscure one's vision or at least in an open location where distant terrain features can be seen. Take a topo map to a spot where you can see for perhaps a half mile around you. Sit there, locate yourself on the map, turn the map around until it is oriented true to the ground and then look at the terrain. Note some visible feature, say a small hill forward of your position. Then look at the map and locate that hill. Note the distance to it both visually and on the map scale. Note as well the distance between contours and relate it to the visible distance from the foot of the hill and its summit. Do the same with other terrain features until the map and ground become one in your mind. Now move to another spot and do the same thing until, once again, your mind sees the agreement between map and ground. After doing this exercise in open terrain, try it in denser growth where you can see only perhaps a couple of hundred meters in any direction. This will require you to see how smaller features are represented on the map, if they are represented at all. Gradually, these exercises will help you not only relate ground to map and vice-versa, but they will also give you sense of how a map-maker represents or ignores terrain features of varying prominence. A map, after all, is not a photograph of terrain, but rather a mapper's rendition of it and thus subject to his idiosyncrasies. Only after you have become skilled at map inspection should you turn to the dials, grids and azimuths of your map-reading tools.

Finally, one should not rely on the beam of a flashlight or headlamp for finding your way in the dark. These are all right for ground that you are already familiar with, such as finding one's way from the car to the house, but finding one's way in unknown terrain by flashlight is a very challenging art as any night orienteerer will tell you. First, it is difficult to get a sense of terrain context from a narrow cone of light. Without the ability to see distant features, one can only view nearby objects, isolated from their relationship with one another. Off-trail in a woodland in a narrow beam of light, one tree looks much the same as the next.

Emergency Shelter

It seems to me that information available in books about emergency procedures in the wild is often highly theoretical, sounding like something written from the comfort of a desk without real experience with the recommendations. For instance, one often is advised to stop well before dark in order to have enough time to make preparations

547

for forced overnighting. The problem with this advice is that few people will do so. If you are lost, it is likely that you will continue to wander around hoping to find your way out until close to dark and will begin making preparations only when imminent nightfall forces you to. At that point speed and ease of construction are the paramount considerations.

Of the proposed shelters for winter survival that I have seen in various handbooks, the easiest is a light, nylon tarp or poncho rigged overhead with lines to nearby trees. This certainly has simplicity and speed of construction to recommend it. However, it is unlikely to guarantee your survival in sub-freezing temperatures for long. While such a fabric covering will defeat radiational cooling to the sky, wind is at least as much your enemy as cold, and any breeze or even draft of air will infiltrate underneath the sheet and steal convective heat. This air movement may even be caused by rising heat from your own body. Other designs that I have read about are more elaborate. They might make sense for a planned overnight, but are too complicated and time-consuming for quick emergency shelter.

Building a Quin-zhee is sometimes advised. This is a mound of snow similar to the snow hut you may have made as a child. However, great care must be used in hollowing out this mound to avoid its collapse, a care that is difficult in a fatigued state and with mounting anxiety due to the gathering cold of impending nightfall. Building an igloo, as is sometimes suggested in the more unrealistic literature, is really impossible. Such structures require several people to build and are only feasible in the high arctic where there is abundant wind-packed snow typical of that environment. Trying to make snow blocks out of the soft or sugary snow typical of lower latitudes ranges from difficult to impossible. A fellow I know was determined to build one of these on a ski trek in Finland. After he crawled in to spend the night, it collapsed on him.

As I have suggested, wind is at least as much of an enemy as is cold. The quickest solution for surviving the night is to wrap yourself up in a poncho and bury yourself in the snow with only your face exposed. Stories about then being entombed in ice formed by melt from you body heat seem far-fetched, but I admit I have never tried this method of survival. With only a little more effort, the wind may be stopped and the cold kept at bay by building a simply constructed and much more comfortable slot-shelter. To do this, pile up snow in a long mound a few feet longer than your own length and about 2 ½ feet high, positioning the long axis of the mound in the direction of the prevailing wind. Then cut a slot of your own length into the mound starting on the leeward end. As you dig out the middle of the mound, pile each shovelful on the walls to build them higher, up to about 3 feet, packing down the added snow as you go. Then cut conifer boughs for the floor in order to keep yourself above the melt water that will accumulate in the bottom. Fir is best for this, since the needles are very springy, with spruce a second choice and hemlock for a third. If you have skis, lay them and your poles across the top. If you still have the energy, lay more boughs over the skis for insulation. Finally, rig a tarp or poncho over the whole. Tie down the corners with "deadboys" — sticks buried perpendicular to the direction of strain — and a slip knot for easy disassembly the next morning. You can build a fire in front of the entrance and reflect its warmth toward the shelter opening by forming a snow wall behind it. Gather some dead wood and kindling for later and crawl into the slot, keeping most of your equipment under you. Insert your feet into your pack for added warmth. A lighted candle gives off a surprising amount of heat in a small, enclosed space.

I have constructed this sort of shelter in about half an hour, with another fifteen minutes needed to prepare and start the fire. Obviously, a few items are required to build the shelter: a pack shovel, a poncho or tarp, a length of cord and a sharp knife or foldable brush saw for cutting bows. A jack knife may do in a pinch, but remember that fir and spruce limbs are very wiry, being designed to bear snow load without breaking. All three of these tools should be part of your kit for back-country tracking on short winter days. A snowshoe can be used for a shovel in a pinch, although it will slow down the process. Also, cut boughs can be used alone for a covering without adding a sheet over the top, although this will allow more heat loss overhead. A headlamp for performing these preparations in low light is a great help as it leaves both hands free to work. Once again it should be powered by fresh lithium batteries. For fire starters, birch bark is good for drying out other small kindling. It is waterproof — that's why Indians used it for canoes — and will always light. Other fire starters that work are steel wool and even cotton balls. A butane cigarette lighter provides a more reliable ignition source than matches, which can get damp. However, it must be kept warmer than freezing temperature in an inside pocket in order for the liquid butane to vaporize. I suggest carrying both in your emergency kit.

If there is no snow, simply pile up leaves and brush into a debris pile the same shape as the slot shelter described above. Lie down in the slot and then pull the debris over you, starting at your feet and ending at your head. Leaves and wood are largely non-conductive; they and the air trapped in the pile will insulate you well enough to allow you to survive the night.

Bibliography

While the information presented in this book on the interpretation of tracks, trails and other sign is original, inevitably the work of others that I have consulted over the years has filtered into the content. While the listed books have not been quoted directly, all have contributed to the background knowledge for *The Next Step*. I am greatly indebted to the experience, scholarship and wisdom shared by their authors.

Brown, David. *Trackards for North American Mammals*. Granville, OH: McDonald and Woodward Publishing Company, 2012.

_____ *The Companion Guide to Trackards for North American Mammals*. Granville, OH: McDonald and Woodward Publishing Company, 2013.

Burt, William, and Richard Grossenheider. *A Field Guide to the Mammals of America North of Mexico*. Boston, MA: Houghton Mifflin, 1976.

Cronon, William. *Changes in the Land*. New York, NY: Hill & Wang, 1983.

DeBruyn, Terry. *Walking with Bears*. New York, NY: Lyons Press, 1999.

DeGraaf, Richard, and Mariko Yamasaki. *New England Wildlife*. Hanover, NH: University Press of New England, 2001.

Elbroch, Mark. *Bird Tracks & Sign*. Mechanicsburg, PA: Stackpole Books, 2001.

_____ *Mammal Tracks & Sign*. Mechanicsburg, PA: Stackpole Books, 2003.

_____ *Animal Skulls*. Mechanicsburg, PA: Stackpole Books, 2006.

Errington, Paul L. *Muskrats and Marsh Management*. Lincoln, NE: University of Nebraska Press, 1961.

_____ *Of Predation and Life*. Ames, IA: Iowa University Press, 1967.

Department of the Army. *U.S. Army Map Reading and Land Navigation Handbook*. Guilford, CT: Lyons Press, 2004.

Halfpenny, James. *A Field Guide to Mammal Tracking in North America*. Boulder, CO: Johnson Books, 1986.

Jorgensen, Neil. *A Sierra Club Naturalist's Guide to Southern New England*. San Francisco, CA: Sierra Club Books, 1978.

Kals, W. S. *Land Navigation Handbook: The Sierra Club Guide to Map, Compass and GPS*. San Francisco, CA: Sierra Club Books, 2005.

King, Carolyn. *Natural History of Weasels and Stoats*. Ithaca, NY: Comstock Publishing Associates, 1989

Kricher, John C. *A Field Guide to Eastern Forests*. Boston, MA: Houghton Mifflin, 1988.

Macdonald, David. *Running with the Fox*. New York, NY: Facts on File Publications, 1987.

Manning, Harvey, et al. *Mountaineering, The Freedom of the Hills*. Seattle, WA: Seattle Mountaineers, 1967.

Marchand, Peter. *Life in the Cold*. Hanover, NH: University Press of New England, 1987.

Murie, Olaus. *A Field Guide to Animal Tracks*. Boston, MA: Houghton Mifflin, 1954.

Powell, Roger A. *The Fisher: Life History, Ecology and Behavior*. Minneapolis, MN: University of Minnesota Press, 1982.

Reid, Fiona. *A Field Guide to Mammals of North America*. Boston, MA: Houghton Mifflin, 2006.

Rezendes, Paul. *Tracking and the Art of Seeing*. Charlotte, VT: Camden House Publishing, 1992.

Whitaker, John, and William Hamilton. *Mammals of the Eastern United States*. Ithaca, NY: Comstock Publishing Associates, 1998.

Williams, Roger. *A Key into the Language of America*. Providence, RI: The Rhode Island and Providence Plantations Tercentenary Committee, 1936.

ᓚ

Index